Travel Knowledge

Travel Knowledge

European "Discoveries" in the Early Modern Period

Edited by

Ivo Kamps and Jyotsna G. Singh

palgrave

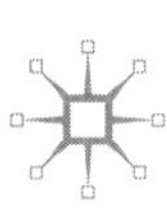

TRAVEL KNOWLEDGE

First published 2001 by
PALGRAVE™
175 Fifth Avenue, New York, N.Y. 10010 and
Houndmills, Basingstoke, Hampshire, England RG21 6XS.
Companies and representatives throughout the world.

PALGRAVE™ is the new global publishing imprint of St. Martin's Press LLC Scholarly and Reference Division and Palgrave Publishers Ltd (formerly Macmillan Press Ltd).

ISBN 0-312-22270-X hardback
ISBN 0-312-22299-8 paperback

Library of Congress Cataloging-in-Publication Data can be found at the Library of Congress

A catalogue record for this book is available from the British Library.

Design by Westchester Book Composition

First edition: January, 2001
10 9 8 7 6 5 4 3 2 1

Printed in the United States of America.

CONTENTS

NOTE ON ORGANIZATION

Travel Knowledge is divided into nine sections consisting of primary material followed by an essay. Each contributor selected, edited, and in some cases annotated their primary material. All the notes from contributors within primary material sections appear in italics.

ACKNOWLEDGMENTS

As editors of *Travel Knowledge,* we have been enriched by the project of putting together this collection and thank our contributors for their efforts and diligence. Our thanks also to Kristi Long, our editor at Palgrave, for her guidance, patience, and geniality throughout this process. We are grateful to the University of Mississippi for a summer research grant, and to the Interlibrary Loan Office of the J. D. Williams Library. Carel F. Kamps, Jr., has our gratitude for his enthusiastic support for the project and for helping us unearth research materials in the Netherlands. And finally, our thanks to Robin Christensen, a research assistant at Michigan State University, whose diligent and cheerful assistance at all stages made our work easier.

—Ivo Kamps and Jyotsna G. Singh

INTRODUCTION

Ivo Kamps and Jyotsna G. Singh

I

Geographers present before men's eyes
How every Land seated and bounded lies.
But the Historian, wise Traveller
[Describes] what mindes and manners sojourn there,
The Common Merchant brings thee home such ware
As makes thy Garment wanton, or . . . fare.
But this hath Traffick in a better kinde
To please and profit both thy virtuous minde.
. . . Read it and thou wilt make this gain at least,
To love thy one true God, and Countrey best.

By Henry Ashwood in dedication:
"To my ancient friend Mr. Edward Terry,
On his Indian Voyage."[1]

This short poem of praise appended to Edward Terry's *A Voyage to East-India* (1655 edition) clearly speaks for all the varied concerns and motivations of European travelers in the early modern period. Geographers, historians, merchants, or moralizing churchmen like Terry himself were all a part of the movement of European exploration and "discovery" from the fifteenth century onwards, setting into motion a process of globalization and transculturation that is still with us today. In stating this, our purpose is not to identify an originary moment in the history of globalization or to suggest for this history a specific teleology. Rather, our point is that travels during the early modern period, undertaken by the Portuguese, the English, the Dutch, and others, to the Ottoman Empire, the Far East, Africa, and the Americas initiated a series of cultural (as well as economic and military) encounters, exchanges, and confrontations that are the dynamic precursor to two centuries of colonialism, as well as its aftermath, postcolonialism. This anthology seeks to do two things: (1) to offer the reader, through primary materials (travel narratives), a contemporary record of these early moments of transculturation, and (2) to provide the reader with a set of new critical essays that, from within our current and different ideological and theoretical horizon, contextualizes and assesses these travel accounts.

One of the things that will become clear from the "dialogue" between primary texts and critical essays is that it is impossible to arrive at a generic characterization of transculturation that satisfies the wide variety of cross-cultural encounters in the early modern period. What we can say, however, is that the exchanges between western travelers and native peoples were reciprocal but often unequal. The power struggle ensuing from these exchanges, which is manifested in economic, military, and religious terms primarily (and which the travel narratives themselves cannot disguise), does not by itself constitute the social formation we now refer to as colonialism, but it does identify some of the very conditions that make colonialism possible.

Travel Knowledge, therefore, enters this scene of travel writing by acknowledging the problematic, or even futile, task of locating any originary moment of colonialism. The discursive field of postcolonial theory, from Edward Said's *Orientalism* (1978) onwards, has often questioned the Eurocentric premises of "discovering" lands, people, and valuable commodities—a process that is recounted in earliest travel narratives through the sixteenth and seventeenth centuries.[2] Mary Louise Pratt, for instance, historicizes European travel writing (from 1750 to around 1980) as an important part of the "text of Euroimperialism," and asks, "How has travel and exploration writing *produced* 'the rest of the world' for European readerships at particular points in Europe's expansionist trajectory?"[3] Like her, David Spurr also views European travel writing—in the modern period of colonization from 1870 to 1960—as a "colonial discourse" or "a form of self-inscription onto the lives of the people who are conceived of as an extension of the landscape. [Thus], for the colonizer as for the writer, it becomes a question of establishing authority through the demarcation of identity and difference."[4] Most postcolonial critics examine this "metaphoric relation between the writer and colonizer" as it applies to European travel writing during the later periods of high imperialism.[5]

Theories set forth by Pratt and Spurr focus on travel writing as a colonial/imperial post-mid-eighteenth century genre, and hence not clearly applicable to the earlier narratives of exploration. Thus, we are sensitive to the dangers of yoking disparate journeys and encounters within an ideological straitjacket of colonial intent—as is sometimes the case in the postcolonial approaches mentioned above. Having made this qualification, however, we do not want to occlude the emerging imperial designs underpinning the new mercantile economy. Thus, even though the essays in this volume record journeys and ventures *prior* to the consolidation of European imperial power in the late eighteenth/early nineteenth century, they nonetheless (in varying forms and degrees) bear the marks of a "colonizing imagination"—tropes, fantasies, rhetorical structures—whereby the writers/travelers frequently fall back on defining the cultural others they encounter in terms of binaries that later consolidate and justify full-blown colonialism: civilization versus barbarism, and pious Christianity versus impious Islam, among numerous others.[6] Frequently, in the case of the English and the Dutch, for instance, Protestantism was the force that maintained these binaries and, more importantly, unified the interests of both nationalism and commercialism—as is evident in several of the essays. Thus, for instance, if we consider Edward Terry's claims to Christian moral superiority over the Indian natives, or Dallam's and Sandys's exoticizing, eroticizing, and demonizing impulses toward the Turks, we can see how these various modes of othering may have helped to define a Protestant, English national identity, within an emerging proto-imperialist formation (see Singh, Vitkus, and Fuller in this volume).[7]

Yet these tropes of otherness are hardly consistent or uniform; even while defining the natives as heathens or barbaric, the Europeans expressed "imperial envy" toward the powerful Ottoman Empire as well as awe and fear of the Mogul emperor. In fact, in most accounts of individual encounters, the binaries are often in play, calling for complicated responses from their readers and audiences, as the cultural, racial, and (in some instances) sexual identities of the writers/travelers shift under the allure or the duress of the varied interactions. Mary Fuller's recapitulation (in this volume) of Dallam's gaze on the harem reveals both the voyeuristic potential of sexual encounters and the attendant fears emerging from an intermingling of heterosexual and homoerotic desire.

All the essays in *Travel Knowledge* reveal (in varying forms and degrees) the ways in which the discursive practices and material realities of voyages and journeys intersect and affect each other; and in doing so, they seek to illuminate the *specificity* of the travelers' lives as *historical subjects,* capturing both the *localized* dimensions of their cross-cultural encounters in alien landscapes and the discourses of *difference* that constitute their shared cultural/racial assumptions. For instance, when the travelers enter local spaces such as the Seraglio, the baths, the Mogul Court, or the Hindu temple, the voices we hear of the Europeans and the natives are dissonant and contradictory.

The Turkish materials reflect a frequent fear of "turning Turke"—which means either deteriorating in character from a supposedly Christian moral rectitude, or being forcibly converted to Islam, as was the common fate of European captives of the Ottomans. In this way, we show that these early modern travel narratives do not produce Said's Orientalism, but instead recount cultural encounters in which self and other are not fixed in opposing positions but are rewritten through discursive and social interventions. Not surprisingly, then, a propagandist and travel-narrative anthologizer such as Richard Hakluyt may ideologically promote the unified ideal of a merchant-colonist—the heroic, "high-minded, practical trade-minded Englishman"—but the historical subjects (the English and others) who emerge in these narratives speak in many voices, struggling to distinguish between identity and difference—in terms of religion, sexuality, nationality, among other factors.

In this dynamic, as several essays in this volume note (see Fuller, Taylor, Kamps, MacLean, and Singh), the *gaze* or simply the *act of seeing* functions as an important structural device, whereby our travelers interpret and record European cultural encounters. In fact, seeing, with its assumptions of *knowing* and *possessing,* becomes an act fraught with meaning and power. Repeatedly, the travelers remind their readers that they are "eyewitnesses" giving us an empirical rendition of unfolding events; but this process is more complicated since, as recent critics remind us, "the European dream of possession rests on witnessing understood as a form of significant and representative seeing," and frequently in travel accounts, the early modern European (male) traveler functions as a "seeing-man," the "landscanning" producer of information whose "imperial eyes passively look out and possess."[8]

In this context, the epistemological question, in particular, is vexing. The shaping power of western languages, religions, beliefs, customs, and practices mark all early modern travel accounts of the other as crucially derivative and always already framed by prior generic and cultural expectations. As Gayatri Spivak and others have argued,[9] the western traveler, because of his subject position, is placed at a structural (and absolute) remove from the object of his interest—the other. The travel *writer* calls on language to bridge the cultural divide, to explain the other. But that very instrument—language—is also an essential reason that the distance exists in the first place. It is both the source of the "problem," and its "solution."[10] Thus, *seeing, witnessing,* and *knowing* all usually involve varying forms of *misrecognition,* whereby Europeans struggle to claim a coherent subjectivity, mostly in opposition to varying manifestations of otherness.

Overall, while acknowledging the long, retrospective reach of the colonial imaginary—whether it defines the period between 1500 and 1720 as proto-colonial or early colonial—the essays in this volume do not read these travel narratives as expressions of a unified Eurocentrism. This term tends to erase or downplay both the notable cultural, religious, political, and economic differences that existed between the English, the Spanish, the Dutch, and the Portuguese, as well as the cultural and geographical differences of the places to which they traveled. It also glosses over the different colonizing "styles" exhibited by these nations. What is more, while we understand that human subjects are ideologically produced (in the Althusserian sense), the term Eurocentrism nonetheless obscures the particular motives and circumstances of the travelers themselves. The Dutchman Jan Huyghen van Linschoten, for instance, sought commercial experience among the Portuguese colonizers on the East Indian coast; Edward Terry went as a chaplain on a commercial voyage to East India and later joined Sir Thomas Roe, King James's ambassador to the Mogul Court; Thomas Dallam went to deliver an organ to the Ottoman sultan, and Mary Wortley Montagu accompanied her husband, the English ambassador to Turkey. Their experiences, backgrounds, and reasons for traveling were diverse and find expression in both the form and content of their written records.

Therefore, it is not surprising that travel writing as a genre seems to be eclectic and all-encompassing in its scope and methodology, ranging (in this anthology) from Sir Henry King's poem to Terry's quasi-ethnography to Leo Africanus's descriptive geography to Montagu's (revised) epistles, and to

Camões's epic poem. In order to preserve historical specificity, therefore, we endorse the critique offered by Steve Clark of a strain in postcolonial criticism that propounds a "model of travel writing and empire [that] would insist that [European] texts promote, confirm, and lament the exercise of imperial power; and that this ideology pervades their representational practices at every level."[11] To do otherwise would be to erase the subtleties and manifold idiosyncrasies of the transculturation process itself. In fact, the epistemological and subjective struggles recorded in these essays show that the traveler can never fully realize the "dream of possession" since the travel text's descriptive and taxonomic frame is also marked by frustration of desire, whereby *knowing* and *possessing* lack certitude within shifting and often fantasmic frames of reference.[12]

II

Despite the discursive struggles emerging in these essays, what is common to the accounts of these and other travelers is their claim to provide *knowledge,* frequently premised on *pleasure,* to a readership back home, while in many cases legitimating and promoting English/European mercantile expansion and cultural/religious influence. Travel, as we claim in our title, was indeed frequently a mode of knowledge-gathering based on the promise of an access to and understanding of new lands, people, and cultures. It has also been observed that European travel knowledge served a practical end in providing necessary information to subsequent waves of merchants and proto-colonists; according to Sir Thomas Palmer in *How to Make Our Travailes Profitable* (1606), for instance, the aim of every traveler should be to get "knowledge for the bettering of himselfe and his Countrie."[13] Examples of this abound. In the case of East India, as has often been noted, descriptive, ethnographic narratives, mostly by those associated with the East India Company, revealed the early forms of economic and commercial penetration into Mogul territories. And, retrospectively, these accounts are useful to historians charting the beginnings of British colonization of India. The travel records of the seventeenth and eighteenth centuries, as Bernard Cohn points out, "reflect the [East India] Company's central concerns with trade and commerce. [Thus, in these sources] we can trace the changes in forms of knowledge which the conquerors defined as useful to their own ends."[14]

In another example, an address "To the Reader," prefixed to the 1598 English translation of van Linschoten's *Itinerario,* boasts that "auncient Trauellers had indeed a certain kind of knowledge of this *Countrey* [the Indies] and *People;* but it was very vncertaine and vnperfect: Whereas we in our times are thoroughly learned and instructed by our owne experience of the *Prouinces, Cities, Riuers, Hauens,* and *Traffficks* of them all: So that nowe it is become knowne to the whole world."[15] How, according to the author of the address, will the *Itinerario* deliver on this promise of knowledge and firsthand experience? It tells of the "The *Voyage* of *Iourney* by sea of the sayde *Hugh Linschote* the *Author,* into the *East* or *Portingall Indies,* together with the *Sea-Coasts, Hauens, Riuers* and *Creekes* of the same, their *Customes* and *Religion,* their *Policie* and *Gouernment,* their *Marchandises, Drugges, Spyces, Hearbs* and *Plants.*"[16] All these things have been recorded by van Linschoten who, most "diligently and considerately obserue[d] and collect[ed] together all occurrents and accidents that happened in his memory and knowledge," and which brought great "benefit" to van Linschoten's native Netherlands.

And why translate the *Itinerario* into English? Because this was recommended by "Maister *Richard Hackluyt,* a man that laboureth greatly to aduance our English Name and Nation." And what constitutes such "aduance"? What are the translator's (and publisher's) hopes for this book?

> *I doo most hartely pray and wish, that this poore Translation may worke in our English Nation a further desire and increase of Honour ouer all Countreys of the World, and as it hath hitherto mightily aduanced the Credite of the Realm by defending the same with our Wodden Walles (as Themistocles called the Ships of Athens): So it would employ the same forraine partes, as well for the dispersing and planting true Religion and Ciuill Conuersation therein: As also for the further benefite and commodity of this Land by exportation of such thinges wherein we doe abound, and importation of those Necessities whereof we stand in Neede: as Hercules did, when hee fetched away the Golden Apples out of the Garden of the Hesperides; and Iason, when with his lustie troupe of courageous Argonautes hee atchieued the Golden Fleece in Colchos.*[17]

There is no space here to tease out the manifold implications of linking English exploration to instances of mythological "theft." All we wish to stress here is the translator's obvious confidence in the capacity of travel narratives to provide knowledge that would promote English interests and influence, as well as create opportunities to spread the true religion (Oumelbanine Zhiri's contribution to this volume confronts the translation issue head-on in her treatment of Leo Africanus's *Description of Africa*). Overall, as most of the essays reveal, cross-cultural and commercial exchanges fed off of one another. Thus, repeatedly and in different forms travel writings reveal complex interactions between the categories of *knowledge, pleasure, profit, and national pride.* It is not accidental that the formation of joint stock-trading companies like the Levant Company (1581) and the East India Company (1600) coincided with the profusion of travel writing. A testimony to this relationship can be found in Hakluyt's *The Principall Navigations* (1589-1600), a project, which scholars note, performed both an ideological and practical function: in promoting a newly emerging nationalism—in which Islam and other creeds helped define "Englishness"—as well as in providing valuable local information for merchants planning commercial ventures.[18]

Travel knowledge, however, is hardly unmediated insofar as it is shaped by political factors,[19] subject to authorial intervention, and plagued by general epistemological problems that attend the movement of information from one culture to another. This does not, in our view, mean that there is no possibility of knowledge at all, but it does mean that we have to understand "knowledge" as a culturally specific and ideologically charged category that relies heavily on "translating" "otherness" in terms that are accessible, in the case of travel literature, to an audience back home. Or, to adapt Hayden White's analysis of the historian's practice, the travel writer makes sense of—"encodes"—his experiences much as a "poet or novelist" does.[20] The travel writer endows "what originally appears to be problematical and mysterious with the aspect of a recognizable, because it is a familiar, form. It does not matter whether the world is conceived to be real or only imagined; the manner of making sense of it is the same."[21] Knowledge, we might say, is what they thought they knew.

This "literary" quality of travel writing is generally linked inextricably, though unevenly, with pleasure and knowledge. The pursuit of pleasure had a concrete impact on the actual form of the narrative and the amount and kinds of "knowledge" it could contain. In 1625, for instance, Samuel Purchas, the important anthologizer of travel accounts, sums up the entertainment goals in his preface, "for those who cannot travel farre I offer a world of travellers for their domestic entertainment."[22] Furthermore, he admits that "vast volumes" of information are "contracted and epitomized," so the "nicer reader might not be cloyed."[23] Purchas's admission typifies many moments in the vast European travel archive when the authors themselves intentionally or inadvertently let slip the *strategies* of representation that go into the *making* of their text, such as embellishment, compression, censorship, editorializing—all aimed at pleasing and edifying their audiences. Shankar Raman's analysis of the relation between historiography and epic in this volume implicitly reveals how pleasure becomes an important motive when modulating history into myth (epic). Thus, overall, as the early modern European travelers record their cultural encounters in alien landscapes—as we emphasize in our selections of primary texts—their representational practices seem to be Janus-like—assuming a transparency of representation and reliability of information even while frequently revealing their own *derivative* and *self-reflexive* attributes.

III

In various ways, the essays in this volume speak to the issues raised so far. The contributions that deal with English journeys to the Levant (with which we open this volume) share, aside from a focus on English encounters with Turks, Islam, and the Ottoman Empire, a strong sense that Englishness and English identity were shaped in crucial ways by these encounters. While English travel narratives describing the Levant offered readers a thrilling and voyeuristic view of the exotic other, their concrete importance lies in the knowledge they contained that aided both in the development of foreign trade and the construction of an English identity. These essays (in varying degrees and forms) find points of convergence with the work of Richard Helgerson, who he demonstrates that the accounts of Englishmen's experiences in the Ottoman Empire bear witness to new forms of knowledge and to emergent identity formation in their varying responses to non-European Others.

Daniel Vitkus's essay on English travelers in the Ottoman Empire reveals travel writing as an early modern genre that places little stress on the importance of converting the infidel (which since medieval times had dwindled into a stale fantasy) or on imperialist confrontation, and instead emphasizes commercial exchange above all else. Like most of the authors in this volume, Vitkus suggests that we can learn about early modern Englishmen by scrutinizing their commercial, religious, and cultural contacts with the other. He tracks both the strange attraction and the strong repulsion the English felt when coming into contact with Turkish culture. George Sandys, for instance, who was gentleman of the privy chamber at the court of Charles I, traveled in the highest circles in the Ottoman Empire, and his narrative details the sultan's court and harem. Vitkus locates Sandys as an English subject in relation to the disciplinary and masculine order of the court, as well as to effeminate luxury, hidden sexuality, and sexual excess of the sultan's harem. In Sandys's narrative, Vitkus discerns a superior tone that thinly veils a deeper sense of imperial envy, sometimes exacerbated by an anxious feeling of competition with the other Europeans in the Levant.

By contrast, William Lithgow, a Scotsman who traveled widely (by his own estimation more than 36,000 miles—mostly on foot) and who committed his remarkable experiences to paper in his *Totall Discourse of the rare Adventures . . . to the most Famous Kingdomes in Europe, Asia and Affrica* (1632), moved in less exalted circles. Although Lithgow finds a great deal to admire in the holy sites to which he travels and the people he meets there, it is clear that his commitment to English Protestantism prevents him from meeting them on their own terms. At one point, Vitkus argues in this volume, Lithgow blames the heresies of Roman Catholicism and Levantine Christianity on corruptions brought on by the Turks. Indeed, Lithgow's narrative is "a Protestant justification of faith, measured against the alleged folly, sensuality and crookedness of all other religions" (Vitkus essay).

Mary Fuller's essay examines two autobiographical narratives—an unpublished diary kept by the organmaker Thomas Dallam, and a twice-printed account by a mariner named John Rawlins—and follows their complex journeys into and away from the Islamic world. The circumstances of their travel differ radically: Dallam, in 1599, spent several months at the court of Mahomet III in Istanbul, where he accompanied a mechanical organ sent as a gift to the sultan by Queen Elizabeth. Twenty years later, Rawlins was taken prisoner by corsairs and sold as a slave to "Turks" who were English converts. Despite the differing trajectories of their entry into the Ottoman/Islamic world, Fuller shows us, via several local encounters, that "both men possessed skills which made them desirable subjects for recruitment and assimilation." Their narratives open perspectives into the metropolitan center and margins of the Ottoman Empire, and onto a landscape peopled with "converts." As Fuller reminds us, the "voluntary conversion of Englishmen to Islam was not uncommon," the possibility of conversion looms large over the narratives, even though both narrators reject that path in the end.

Assimilation and conversion, whether voluntary or forced, in the early modern Islamic world are represented in eroticized terms by both Dallam and Rawlins: the male body becomes the recipient of attention from other men, ranging along a continuum from the sexual predation on boys alleged by Rawlins to the affectionate courtship of Dallam by his guards. As both narratives make clear, the zone of contacts and attempts at recruitment were charged with a range of homoerotic energies; in it, Englishmen experienced themselves as both desiring and desired, whatever their eventual responses to the seductions—benign or coercive—of the Islamic world.

Gerald MacLean's essay is explicitly concerned with the process of self-construction and self-representation that is revealed in such English travel narratives as Henry Blount's *Voyage in the Levant* (1636). Blount, MacLean argues, learned from observing the Turks how to imagine the British Empire into being. "Ottomanism"—a set of tropes used by the English to make the other knowable—became a mode of knowledge production that, driven by both lack and desire, intertwined commercial and cultural interests into a strategy of engagement that tells us rather more about the desiring subject than about the object of knowledge. MacLean points out, however, that this idea of Ottomanism precedes and differs from what has become known as Orientalism. Without a doubt, both terms describe an activity or mode of thought engaged in by Westerners for the purpose of defining, shaping, and exerting a measure of control over a newly discovered region of the world. Yet, the early English travelers—those who came before Mary Wortley Montagu—who had an Ottomanist outlook, never failed to understand that the Ottoman Empire existed, and that it had already achieved the great power and fantastic wealth of which the English were only just beginning to dream. Orientalism, on the other hand, was a product of a later historical moment—a moment subsequent to Islamic imperialism during which western power was in ascendancy, and that "set about understanding and eventually controlling those cultures and nations by designating them the 'Orient.'" MacLean is especially interested in the transition from Ottomanism to Orientalism: How, he asks, might Ottomanist discourse be said to have produced an imperial subject such as Lady Mary Wortley Montagu?

Rebecca Chung follows Lady Mary Wortley Montagu's journey to the Ottoman Empire via Montagu's nonfiction prose account, based on her letters, entitled *Travels of an English Lady in Europe, Asia, and Africa* (1724). Since travel writings generally encode the traveler as male, mainly because men traveled more and left more records of their journeys, Lady Mary's account has great historical significance. Women's access to the more explicitly feminine spaces of the *hamam* or bath, for instance, and their interactions within a woman-only environment produce very different gender inflections from those found in male accounts. Chung's essay is sensitive to these gender inflections, not only in Montagu's actual experiences in Turkey but also in the reception to and interpretations of her work, especially by some male readers/critics. At issue in this essay is Montagu's visit to the Turkish baths and the extent to which she undressed before the native women, which Chung examines via some confusions in the textual history of the words *skirt* or *shirt*. To whatever extent she undressed, the Turkish ladies were shocked to see her corset, which made them believe that she was "lock'd up" in the "contrivance they attributed to [her] husband." But Montagu herself was equally shocked to see uncorseted women who drank coffee and conversed unselfconsciously in public even though they were naked. This genuine cross-cultural shock is the most repressed moment in her representation of the Turkish bath scene because it exposes to Montagu herself how patriarchal sexual fantasies function as a rationalization for the English patriarchal practices of corseting and isolating its women. Thus, following Chung's complex account into the textual history of this episode, we get a glimpse of how European sexual and cultural stereotypes about Turks and vice versa come into play. What we witness here is an encounter in which Europeans and Turks confront sexual and cultural stereotypes about each other, and use the moment of contact to consider what constitutes female freedom and agency in a cross-cultural contact.

Shankar Raman explores the very early stages of "colonialism," namely, Vasco da Gama's voyage to India (1497-1498), by historicizing one of the earliest literary works to deal with the Portuguese colonial voyages eastward: Luis de Vaz de Camões's famous epic, *Os Lusíadas* (1572). While reflecting on how Camões tries to balance the "generic demands of the classical epic" with the "burden of recounting a contemporaneous history," Raman compares the epic narrative to Lopes de Castanheda's *História* (1551-1554), which ostensibly narrates the same history of the conquest of India. Casting this comparison in the context of the Catholic Counter-Reformation, Raman's essay underscores how Camões's epic revises Castanheda's account to strike a blow against the idolatry that follows the misinterpretation of false icons. In presenting two different versions of a potentially idolatrous and dangerous encounter between the Portuguese voyagers and the natives in Mombasa, Raman stresses the epic's fixation with deception, as these early colonizers must differentiate "Us from Them, the European colonizers from those who are to be made their subjects."

Camões's concern in *Os Lusíadas* with distinguishing the "true representation from the counterfeit" reflects deeper tensions within Renaissance historiography over the problems of "origin and meaning." Raman examines the dynamic of this tension by further exploring the opposition between the original and counterfeit, this time as it involves a false prophecy found in another "travel" text—Abraham Ortelius's atlas, *Theatrum Orbis Terrarum* (1606); further comparing this moment to Camoes's treatment of King Manuel's true dream of Portuguese imperial enterprise in *Os Lusíadas,* Raman illustrates the process whereby the history of Portuguese colonial expansion could be rewritten as prophecy and myth (epic).

Ivo Kamps examines portions of the earliest Dutch narrative account of travel to the East, Jan Huyghen van Linschoten's *Itinerario* (1596). Van Linschoten, a Catholic from the Protestant town of Enkhuizen, and a citizen of a country engaged in an exhausting 80-year war with Spain, boldly traveled to Spain and Portugal to seek his fortune. In Portugal he entered the employ of João Vincente da Fonseca, who was appointed Archbishop of the Indies by Philip II of Spain. Van Linschoten traveled with the Archbishop to Goa, India, where he was uniquely positioned to scrutinize the Portuguese colonizers and to document (for subsequent publication) their practices and weaknesses. Van Linschoten's account of the Portuguese colonizers itself can therefore be read as a colonizing act—as an act that was of course about to be repeated by the Dutch, who were about to displace the Portuguese as the dominant colonizing power in India and the surrounding areas. A witness to economic, cultural, and biological hybridity in *Asia Portuguesa,* van Linschoten appears convinced that ideally a colonizing act should be a Manichean act that insists on radical differences between colonizer and colonized. The failure of the Portuguese to maintain strict differences, the *Itinerario* proffers, hastens their decline. So might well have, incidentally, the publication of the *Itinerario.*

Jyotsna Singh's essay follows the trajectory of Edward Terry's travels to India where he stayed from 1616 to 1619 as a chaplain to Sir Thomas Roe, King James's ambassador to the Court of the Great Mogul, Jehangir. This essay examines the way in which moral and religious concerns shape his detailed quasi-ethnographic narrative, *A Voyage to East-India (1655 & 1666),* and his *Sermon* (1649) to the East India Company, while legitimating a providential view of English commercial and proto-colonial expansion. While Terry's account offers fairly accurate information about Mogul India, it also demonstrates familiar representational practices of early modern European travel writing; most notably, it emphasizes eyewitness authenticity and objectivity, while falling back on familiar proto-colonial binaries: Christianity versus barbarism, for instance, as a way of distinguishing between the English and the Indian natives.

Thus, typically, Terry's detailed descriptions of the geographical features, social and religious customs, and natural resources and commodities, among other things, are permeated by religious and moral homilies. It is obvious that Protestant Christianity is the prism through which the chaplain interprets the world he encounters. While his classificatory impulses bear the marks of a characteristically European colonizing imagination, his narrative also reveals a discursive and ideological uncertainty when he is faced with the grandeur and power of the Mogul King. Thus, *A Voyage to East-India* in many ways can be considered a fairly representative text in the early modern scene of travel writing, reflecting all the tensions and contradictions of European "discoveries" in that period.

Shifting continents, Gary Taylor's essay reconstructs for us the first known performance of *Hamlet* in Africa, in Sierra Leone in the year 1607. For us at the dawn of the twenty-first century, there is no such thing as a pristine *Hamlet*—a *Hamlet* that is not inextricably woven into centuries of criticism, theatrical and cinematic performances, classroom instruction, commercialism, and widespread cultural veneration if not fetishization. But Taylor wants to take us back to a moment when the play must have been viewed with a sense of astonishment and, for the African members of the audience, a likely sense of unmitigated bewilderment. Because of the blurring effects of historical distance and the relative paucity of the historical record, Taylor admits that this performance can only take place in the mind's eye; but these impediments, Taylor's essay amply demonstrates, need not bar us from a seat in the audience (albeit in the back row where our view might at times be less than perfect) on the deck of the English merchant vessel *Red Dragon* as crew members stage the tragedy of Denmark's prince. Taylor wonders—and tries to reconstruct—how the play might have struck the four Africans in attendance, who heard the play's language mediated through a running translation into Portuguese, and perhaps another translation into Temne, the local African language. The primary text included by Gary Taylor is the diary of John Hearne and William Finch, two merchants aboard the *Red Dragon.*

The issue of translation (both of translating one language into another and of one culture into another) is of central concern in Oumelbanine Zhiri's essay on Leo Africanus. If Gary Taylor's essay meditates on the textual traces of Shakespeare's arrival in Africa, Oumelbanine Zhiri's essay tells the story of how Leo Africanus's *Description of Africa* transmitted some of the earliest images of Africa to a largely uninformed western European audience. However, Zhiri stresses that this act of transmission was, in demonstrable ways, hardly an unmediated event. In fact, it resembles more closely a translation—and a flawed one at that—of one culture into a mode of expression accessible to another, than a direct transmission. Part of the story of Leo Africanus's "transmission" of things African to European readers lies in the irreducible presence of the author's own complex personal history in his *Description:* Leo was an Arab who, while living in Europe, wrote his text about his time in Africa in less than flawless Italian for western readers. Even though, as Zhiri points out, Leo's limited knowledge of Italian and his desire to have Europeans understand African customs and laws (which led him to "translate" ideas) gave rise to significant misrepresentations, it is significant that the author was acutely aware of how the very nature of his project and his unique subject position affected his text and the knowledge it sought to transmit. Zhiri rightly suggests, therefore, that Leo's highly overdetermined text can be studied as a densely woven metaphor for a certain trend in cultural studies today, a trend that is increasingly inclined to search out connections between local cultural events and their globalization. Leo's text stands as an early instance of a self-conscious, multicultural textual production.

IV

Necessarily, the primary texts and essays collected here draw our attention to the historical and geographical *distance* between the events described in them and ourselves and our historical moment. But we do not want that distance to mask the palpable connections that many of us, including the editors and authors of this volume, have to the events depicted in these travel accounts. Our diverse group of contributors bring to this volume their rich experiences drawn from their wide-ranging journeys and locations: from Egypt, Britain, Morocco, India, and the United States. Given the complex and broad range of experiences faced by all of us and by our subjects, the early modern European travelers, it seems that the very contemporary themes of travel, disapora, border crossings, and self-location have been with us a very long time.[24]

I (Ivo Kamps) for one see my work on van Linschoten in part as a journey that returns me to a phase in my family history that has been neglected and even suppressed. Growing up in the Netherlands, I always considered myself Dutch (and white), but van Linschoten's strong urging that the Dutch government establish an economic presence on the Indonesian isle of Java may well have made it possible for a German named Abraham Israël to settle on Java in 1807 and have several children with a native woman named Klawit.[25] My maternal grandfather is one of their many descendents. When I look at an old photograph of my grandfather, I don't see a Dutchman but a distinctly Indonesian gentleman in a smart, white, tropical uniform. He was a translator of Javanese languages in the government translator's office in Soerakarta. My Dutch grandmother—the Swiss-boarding-school educated daughter of a wealthy manufacturer of carriages, cars, and eventually airplanes—was evidently disinherited for marrying this man of color. After living on Java together for two years, the couple was pressured by the Dutch side of the family to "come home." There may have been a fear that my grandmother, like too many Dutch people who lived in Indonesia, would "go native," an apprehension perhaps intensified by the example of her uncle who spend long stretches of time on Java, where he lived with a native woman.

It seems that my family's anxious insistence that my grandmother return to her Dutch roots was an attempt to reverse the colonial trajectory that sends white people away to represent the colonizing nation, only to "lose" them to cultural and biological miscegenation. In Amsterdam, my grandfather became the personal secretary of a rich and racist aunt, and was kept on a short financial leash. There was nothing Indonesian about the brick house in which my grandparents lived. The portraits on the walls were of Dutch and Norwegian ancestors. The furniture and art objects were mostly European, as was the cooking. As an aunt of mine used to say, our family heritage is Dutch colonial; she never said Dutch-Indonesian. This kind of race-erasure works. As I said, I always thought (and still think) of myself as Dutch. Race, when it was acknowledged in my childhood, was admitted to as a historical fact but never as a racial reality: Indonesian background, white race. The only exception was during summer holidays in Austria and Switzerland when my mother would explain to me that my skin tanned easily because of my Indonesian blood. So much for race. After working on this book, I regret to admit that I don't *feel* any more Indonesian than I did before, but I'm more aware that I *am* part Indonesian.

Until I (Jyotsna Singh) came to the United States, to go to graduate school in the early 1980s, I was a typical naturalized product of the aftermath of colonialism in India. While being required to study the British romantic poets in great detail during my undergraduate education in India, I never questioned the relevance of reading poetry about the Lake District and daffodils, the skylark, and the myth of King Arthur, among other subjects. I had never seen daffodils in India! Neither did I wonder why we needed to continue a study of English canonical texts and how they had been a legacy of empire. It was only on leaving home that I could understand my own cultural formation and why, for instance, my father—a product of colonial education—utterly venerates T. S. Eliot and Shakespeare.

Today, as a result of the rich, though often contentious, cultural struggles of the past decade—both within the Anglo-American academy and in Indian intellectual and social life—we are more aware of ourselves as products of "the historical process to date," which has deposited in us "an infinity of traces, without leaving an inventory."[26] To inventory the "traces" of English culture in my personal formation as a postcolonial subject would be a complicated task. But since this anthology deals with some of the earliest English travelers to the east, I want to locate myself imaginatively within the long shadow of their former presence. For me, the British presence in India literally meant "their ghosts," which, I used to believe, haunted the crumbling cemeteries and elegant old bungalows in the Indian Himalayan towns of my childhood: Simla, Mussoorie, and Kasauli. The British names—or mostly Irish and Scottish names—on gray and mold-covered gravestones typically told stories of military and bureaucratic service, and often of premature deaths of young women and children.

Thus, while watching the Raj nostalgia in the 1980s, my responses were mixed: on one hand, I imagined the world in which my parents (in their youth) and my grandparents lived under British presence; but on the other hand, I also felt a nostalgia for the colonial architecture and the lingering English manners and customs that have left traces in my own socialization. I recalled my grandparents' impressions of colonial administrators as being "fair," but also of British social arrogance, which was evident in the signs on certain streets that proclaimed "Indians and dogs not allowed." I also remembered my father's narrative of joining the Indian independence movement in its last stages, even while he greatly admired British values and culture. Writing (in the twenty-first century) about Edward Terry's travels to India in the early seventeenth century revives memories of the crumbling British cemeteries in India—where he could have been buried if he had died there (as so many did), in the tropical land. But he, as it were, "lived to tell the tale"—a tale that gradually became a story of domination of one people over another—and I am left to struggle with my (perhaps?) misplaced nostalgia!

But regardless of the "personal" connections many of us have to the issues discussed in this book, there is room to argue—as some have done—that this kind of criticism does not really seek to give the oppressed of history a voice but is "in fact complicit in the disposition and operations of the current, neocolonial world order."[27] It is certainly true that with the exception of Leo Africanus's *Description of Africa* all the primary materials included in this volume were written by Europeans, and that the alien voices we hear in them are ventriloquized and therefore inevitably muffled and distorted. Gary Taylor's essay is the only one that in a sense "reverses the gaze" of the colonizers, while revealing non-European reactions to a European self-representation. Overall, unlike early modern travel writers, at least we are able to recognize that the voices are muffled and distorted, and that we have to listen with the utmost attention to hear them at all. Just as Kamps argues in his essay that the Dutchman van Linschoten finds a way to colonize the Portuguese colonizers in India (and is able to do so by virtue of the *differences* between their respective colonizing perspectives), so do the essayists here "colonize" the writings of early modern colonizers, but do so—crucially—from a significantly different ideological vantage point than did the early modern travelers, allowing for readings that the travel writers could hardly have imagined themselves. While that may not get us out of the "appropriation" conundrum, it seems to us far preferable to the utter silence on these matters that characterized American academia for most of the last century.

That said, while working on this volume it became increasingly clear to its editors that literary and cultural studies scholars in the West need to follow the example of anthropologists and do more to retrieve the voices of indigenous people who lived in the precolonial era and who gave expression to their experiences and observations in writing, in art, or otherwise. Only if we grant these voices an effective forum can we begin to paint a more satisfying picture on the canvas of precolonial transculteration. Not that these indigenous voices will bring us the unmitigated historical truth but they may give us the traces of a record of the lived relationship to the real (to borrow another phrase from Althusser) that at one point in history passed for reality. Without that other version of the real, our analysis of precolonial history will ultimately remain an analysis of our own history—which, as we all know, is only a part of the story. As Raja Rao has written, "There is no village in India, however mean, that has not a rich *sthalapurana,* or legendary history, of its own. . . . [A] grandmother might have told you, newcomer, the sad tale of her village."[28]

Notes

1. The poem prefaces *A Voyage to East-India* (1655 edition) by Edward Terry. In it the writer praises Terry's account for its moralizing vision, that will "please and profit the virtuous mind."
2. Edward Said's *Orientalism* (New York: Vintage Books, 1979) acknowledges the gradual development of colonization, noting the "period of extraordinary European ascendancy from the late Renaissance to the present" (7), but his focus is mainly on the period of high imperialism from the end of the eighteenth century to the early twentieth century. More importantly, he shows that colonization was not achieved by mere military might, but rather by a set of discursive practices, which he defined as "Orientalism." This term describes a process of signification in which western observers and authors of literary and non-literary texts construct images—often stereotypes—of eastern and Middle Eastern peoples. Travel books of the period, according to Said, are among the significant genres that constitute "colonial discourse."
3. In *Imperial Eyes: Travel Writing and Transculturation* (London: Routledge, 1992)—a study that spans from 1750 to the mid-1980s—Mary Louise Pratt considers travel accounts in terms of the "space of colonial encounters"(6). See pp. 1-11 for a detailed conceptualization of her approach to European travel writing. We use her term "transculturation" in a more generalized sense to refer to varied aspects of cross-cultural encounters.
4. We use the term "colonial discourse" as drawn from David Spurr, *The Rhetoric of Empire: Colonial Discourse in Journalism, Travel Writing, and Imperial Administration* (Durham: Duke University Press, 1993), p. 7. For a full discussion of his analogy between colonizing and writing, see pp. 1-12.
5. See Spurr, *Rhetoric of Empire,* p. 7.
6. See Richard Fardon, "Introduction," in *Localizing Strategies: Regional Traditions of Ethnographic Writing,* ed. Fardon (Edinburgh: Scottish Academic Press, 1990), p. 6.
7. One could go so far as to argue, as Indira Ghose does, that "the transcendent traveler's gaze *is* the colonial gaze. . . . [T]he opposition between self and other set up in travel writing is, like all binary oppositions, grounded on an implicit (and sometimes explicit) hierarchy" (emphasis added [*Women Travellers in Colonial India: The Power of the Female Gaze* (Delhi: Oxford University Press, 1998), p. 9]). On this question, also see Kamps's essay in this volume.
8. The first observation is a part of Stephen Greenblatt's detailed commentary on "witnessing" in *Marvelous Possessions* (Chicago: University of Chicago Press, 1991), p. 122. The concept of the European traveler as a "seeing-man" is drawn from Mary Louise Pratt's *Imperial Eyes Travel Writing and Transculturation,* pp. 7, 78. Pratt's book about "witnessing" applies more explicitly to post–eighteenth-century European travelers whose production of information under the auspices of natural science was associated with "panoptic apparatuses of the bureaucratic state" (78).
9. Gayatri Chakravorty Spivak, "Can the Subaltern Speak?" in *Marxism and the Interpretation of Culture,* ed. Cary Nelson and Lawrence Grossberg (London: Macmillan, 1988).
10. Greenblatt, in *Marvellous Possessions,* emphasizes the role of language in the voyages of discovery and conquest, showing how the "early modern discourse of discovery . . . is a superbly powerful register of the characteristic claims and limits of European representational practice" (23). Also see Mary Fuller's *Voyages in Print, English Travel to America, 1576-1624* (Cambridge: Cambridge University Press) who argues for the centrality of print culture to early modern travel:

"The voyage narratives came into being not only as after-the-fact accounts for ideological purposes, but as an integral part of the activities they documented" (2).

11. See the "Introduction" by Steve Clark in *Travel Writing and Empire: Postcolonial Theory in Transit* (London and New York: Zed Books, 1999), pp. 1-28.
12. For a suggestive discussion of a traveler "wavering between two worlds": one world produced by a concrete record of possession and the other filled with ghosts of "frustration and derealization," see Brian Musgrove's "Travel and Unsettlement: Freud on Vacation," in *Travel Writing and Empire*, pp. 38-41. Musgrove's reading is based on Megan Morris's essay, "Panorama: The Live, the Dead, and the Living," in *Islands in the Stream: Myths of Place in Australian Culture*, ed. Paul Foss (Sydney: Pluto, 1988), pp. 160-87.
13. For a further discussion of Palmer's work and the dual motives of pleasure and profit that led to the gathering of information about cultural Others, see Vitkus's essay in this volume.
14. For a detailed account of the links between travel knowledge and mercantile and colonial expansion of the English in India, see Bernard Cohn, "The Command of Language and the Language of Command," in *Subaltern Studies: Writings on South Asian History and Society*, vol. 4, ed. Ranajit Guha (Oxford: Oxford University Press, 1985), pp. 276-329. Also relevant here is Percival Spear's study of the decline of Mogul power and the expansion of the East India Company's power in *A History of India*, vol. 2., rev. ed. (London: Penguin, 1978), pp. 61-92.
15. John Wolfe (?), "To the Reader," in *The Voyage of John Huyghen van Linschoten to the East Indies* (1598), vol. 1, ed. Arthur Coke Burnell (London: Hakluyt Society, 1885), p. l
16. Wolfe (?), "To the Reader," p. li.
17. Wolfe (?), "To the Reader," p. lii.
18. See David B. Quinn, ed., *The Hakluyt Handbook* (London: Hakluyt Society, 1974), and Richard Helgerson, *Forms of Nationhood: The Elizabethan Writing of England* (Chicago: University of Chicago Press, 1992), pp. 149-87. Vitkus's essay in this volume also discusses the commercial and ideological goals of travelers/merchants. Also note MacLean's essay, which describes the nationalist and frequently proto-colonialist narratives and self-representations of British visitors to the Ottoman empire from the late sixteenth century to the early eighteenth century (1720).
19. As noted by Andrew Hadfield, "Travel and colonial writings were undoubtedly political genres in a double sense. First, as I have suggested, they often possessed a political content. Second, they were frequently caught up in the turbulent political history within which they were produced" (*Literature, Travel, and Colonial Writing in the English Renaissance, 1545–1625* [Oxford: Clarendon Press, 1998], p. 2).
20. Hayden White, *Tropics of Discourse: Essays in Cultural Criticism* (Baltimore: Johns Hopkins University. Press, 1978), p. 98.
21. White, *Tropics*, p. 98.
22. For Purchas's definition of the genre, see his preface "To the Reader," in *Hakluytus Posthumous or Purchas His Pilgrimes: Contayning a History of the World in Sea Voyages and Lande Travells by Englishmen and others* (1625), vol. 1. (Glasgow: James MacLehose and Sons, 1905), pp. xxxix-xlvii.
23. Here, it is important to note Purchas's justification of editorial changes to please a potential readership.
24. We are indebted for these terms to James Clifford's *Routes: Travel and Translation in the Late Twentieth Century* (Cambridge: Harvard University Press, 1997).
25. Han van der Lelie, *Carel Theodore Israël and Clara Petronella van der Zee. Genealogy*, privately published in the Netherlands, 1994 (revised 1997), p. 7.
26. Edward Said, in *Orientalism*, richly delineates his attempt to "inventory the traces" of western influences in his life, pp.25-26. I am indebted to his model of analysis.
27. Bart Moore-Gilbert, *Postcolonial Theory: Contexts, Practices, Politics* (New York: Verso, 1997), p. 153. Moore-Gilbert does not himself make this claim. See Aijaz Ahmad, *In Theory: Classes, Nations, Literatures* (New York: Verso, 1992).
28. Raja Rao, "Author's Foreword," *Kanthapura* (1938) (Bombay: New Directions, 1963).

LIST OF SUGGESTED READINGS

Ashcroft, Bill, Gareth Griffith, and Helen Tiffin, eds. *The Post-Colonial Studies Reader.* New York: Routledge, 1995.

Bhabha, Homi. *Nation and Narration.* New York: Routledge, 1990.

———. *The Location of Culture.* New York: Routledge, 1994.

Brenner, Robert. *Merchants and Revolution: Commercial Exchange, Political Conflict and London's Overseas Traders, 1550-1653.* Princeton: Princeton University Press, 1993.

Chard, Chloe. *Pleasure and Guilt on the Grand Tour: Travel Writing and Imaginative Geography, 1600–1830.* New York: Manchester University Press, 1999.

Chaudhuri, K.N. *Asia before Europe: Economy and Civilization of the Indian Ocean from the Rise of Islam to 1750.* Cambridge: Cambridge University Press, 1990.

Clark, Steve, ed. *Travel Writing and Empire: Postcolonial Theory in Transit.* London and New York: Zed Books, 1999.

Clifford, James. *Routes: Travel and Translation in the Late Twentieth Century.* Cambridge: Cambridge University Press, 1997.

Daunton, Martin, and Rick Halpern, eds. *Empire and Others: British Encounters with Indigenous Peoples, 1600-1850.* Philadelphia: University of Pennsylvania Press, 1999.

Fuller, Mary C. *Voyages in Print: English Travel to America 1576–1624.* Cambridge: Cambridge University Press, 1995.

Ghose, Indira. *Women Travellers in Colonial India: The Power of the Female Gaze.* Delhi: Oxford University Press, 1998.

Greenblatt, Stephen, ed. *Marvelous Possessions: The Wonder of the New World.* Chicago: University of Chicago Press, 1991.

———. *New World Encounters.* Berkeley: University of California Press, 1993.

Grossberg, Lawrence, Cary Nelson, and Paula Treichler, eds. *Cultural Studies.* New York: Routledge, 1992.

Hadfield, Andrew. *Literature, Travel, and Colonial Writing in the English Renaissance, 1524-1625.* Oxford: Clarendon Press, 1998.

Helgerson, Richard. *Forms of Nationhood: The Elizabethan Writing of England.* Chicago: University of Chicago Press, 1992.

Lewis, Bernard. *Islam and the West.* Oxford: Oxford University Press, 1993.

Mater, Nabil. *Islam in Britain—1558-1685.* Cambridge: Cambridge University Press, 1998.

———. *Turks, Moors, and Englishmen in the Age of Discovery.* New York: Columbia University Press, 1999.

Mills, Sara. *Discourses of Difference: An Analysis of Women's Travel Writing and Colonialism.* London: Routledge, 1991.

Montrose, Louis. "The Work of Gender in the Discourse of Discovery." *New World Encounters.* Ed. Stephen Greenblatt. Berkeley: University of California Press, 1993: 177-217.

Nussbaum, Felicity. *Torrid Zones: Maternity, Sexuality, and Empire in Eighteenth-Century Narrative.* Baltimore: Johns Hopkins University Press, 1995.

Pagden, Anthony. *European Encounters with the New World: From Renaissance to Romantics.* New Haven: Yale University Press, 1993.

Peirce, Leslie Penn. *The Imperial Harem: Women and Sovereignty in the Ottoman Empire.* New York: Oxford University Press, 1993.

Pratt, Mary Louise. *Imperial Eyes: Travel Writing and Transculturation.* London: Routledge, 1992.

Said, Edward. *Orientalism.* New York: Vintage, 1979.

———. *Culture and Imperialism*. London: Chatto and Windus, 1993.

Singh, Jyotsna G. *Colonial Narratives/Cultural Dialogues: "Discoveries" of India in the Language of Colonialism*. London: Routledge, 1996.

Spurr, David. *The Rhetoric of Empire: Colonial Discourse in Journalism, Travel Writing, and Imperial Administration*. Durham, N.C.: Duke University Press, 1994.

Stratton, F. *African Literature and the Politics of Gender.* New York: Routledge, 1994.

Tiffin, Chris, and Alan Lawson, eds. *De-scribing Empire: Postcolonialism and Textuality.* London: Routledge, 1994.

Viswanathan, Gauri. *Masks of Conquest: Literary Study and British Rule in India.* New York: Columbia University Press, 1989.

Young, Robert J. C. *Colonial Desire: Hybridity in Theory, Culture, and Race.* London: Routledge, 1995.

PART I:

Travelers into the Levant

SECTION ONE

GEORGE SANDYS

George Sandys is the author of *A Relation of a Journey begun An. Dom. 1610.* The earliest extant version of this text was printed in 1615. Further editions appeared in 1621, 1627, 1637, 1652, and 1673. The book, which is elegantly illustrated with copperplate engraved maps and images, is an account of Sandys's exploits and observations during a trip that he began in 1610. His journal commences its narrative in Venice and describes Ionia, the Aegean, Smyrna, and Constantinople, where he stayed as the guest of the English ambassador, Sir Thomas Glover. He then traveled to Egypt and Palestine and returned to England via Italy and France, arriving home in 1611. Sandys, an aristocratic humanist and translator of Ovid, sprinkles his report with allusions to classical history, literature, and religion. He also has a great deal to say about Ottoman society, its customs, religion, wealth, and power. The excerpt included here, describing the Ottoman sultan Ahmet I and his seraglio, is taken from the 1615 text (George Sandys, *A Relation of a Journey begun An. Dom. 1610. Foure Bookes. Containing a description of the Turkish Empire, of Aegypt, of the Holy Land, of the Remote parts of Italy, and Ilands adjoyning* [London, 1615]).

—Daniel J. Vitkus

[A description of Sultan Ahmet I and his seraglio at Topkapi palace in Istanbul]

I have spoken sufficiently, at least what I can, of this Nation in generall: now convert we to the Person and Court of this Sultan.

He is, in this yeare 1610, about the age of three and twenty, strongly limd, & of a just stature, yet greatly inclining to be fat: insomuch as sometimes he is ready to choke as he feeds, and some do purposely attend to free him from that danger. His face is full and duely proportioned: onely his eyes are extraordinary great, by them esteemed (as is said before) an excellency in beauty. Fleame[1] hath the predominancy in his complexion. He hath a litle haire on his upper lip, but lesse on his chin, of a darksome colour. His aspect is as hauty as his Empire is large, he beginneth already to abstaine from exercise; yet are there pillars with inscriptions in his Serraglio, betweene which he threw a great iron mace, that memorise both his strength, and activity. Being on a time rebuked by his father Mahomet that he neglected so much his exercises and studies, he made this reply: that, Now he was too old to begin to learne; intimating thereby that his life was to determine with his fathers: whereat the Sultan wept bitterly. For he then had two elder brothers, of whom the eldest was strangled in the presence of his father upon a false suspition of treason; and the other by a naturall death did open his way to the Empire. Perhaps the consideration thereof hath made him keepe his younger brother alive, contrary to their cruell custome: yet strongly guarded, and kept within his Serraglio. For he is of no bloudy disposition, nor otherwise notoriously vicious, considering the austerity of that government, and immunities of their Religion.[2] Yet is he an unrelenting punisher of offences, even in his owne houshold: having caused eight of his Pages, at my being there, to be throwne into the Sea for Sodomy (an ordinary crime, if esteemed a crime, in that nation) in the night time; being let to know by the report of a Cannon that his will was fulfilled. Amongst whom, it was given out that the Vice-royes naturall sonne of Sicilia was one (a youth lately taken prisoner, and presented unto him) yet but so said to be, to dishearten such as should practise his escape. His valour rests yet untried, having made no warre but by disputation: nor is it thought that he greatly affects it: despairing of long life in regard of his corpulency. Whereupon he is now building a magnificent Mosque, for the health of his soule, all of white marble; at the East end, and South side of the Hippodrom; where he first broke the earth, and wrought three houres in person. The like did the Bassas:[3] bringing with them presents of money, and slaves to further the building. His occupation (for they are all tied to have one) is the making of ivory rings, which they weare on their thumbs when they shoote, whereupon he workes daily. His Turbant is like in shape to a pumpion,[4] but thrice as great. His under and upper garments are lightly of white sattin, or cloth of silver tishued with an eye of greene, and wrought in great branches. He hath not so few as foure thousand persons that feede and live within his Serraglio, besides Capagies[5] of whom there are five hundred attired like Janizaries,[6] but onely that they want the socket in the front of their bonnets,[7] who waite by fifties at every gate. The chiefe officers of his Court are the Maister (as we may terme him) of the Requests, the Treasurer, and Steward of his houshold, his Cupbearer, the Aga of the women,[8] the Controuler of the Jemoglans:[9] who also steereth his barge, and is the principall Gardiner. Divers of these Jemoglans marching before the Grand Signior[10] at solemne shewes, in a vaine ostentation of what they would undergo for their Lord, gathering up the skin of their temples to thrust quils through, and sticke therin feathers for a greater bravery: so wearing them to their no small trouble, untill the place putrifie; some when the old breaks out, cutting new holes close to the broken. Yea the standard-bearers of this crew thrust the staves sometimes of their standards through the skin and fat of their bellies, resting the lower end on a stirrop of leather and so beare them through the Citie. Fiftly, Mutes he hath borne deafe and dumbe, whereof some few be his daily companions; the rest are his Pages. It is a wonderfull thing to see how readily they can apprehend, and relate by signes, even matters of great difficulty. Not to speake of the multitude of Eunuchs, the footmen of his guard, cookes, sherbet-men, (who make the foresaid bevrage) gardeners, & horsekeepers. Relate we now of his women: wherein we will include those as well without as within his Serraglio.

And first begin we with his Virgins, of whom there seldome are so few as five hundred, kept in a Serraglio by themselves, and attended on onely by women, and Eunuchs. They all of them are his slaves; either taken in the warres, or from their Christian parents: and are indeede the choisest beauties of the Empire. They are not to be presented to the Emperour, untill certaine moneths be expired after their entrance, in which time they are purged and dieted, according to the custome of the ancient Persians. When it is his pleasure to have one, they stand ranckt[11] in a gallery; and she prepareth for his bed to whom he giveth his hankercher: who is delivered to the aforesaid Aga of the women (A Negro Eunuch) and conducted by him into the Sultans Serraglio. She that beareth him the first sonne is honoured with the title of Sultana. But for all his multitude of women, he hath yet begotten but two sonnes and three daughters, though he be that way unsatiably given, (perhaps the cause that he hath so few) and all sorts of foods that may inable performance. He cannot make a free woman his concubine: nor have to do with her whom he hath freed, unlesse he do marry her; it being well knowne to the wickedly witty Roxolana[12]: who pretending devotion, and desirous for the health, forsooth, of her soule to erect a Temple, with an hospitall, imparting her mind to the Mufti,[13] was told by him that it would not be acceptable to God, if built by a bondwoman. Whereupon she put on a habite of a counterfet sorrow; which possest the doting Solyman with such a compassion, that he forthwith gave her her freedome that she might pursue her intention. But having after a while sent for her by an Eunuch, she cunningly excused her not coming, as touched in conscience with the unlawfulnesse of the fact; now being free, and therefore not to consent unto his pleasure. So he whose soule did abide in her, and not able to live without her, was constrained to marry her. The onely marke that she aimed at, and whereon she grounded her succeeding tragedies. This also hath married his concubine, the mother of his yonger sonne, (she being dead by whom he had the eldest) who with all the practises of a politicke stepdame endevours to settle the succession on her owne, adding, as it is thought, the power of witch-craft to that of her beauty, she being passionately beloved of the Sultan. Yet is she called Casek Cadoun,[14] which is, the Lady without haire: by Nature her selfe, both graced, and shamed. Now when one Sultan dieth, all his women are carried into another Serraglio; where those remaine that were his predecessors: being there both strictly lookt unto, and liberally provided for. The Grand Signior not seldome bestowing some of them (as of his Virgins, and the women of his owne Serraglio) upon his great Bassas and others; which is accounted a principall honour. But for his daughters, sisters, and aunts they have the Bassas given them for their husbands: the Sultan saying thus, "Here sister, I give thee this man to thy slave, together with this dagger, that if he please thee not thou maist kill him." Their husbands come not unto them untill they be called: if but for speech onely, their shooes which they put off at the doore are there suffered to remaine: but if to lie with them, they are layd over the bed by an Eunuch, a signe for them to approch; who creepe in unto the at the beds feete. Mustapha and Hadir, (two of the Vizers of the Port)[15] have married this Sultans sister, and neece; and Mahomet Bassa of Cairo, his daughter, a child of sixe yeares old, and he about fiftie, having had presents sent him according to the Turkish solemnities: who giveth two hundred thousand Sultanies[16] in dowry. Not much in habite do the women of the Serraglio differ from other, but that the Favorite weares the ornament of her head more high, and of a particular fashion, of beaten gold, and inchaced[17] with gems; from the top whereof there hangeth a veile that reacheth to her ancles, the rest have their bonnets more depressed, yet rich; with their haire disheveled.[18]

When the Sultan entertaineth Embassadours, he sitteth in a roome of white marble, glistring with gold and stones, upon a low throne, spred with curious carpets, and accommodated with cushions of admirable workmanship; the Bassas of the Bench[19] being by, who stand like so many statues without speech or motion. It is now a custome that none do come into his presence without presents, first fastned upon his Bassas, as they say, by a Persian Embassadour: who thereupon sent word to the Sophy his maister that he had conquered Turkie. The stranger that approcheth him is led betweene two: a custome observed ever since the first Amurath was slaine by the Servian Cobelitz, a common souldier; who in the overthrow of Cossova,[20] rising from amongst the dead bodies, and reeling with his wounds, made towards the Sultan then taking a view of the slaine, as if he had something to say: by whom admitted to speech, he forthwith stabd him with a dagger, hid under his cassocke for that purpose. They go backward from him, & never put off their hats: the shewing of the head being held by the Turke to be an opprobious indecency. Now when he goeth abroad, which is lightly every other Friday (besides at other times upon other occasions) unto the Mosque: and when in state, there is not in the world to be seene a greater spectacle of humane glory, and (if so I may speake) of sublimated manhood. For although (as hath bene said) the Temple of Sancta Sophia,[21] which he most usuallly frequenteth, is not above a stones cast from the out-most gate of the Serraglio, yet hath he not so few as a thousand horse (besides the archers of his guard and other footmen) in that short procession: the way on each side inclosed as well within as without, with Capagies and Janizaries, in their scarlet gownes, and particular head ornaments. The Chauses[22] ride formost with their gilded maces; then the Captaines of the Janizaries with their Aga, next to the Chiefetaines of the Spachies,[23] after them their Sanziaks; those of the souldery wearing in the fronts of their bonnets the feathers of the birds of Paradise, brought out of Arabia, and by some esteemed the Phoenix. Then follow the Bassas and Beglerbegs[24]: after them the Pretorian footmen called the Solacchi,[25] whereof there be in number three hundred; these are attired in calsouns[26] and smocks of callico,[27] wearing no more over them then halfe-sleeved coates of crimson damask, the skirts tuckt under their girdles: having plumes of feathers in the top of their copped[28] bonnets; bearing quivers at their backs, with bowes ready bent in their left hands, and arrowes in their right: gliding along with a marvellous celerity. After them seaven or nine goodly horses are led, having caparisons and trappings of inestimable value; followed by the idolized Sultan gallantly mounted. About whome there runne fortie Peichi (so called in that they are naturally Persians) in high-crowned brimlesse caps of beaten gold, with coats of cloth of gold girt to them with a girdle called Chochiach[29]: the Pages following in the reare, and other officers of the houshold. But what most deserveth admiration amongst so great a concourse of people, is their generall silence: in so much as had you but onely eares, you might suppose (except when they salute him with a soft and short murmur) that men were then folded in sleepe, and the world in midnight. He that brings him good newes (as unto others of inferior condition) receiveth his reward, which they call Mustolooke. But this Sultan to avoid abuses in that kind, doth forthwith commit them to prison, untill their reports be found true or false; and then rewards or punisheth accordingly. Although he spends most of his time with his women, yet sometimes he recreates himselfe in hauking: who for that purpose hath (I dare not name) how many thousand Faulkners[30] in pension, dispersed thoughout his dominions: and many of them ever attendant. . . . Although he affects not hunting, yet entertaines he a number of huntsmen. Their dogs they let go out of slips in pursute of the Wolfe, the Stag, the Bore, the Leopard, &c. Those that serve for that purpose are stickle haired,[31] and not unlike to the Irish grayhounds.

Now the yearely revenew which he hath to defray his excessive disbursments, such a world of people depending upon him, amounts not to above fifteene millions of Sultanies, (besides the entertainment for his Timariots[32]) which is no great matter, considering the amplitude of his dominions: being possest of two Empires, above twenty kingdomes, beside divers rich and populous Cities; together with the Red, most of the Mid-land, the AEgean, Euxine, and Proponticke seas. But it may be imputed to the barbarous wastes of the Turkish conquests: who depopulate whole countries, and never reedifie what they ruine. So that a great part of his Empire is but thinly inhabited, (I except the Cities) and that for the most part by Christians: whose poverty is their onely safety and protectresse. But his casuall incomes do give a maine accession to his treasury: as taxes, customes, spoiles, and extortions. For as in the Sea the greater fishes do feede on the lesse, so do the Great ones here on their inferiours, and he on them all: being, as aforesaid, the commander of their lives, and generall heire of their substances. He hath divers mines of gold and silver within his dominion: that of Siderocapsa in Macedon having bene as beneficiall unto him as the largest Citie of his Empire, called anciently Chrysites: and not unknowne to Philip the father of Alexander; who had the gold from thence wherewith he coyned his Philips. . . .

Notes

1. Fleame: phlegm, indicating a "phlegmatic" disposition.
2. immunities of their Religion: freedoms allowed under Islam.
3. Bassas: pashas; high-ranking officials in the Ottoman empire.
4. pumpion: pumpkin.
5. Capagies: gatekeepers.
6. Janizaries: originally slaves, they were trained from childhood to serve in the elite corps of the Ottoman military.

7. socket . . . bonnets: insignia worn on the turban to indicate janissary status.
8. Aga of the women: eunuch charged with overseeing the women of the seraglio.
9. Jemoglans: janissary recruits.
10. Grand Signior: the Ottoman sultan or "Great Turk."
11. ranckt: in ranks or rows.
12. Roxolana: Hurrem Sultan (1500-1558), who became the wife of "Solyman" (Suleiman I, reigned 1520-1566).
13. Mufti: ranking religious official and chief justice of Islamic law.
14. Cadoun: mother of a prince.
15. Vizers of the Port: viziers of the Ottoman court (known as the "Sublime Porte").
16. Sultanies: Turkish gold coin worth about eight shillings.
17. inchaced: encrusted.
18. disheveled: worn down and loose.
19. Bassas of the Bench: high court officials attending on the sultan.
20. Cossova: refers to the Serbian defeat at the battle of Kosova in 1389.
21. Temple of Sancta Sophia: the Hagia Sophia.
22. Chauses: ambassadors or messengers.
23. Spachies: cavalrymen.
24. Beglerbegs: governors or viceroys.
25. Solacchi: royal guards.
26. calsouns: hose or trousers.
27. smocks of callico: fine shirts of linen and silk.
28. copped: high-peaked.
29. Chochiach: Turkish "kusak" meaning "broad belt."
30. Faulkners: falconers.
31. stickle haired: bristly.
32. Timariots: feudal landholders who were obliged to supply the Ottoman army with troops and supplies for military campaigns.

William Lithgow

William Lithgow began his restless career as a wanderer with walking tours of his native Scotland, and by 1609 he had already made two voyages to the Orkney and Shetland isles. Later that year, Lithgow set out for Paris on the first of the three long journeys that are recounted in his *Totall Discourse of the Rare Adventures, and painefull Peregrinations of long nineteene Yeares Travayles, from Scotland, to the most Famous Kingdomes in Europe, Asia and Affrica,* printed in 1632 (the excerpt included here is from this version of the text). From 1610 to 1613, Lithgow's first great journey took him to Constantinople, the islands of the eastern Mediterranean, Syria, Palestine, Egypt, and then home. These wanderings were recounted in *A Most Delectable and True Discourse, of an admired and painefull peregrination from Scotland, to the most famous Kingdomes in Europe, Asia and Affricke*, published first in 1612 and then again, "corrected and enlarged," in 1614. A "second Impression, Corrected and enlarged by the authour" appeared in 1616, followed by a third edition in 1623 with further additions. Lithgow's next trip, which lasted from 1613 to 1616, took him to Algiers, Fez, Tunis, and the Libyan desert. He returned to England and set out once more, intending to reach Ethiopia, but he was seized by the Spanish authorities in Malaga, where he was tortured by the Spanish governor, who thought he was a spy. After months of imprisonment and much suffering, Lithgow returned to the English court, where he exhibited his wounds and petitioned for redress and compensation from James I and from Spain. When the Spanish ambassador, Gondamar, promised to compensate Lithgow for his sufferings and then failed to keep his word, Lithgow grew impatient and assaulted him in the King's presence chamber. For this, Lithgow was sent to the Marshalsea prison, where he remained for many months. After his release from prison, Lithgow began to attend and petition Parliament during a period of 17 weeks in 1624. *The Totall Discourse*, published in 1632, brought renewed persecutions from the court, where Spanish influence was powerful at that time. *The Totall Discourse* was reprinted in 1640 and 1682. Lithgow claimed to have traveled, in total, more than 36,000 miles during the second decade of the seventeenth century, and most of this was accomplished on foot, without the privileges and comforts enjoyed by his contemporary, George Sandys. The excerpt that appears here describes Lithgow's arrival in Jerusalem on Palm Sunday, 1612, and his observations on both the reception of European pilgrims and the behavior of "oriental" Christians during the Easter festival there.

—Daniel J. Vitkus

[Description of Lithgow's visit to the Holy Sepulcher in Jerusalem]

And now the Souldiers and wee being advanced in our Way, as wee returned to Jerusalem, wee marched by an olde Ruinous Abbey, where (say they) Saint Jerome dwelt, and was fed there by wilde Lyons: Having travailed sore and hard that afternoone, wee arrived at Jerusalem an houre within night, for the Gate was kept open a purpose for us and our Guard: and entring our Monastery, wee supped, and rested our selves till midnight; having marched that halfe Day, more as 34. miles. A little before midnight, the Guardian and the Friers, were making themselves ready to goe with us to the Church of the Holy Sepulcher, called Sancto Salvatore; where wee were to stay Good-friday and Satturday, and Easter-Sunday till mid-night: They tooke their Cooke with them also to dresse our Dyet, carrying Wine, Bread, Fishes, and Fruites hither in abundance. Meane while, a Jew, the Trench-man[1] of the Turkies Sanzacke,[2] came to the Monastery, and received from every one of us Pilgrimes, first two Chickens[3] of Gold, for our severall heads, and entrey at Jerusalem: and then nine Chickens a peece for our in going to the Holy Grave; and a Chicken of golde a man, to himselfe the Jew, as beeing due to his place.

Thus was there twelve Chickens from each of us dispatched for the Turke: And last one, and all of us, behoved[4] to give to the Guardian two Chickens also for the Waxe Candles and fooleries hee was to spend, in their idle and superstitious Ceremonies, these three aforesayd nights, which amounted in all to every one of us, to foureteene Chickens of gold, sixe pounds sixe shillings starling. So that in the whole from the sixe Germanes, foure French men, and nine Commercing Franks in Cyprus and Syria, Venetians, and Ragusans, and from my selfe, the summe arose for this nights labour to a hundred and twenty sixe pounds starling.

This done, and at full mid-night wee came to the Church where wee found twelve Venerable like Turkes, ready to receive us, sitting in the Porch without the Doore; who foorthwith opened at randone the two great Brazen halfes of the Doore, and received us very respectively: We being within the doore made fast, and the Turkes returned to the Castle, the first place of any note we saw, was the place of Unction, which is a foure squared stone; inclosed about with an yron Revele,[5] on which (say they) the dead body of our Saviour lay, and was imbalmed; after hee was taken from the Crosse, whiles Joseph of Arimathea, was preparing that new Sepulcher for him wherein never man lay: from thence we came to the holy Grave. Leaving Mount Calvary on our right hand toward the East end of the Church; for they are both contained within this glorious edifice.

The Holy Grave is covered with a little Chappell, standing within a round Quiere,[6] in the west ende of the Church: It hath two low and narrow entries: As we entred the first doore, three after three, and our shoes cast off, for these two roomes are wondrous little, the Guardiano fell downe, ingenochiato,[7] and kissed a stone, whereupon (he sayd) the Angell stood, when Mary Magdalen came to the Sepulchre, to know if Christ was risen, on the third day as he promised: And within the entry of the second doore, we saw the place where Christ our Messias was buried, and prostrating our selves in great humility, every man according to his Religion, offered up his prayers to God.

The Sepulchre it selfe, is eight foote and a halfe in length, and advanced about three foote in height from the ground, and three foote five inches broad, being covered with a faire Marble stone of white colour.

In this Chappell, and about it, I meane without the utter sides of it, and the inward incirclings of the compassing Quiere, there are alwayes burning above fifty Lampes of oyle, maintained by Christian Princes, who stand most of them within incircling bandes of pure Gold, which is exceeding sumptuous, having the names of those, who sent or gave them, ingraven upon the upper edges of the round circles: each of them having three degrees, and each degree depending upon another, with supporters of pure Gold, rich and glorious. The fairest whereof was sent thither by King John of England, whereon I saw his Name, his Title, and crowne curiously indented, I demanded of the Guardiano if any part of the Tombe was here yet extant, who replied, there was; but because (said he) Chistians resorting thither, being devoutly moved with affection to the place, carried away a good part thereof, which caused S. Helen inclose it under this stone; whereby some relicts of it should alwaies remaine. I make no doubt but that same place is Golgotha, where the holy Grave was, as may appeare by the distance, betweene Mount Calvary and this sacred Monument; which extendeth to forty of my pases: This Chappell is outwardly decored, with 15. couple of Marble Pillars, and of 22. foote high; and above the upper coverture of the same Chappell, there is a little sixe-angled Turret made of Cedar wood, covered with Lead, and beautified with sixe small Columnes of the same tree. The Chappell it selfe standeth in a demicircle or halfe Moone, having the little doore or entry looking East: to the great body of the Church, and to Mount Calvary, being opposite to many other venerable monuments of memorable majesties.

The forme of the Quiere wherein it standeth, is like unto that auncient Rotundo in Rome, but a great deale higher and larger, having two gorgeous Galleries; one above another, and adorned with magnificent Columnes being open at the top, with a large round; which yeeldeth to the heavens the prospect of that most sacred place.

In which second Gallery we strangers reposed all these three nights we remained there: whence we had the full prospect of all the spacious Church, and all the Orientall people were there at this great feast of Easter day, being about 6000. persons: from this curious carved Chappell we returned through the Church to Mount Calvary; To which we ascended by twenty one steps, eighteene of them were of Marble, and three of Cedar-wood: where, when we came I saw a most glorious & magnifick roome, whose covert[8] was supported all about with rich columnes of the Porphyre stone, and the oversilings[9] loaden with Mosaick worke, & overgilded with gold, the floore being curiously indented with intermingled Alabaster and black shining Parangone:[10] On my left hand I saw a platformd rocke, all covered with thicke and ingraven boords of silver; and in it a hole of a cubits deepe, in which (say they) the Crosse stood whereon our Saviour was crucified: And on every side thereof a hole for the good & bad theeves, were then put to death with him. Discending from Mount Calvarie, we came to the Tombe of Godfrey du Bulloine,[11] who was the first proclaimed Christian King of Jerusalem, and refused to be crowned there, saying; It was not decent, the Servants head should be crowned with gold, where the Maisters head had beene crowned with thornes; having this Inscription ingraven on the one side:

> Hic jacet inclytus Godfridus de Bullion, quitotam hanc terram acquisivit cultui divino, cujus anima requiescat in pace.[12]

And over against it, is the Tombe of King Baldwine his brother, which hath these Verses in golden Letters curiously indented.

> Rex Baldevinus, Judas alter Machabeus
> Spes patriae, Vigor Ecclesiae, Virtus, utriusque;
> Quem formidabant, cui dona, tributa ferebant.
> Caesar, AEgypti Dan, ac homicida Damascus;
> Proh dolor! in modico clauditur hoc Tumulo.[13]

The other things within the Church they shewed us, were these, a Marble Pillar, whereunto (say they) our Saviour was bound, when he was whipped, and scourged for our sakes: the place in a low Celler, about fourteene stone degrees under the ground, where the Crosse was hid by the Jewes, and found againe by S. Helen: the place where Christ was crowned with thornes, which is reserved by the Abasines,[14] and where the Souldiers cast lots for his Garment; the place where he was imprisoned, whiles they were making of his Crosse, and where the Crosse, being laid along upon the gound, our Saviour was nailed fast to it; the Rocke, which (as they say) rent at his crucifying, which is more likely to be done with hammers, and set one peece a foote from another, for the slit lookes, as if it had beene cleft with wedges and beetles.[15] And yet the sacred Scriptures say that it was not a Rocke, but the Temple that did rent in two from the bottome to the top, wherein these silly soule-sunke Friers are meerely blinded, understanding no more than leying traditions; perfiting this their nationall Proverb;

> Con arte, et con inganno, ci vivnono medzo l' anno
> Con inganno et con arte, ci vivona l' altera parte.
>
> With guile and craft, they live the one halfe yeare
> With craft and guile, the other halfe as cleare.

And lastly, they take upon them below Calvary to shew us where the head of Adam was buried. These and many other things, are so doubtfull, that I doe not register them for trueth (I meane in demonstrating the particular places) but onely relates them as I was informed.

There are seven sorts of Nations, different in Religion, and language, who continually (induring life) remaine within this Church, having incloystered lodgings joyning to the walls thereof: their victuals are brought dayly to them by their familiars, receiving the same at a great hole in the Church-doore; for the Turkes seldome open the entry unlesse it be when Pilgrimes come, save one houres space onely every Saturday in the afternoone, and at some extraordinary Festivall daies: and yet it doth not stand open then, but onely opened to let strangers in and shut againe: For this purpose each family have a Bell fastened at their lodging, with a string reaching from thence to the Church doore, the end whereof hangeth outwardly, By the which commodity, each furnisher ringing the Bell, giveth warning to his friends, to come receive their necessars,[16] for through the body of the Church they must come to the porch-doore, and returne from it, to the cloyster.

The number of those, who are tied to this austere life, are about three hundred and fifty persons, being Italians, Greekes, Armenians, AEthiopians, Jacobines,[17] a sort of circumcised Christians, Nestorians, and Chelfaines[18] of Mesopotamia.

The day before the Resurrection, about the houre of mid-night, the whole Sects and sorts of Christians Orientall (that were come thither in Pilgrimage, and dwelt at Jerusalem) convened together, which were about the number of sixe thousand men, women, and children: for being separated by the Patriarkes in two companies, they compassed the Chappell of the Holy Grave nine times; holding in their hands burning Candles, made in the beginning pittifull, and lamentable regreetings, but in the ending, there were touking[19] of kettle-drummes, sounding of horne-trumpets, and other instruments, dauncing, leaping, and running about the Sepulcher, with an intollerable tumult, as if they had beene all mad or distracted of their wits.

Thus is the prograce of their procession performed in meere simplicitie, wanting civilitie, and government. But the Turkes have a care of that; for in the middest of all this hurley burley, they runne amongst them with long Rods, correcting their misbehaviour with cruell stroakes: and so these slavish people, even at the height of their Ceremonious devotion are strangely abused.

But our Procession begun before theirs, and with a greater regard, because of our tributes: The Turkes meane while guarding us, not suffering the other Christians to be participant in the singular dottage of the Romish folly, being after this manner: First the Guardian, and his Friers brought forth of a Sacrastia,[20] allotted for the same purpose, the wodden Protracture[21] of a dead Corpes, representing our Saviour, having the resemblance of five bloody Wounds, the whole body of which Image, was covered with a Cambricke vale: Where having therewith thrice compassed the Chappell of the Holy Grave, it was carried to mount Calvary, and there they imbalmed the five Timber holes; with Salt, Oyle, Balme, and Odoriferous perfumes.

Then the Guardian, and the other twelve Friers kneeled downe, and kissed each one of the five Suppositive Wounds: the Turkes meanewhile laughing them to scorne in their faces, with miserable derision. Thence they returned, and layd the senselesse blocke uppon the Holy Grave, whence being dismissed, the Papall Ceremony ended.

Truely hereupon, may I say, if the Romane Jesuites, Dominicans, and Franciscans, there Resident in certayne speciall parts of the Turkes Dominions, had onely behaved themselves as their pollliticke charge required, and dismissed from the Paganisme eyes, onely their idolatrous images, veneration of Pictures, Crosses, and the like externall superstitious Rites: These Infidels I say, had long agoe (without any insight of Religion) bene converted to the Christian Faith. For besides all this blindnesse, what infinite abhominable Idolatries commit they in Italy and Spaine; in clothing the Pictures of dead Abbots, Monkes, Priors, Guardians, and the better kind of officiall Friers and Priests, with robes of Sattin, Velvet, Damas, Taffaty, long gownes and coules[22] of cloth, shirts, stockings, and shoes: And what a number of livelesse portrayed Prioresses, motherlesse Nunnes, yet infinite mothers, be erected (like the Maskerata[23] of Morice-dancers) in silver, gold, gilded brasse, yron, stone, tynne, lead, copper, clay, and timber shapes, adorned with double and triple ornaments: over-wrought with silke, silver, and gold-laces, rich bracelets, silke grograine,[24] and cambricke vales, chaines, smockes, ruffes, cuffes, gloves, collers, stockings, garters, pumpes, nosegayes, beeds, and costly head-geire; setting them on their Altars, O spectaculous Images! adoring them for gods, in kneeling, praying, & saying Masses before them: Yet they are none of their avowed, allowed, and canonized pontificall Saints: for although they be bastards & wooden blocks, yet are they better clad, then their lu[m]pish legitimate ones, no, I may say, as the best Kings daughter alive. Which is a sinfull, odious, and damnable idolatry; and I freely confesse at some times, and in some parts I have torne a peeces those rich garments from their senselesse images and blockes, thinking it a greater sinne not to do it than to stand staring on such prodigall prophannesse, with any superstitious respect, or with indifferent forbearance to winke at the wickednesse of Idolaters.

Here the Guardiano offered for ten peeces of gold (although my due be thirty Chickens sayd he) to make me Knight of the holy Grave, or of the order of Jerusalem, which I refused, knowing the condition of that detestable oath I behooved to have sworne; but I saw two of these other Pilgrimes receive that Order of Knighthood.

The manner whereof is thus: First they bind themselves with a solemne vow, to pray (during life) for the Pope, King of Spaine, and the Duke of Venice, from whom the Friers receive their maintenance; and also in speciall, for the French King, by whose meanes they obtaine their liberty of the great Turke, to frequent these monumentall places. Secondly, they are sworne enemies to Protestants, and others, who will not acknowledge the superiority of the Romane Church. Thirdly, they must pay yearely some stipend unto the Order of the Franciscans. These attestations ended, the Frier putteth a gilded spurre on his right heele, causing the yong made Knight stoope downe on his knees, and lay his hands on the holy Grave: after this he taketh a broad sword from under his gray gowne (being privately carried for feare of the Turkes) which is (as he sayd) the Sword, wherewith victorious Godfrey conquered Jerusalem, and giveth this new upstart Cavaliero, nine blowes upon the right shoulder. Loe here the fashion of this Papisticall Knighthood, which I forsooke.

Indeed upon the Knight-hood they have certaine priviledges among the Papists, of which these are two: If a malefactor being condemned and brought to the Gallows, any of these Knights may straight cut the rope and releeve him: The other is, they may carry and buy silkes through all Spaine and Italy, or elsewhere, and pay no Custome, neither in comming nor going, nor for any silke ware, where the Romish Church hath any commandement.

After our Guardiano had ended his superstitious Rites and Ceremonies, upon Easter day, before midnight, we returned to the Monastery, having stayed three dayes within that Church. . . .

Notes

1. Trench-man: translator.
2. Turkies Sanzacke: Ottoman governor in Jerusalem.
3. Chickens: gold coins minted in Italy or Turkey, worth 7-8 shillings each.
4. behoved: required.
5. Revele: railing.
6. Quiere: a space resembling the choir or chancel in an English church.
7. ingenochiato: (It.) ingenuously.
8. covert: roof.
9. oversilings: arches.
10. Parangone: basalt?
11. Godfrey du Bulloine: born ca. 1060, hero of the First Crusade. He was the first crusader to rule over Jerusalem after its capture in 1099.
12. Hic jacet . . . in pace: Here lies Godfrey of Bulloigne who won all this land for the worship of God. May his soul rest in peace.
13. Rex Baldevinus . . . hoc Tumulo: Baldwin the King, another Judas Maccabeus, the Church and Nation's strength and hope, the glory of both; whom Cedar, Egypt's Dan, Damascus, loaded with dead; both feared and tribute brought; O grief! within this little tomb doth lie.
14. Abasines: Abyssinians; Ethiopian Christians.
15. beetles: mallets.
16. necessars: necessaries.
17. Jacobines: members of a monophysite sect from Syria or Mesopotamia.
18. Chelfaines: Chaldeans?
19. touking: beating.
20. Sacrastia: sacristy.
21. Protracture: form; representation.
22. coules: cowls.
23. Maskerata: masques.
24. grograine: grogram (from Fr. *gros grain*), a coarse, stiff silk fabric.

TRAFFICKING WITH THE TURK: ENGLISH TRAVELERS IN THE OTTOMAN EMPIRE DURING THE EARLY SEVENTEENTH CENTURY

Daniel J. Vitkus

In the last three decades of the sixteenth century, Englishmen began traveling to the Mediterranean, including areas controlled by the Ottoman sultanate, in greater numbers. They came as merchants, diplomats, sailors, soldiers, and tourists. Some stayed for long periods of time, including the first ambassadors, consuls, and factors who lived and worked in Ottoman cities. There were also English renegades who had "turned Turk" and served in the corsair ships of North Africa. And there were those who traveled to the Mediterranean aboard English ships that were taken by these pirates, who then sold the captive crew and passengers into slavery.[1] Some of these English travelers and temporary residents wrote narrative accounts of their experiences in the Ottoman Empire, and these texts bear witness to new forms of knowledge and to an emergent identity-formation that was shaped by the religious, political, and economic conditions of the time.

After the Reformation and with the arrival of venture capitalism, English Protestants rejected the devotional theology of pilgrimage, and the notion of a pan-Christian crusade to recover Jerusalem dwindled into a stale fantasy. Many English writers continued to pay lip service to the notion of Christian solidarity against Islam, but the overwhelming reality was the existence of an Islamic power, the Ottoman Turks, who were firmly established in the Holy Land and whose presence was felt throughout the Mediterranean.[2] The Turkish empire was a force that Western Europeans could not hope to overcome. Instead of mounting a crusade, many Christian rulers and merchants sought friendly relations with the Islamic powers, relations that might bring profit and strategic advantage to themselves and their nation in their intensifying competition with other seagoing powers.[3]

English travelers in the Islamic world were neither pilgrims nor crusaders: rather, their motives were defined by the peculiar position of Protestant commercialism in a Mediterranean world that was multicultural and multireligious, but devoid of Protestant rulers. This sense of being the isolated Other, strangers in a strange land where Roman Catholicism, Orthodox Christianity, and Judaism were all tolerated under Islamic rule, positioned English writers as radically different, and yet these English writers often wished to befriend and emulate the powerful Muslims that they encountered. The wealth, order, and discipline of the Ottomans was frequently admired and praised by writers who, at the same time, expressed contempt. The contradiction between condemnation and emulation is strongly apparent when these early modern travelers write about Jerusalem and Constantinople, two sites of great ideological significance for Christianity. From the perspective of Protestant visitors like George Sandys and William Lithgow, the Holy Land and the seat of the Byzantine empire were (like Rome) under the control of what they believed to be a false faith. Pilgrimage, in the medieval sense, is no longer possible for them, but nonetheless English travelers are drawn to sites that retained for them the spectral imprint of a Christian and classical history that they constructed or read against both Islam and Rome. When describing the Middle East, their Protestant travel narratives emphasize the past, and when confronted with the presence of Islamic rule, they attest anxiously to what they see as the spiritual and political excesses permitted by "the Turk" in both these places. English and Scottish visitors to Jerusalem and Constantinople saw themselves, not as pilgrims seeking grace through ritualized and sanctified means, but as "anti-pilgrims" who were there to witness and record their post-Reformation abhorrence of both Catholic "idolatry" and Turkish "tyranny." The reality of Ottoman Jerusalem and Constantinople compelled them to act as iconoclastic anti-pilgrims whose purposes are often more commercial than pious, ethnographic rather than devotional.

At the midpoint of the sixteenth century, England was commercially unsophisticated and isolated, but by the end of the century, English merchants and mariners began to assert themselves as players in the world of international commerce and cross-cultural exchange. This increase in contact with the Mediterranean and Islamic world was brought on by the rapid development of commercial activity in the Mediterranean region, made possible by the deployment of superior nautical technologies that gave English merchants greater mobility and access to Mediterranean ports, markets, and commodities. The "expansionary thrust" of the English into the Mediterranean during the late sixteenth century helped to stimulate commercial movement toward other areas, including the West and East Indies, during the century to come.[4] One prominent economic historian has described this period, 1570-1630, as exhibiting "one of the most striking transformations in economic history," manifested in "New forms of organization, a new breed of merchants and promoters, new sources of capital, a new sense of purpose, and a new vitality in economic enterprise. . . ."[5] The joint stock company was first widely used in England in the late sixteenth century, as various charters and cooperative efforts were initiated, from the Muscovy Company in the 1550s to the East India Company in the early seventeenth century.

Soon after 1570, formal diplomatic relations with Muslim rulers in the Ottoman empire and in the autonomous principalities of North Africa were established.[6] In 1575, two London merchants, Edward Osborne and Richard Staper, initiated a dialogue with the Ottoman sultan that led, first, to the Queen's appointment of William Harborne as English envoy to the sultan (he arrived in the Turkish capital in October of 1578) and, second, to the promulgation of commercial capitulations agreed between the English monarchy and the Ottoman sultanate.[7] As a result of this agreement, the Levant Company was founded in 1581. After commercial capitulations were established in 1580, Harborne began efforts, initiated by Elizabeth and her councilors, to form a political and military alliance with the Ottomans against the Spanish.

The Levant Company's charter of 1581 declared that trade with Turkey had declined to the degree that such commerce was not "in the memory of any man nowe living."[8] Once official sanction was given by the sultan to the Turkey trade, English shipping traffic began to increase throughout the eastern Mediterranean. Before long, there were English "factories" established at Constantinople, Aleppo, Scanderoon, Tripoli (in Syria), Cairo, and Alexandria.[9] In Hakluyt's second edition of *The Principall Navigations,* in the second volume, the "Epistle Dedicatorie" responds to those who "take exception against this our new trade with Turkes and misbeleevers." According to Hakluyt, those who oppose commerce with non-Christians "doth not acknowledge, that either hath travailed the remote parts of the world, or read the Histories of this later age, that the Spaniards and Portugals in Barbarie, in the Indies, and elsewhere, have ordinarie confederacie and traffike with the Moores, and many kindes of Gentiles and Pagans."[10] Trafficking with the Turk and other "misbeleevers" was necessary if English merchants wished to compete in the rapidly evolving context of commercial, proto-colonialist expansion beyond Europe.

Consequently, this was a period of intensive intelligence-gathering. Not only in the Mediterranean but wherever English maritime enterprise was carried out, those who traveled for the sake of trade helped to produce a body of knowledge that would serve the purpose of profit. And yet these descriptions and accounts of the Other were not produced or consumed for the sake of commerce alone: they also satisfied ethnographic curiosity and provided readers with the pleasures of the strange and the exotic.[11] Thomas Palmer, in *How To Make Our Travailes Profitable* (1606), describes the need for travelling "intelligencers" to acquire and disseminate knowledge of foreign lands and peoples for the benefit of the state and the merchant class:

> [T]he very point which every Travailer ought to lay his wittes about [is] To get knowledge for the bettering of himselfe and his Countrie: This, being the object of their Countries defects and the subject of Travailers, in a word containeth Six generall heads, . . . namely, the tongue, the Nature of the People, the Countrey, the Customes; the Government of the State; and the secrets of the same: the which are to be sought out wheresoever these shall come. (52-53)

The kind of intelligence-gathering described by Palmer was undertaken privately, and the accumulated data then circulated in manuscript. As William Sherman has shown, an official program of this kind really began with John Dee and his circle at the English court.[12] Some of this information was then made more widely available in the printed texts edited and compiled by Richard Hakluyt, beginning with Hakluyt's *Divers voyages touching the discoverie of America* (1582) and continuing in Hakluyt's *Principall Navigations* (1589, one volume) and his *Principall Navigations, Voyages, Traffiques and Discoveries* (1598-1600, three volumes). Later, this project would be sustained and supplemented in the texts put together for publication by Samuel Purchas in *Purchas His Pilgrimage* (1613) and *Hakluytus Posthumus or Purchas His Pilgrimes* (1625).[13] While Hakluyt and Purchas carried out their work as compilers and editors, there were other ethnographic and geographic texts being translated and printed by their contemporaries. This period saw a stream of printed "reports" and "descriptions" by other authors who were gathering information about the new plantations in North America and the other areas that were being "discovered."[14]

Hakluyt's *Principall Navigations* project was undertaken for a variety of purposes: it served to document, to entertain, and to encourage patriotic feeling or commercial enthusiasm. Richard Helgerson and others have pointed to its function as nationalistic propaganda, and its pro-English bias helps to construct a certain kind of English traveler.[15] This effect is described by David B. Quinn: "The high-minded, practical, trade-conscious and courageous Englishman who emerges from *The Principall Navigations* as the personification of his time is in some degree the self-projection of the narrators, in some degree also . . . the creation, through selection, of the compilers."[16] Perhaps more than anything else, Hakluyt's books comprised a set of manuals for merchants interested in organizing commercial ventures. According to Quinn, "The object was greater overseas commercial expansion for England; greater plunder of the overseas possessions of Spain and Portugal; greater glory in navigational triumphs, greater responsibilities in the establishment of trading factories, colonies, protectorates perhaps."[17] In the second edition of the *Principall Navigations,* in an "Epistle Dedicatorie" to Walsingham, Hakluyt celebrates England's newly established role in the eastern Mediterranean trade: "Who ever saw before this [Queen Elizabeth's] regiment, an English Ligier in the stately porch of the Grand Signior at Constantinople? who ever found English Consuls & Agents at Tripolis in Syria, at Aleppo, at Babylon, at Balsara . . . ?[18]

Hakluyt's work was both narrowly pragmatic and broadly ideological: it provided a practical guide for the proto-imperialist efforts of the English, and at the same time it helped to construct the image of a resourceful and heroic merchant-colonizer. *The Principall Navigations* provides a variety of local knowledges: lists of weights, measures, currencies, customs levied, which commodities were in demand or available for trade, descriptions of the harbor, the consulates, local contacts, and officials to be dealt with, and everything that a merchant arriving in a foreign port would need to know in order to do business.[19] C. F. Beckingham points out, "The establishment of commercial relations with the Ottoman empire was, for [Hakluyt's] purpose, the most important event in the recent history of the Near East and much of his material is relevant to it."[20]

As new information about Islamic culture flowed back to England, interest grew in travel narratives and descriptions of events, customs, or places in the Ottoman empire and North Africa. The production of knowledge about Islam, Turks, Moors, and Arabs was accelerated and dispersed to various sites of cultural production and consumption, including popular ballads, the visual arts, public pageants, court entertainments, public theater, as well as printed materials such as travel narratives or ethnographies.[21] Commercial energies drove the production of texts about Turkish and Mediterranean culture, but other, more ideological, forces were also at play. Early modern representations of the Islamic Other helped to construct an identity for Protestant England when English identity was developing a proto-imperialist formation. Imperial envy, accompanied by anxiety about religious difference, is often expressed in English texts describing the Turks. Turkish power was a difficult reality to confront at a time when English authors and readers sought to construct a self-image of metropolitan masculinity. Fear and admiration of Turkish culture, and of "the Great Turk" or "Grand Seigneur" as the Ottoman sultan was called, were often mixed with condemnation and loathing. And yet English Protestant animosity for Spanish or Roman Catholic "superstition" was usually stronger than feelings of hostility toward the more distant Ottoman Muslims. Some English narrators describing Turkish society are captivated by the sophistication, order, and strength that they observe, a unified power that they saw as a foil to a divided and corrupted Christendom. In any case, it was commercial exchange, not imperialist confrontation or conflict, which dominated Anglo-Islamic relations during the late sixteenth and seventeenth centuries.

Not all of the knowledge gathered by travelers was delivered to Hakluyt or Purchas. Fynes Moryson, the author of a travel narrative titled *An Itinerary . . . Containing His Ten Yeeres Travell* produced a Latin manuscript that was not printed until after Moryson himself translated it into English in 1617.[22] Moryson, who toured the Ottoman empire from November 1595 to July 1597, visited Constantinople, Asia Minor, Syria, and Palestine before returning home. His *Itinerary* is a densely packed container, full of knowledge about foreign cultures and places: it is brimming with practical advice to other travelers and contains detailed descriptions of politics, religion, and commerce in the Levant.

Moryson traveled to Jerusalem, but before describing his experiences there, he provides this disclaimer:

> by my journy to this City, I had no thought to expiate any least sinne of mine; much lesse did I hope to merit any grace from God. . . . I thought no place more worthy to be viewed in the whole world, then this City, where howsoever I gave all divine worship to God, and thought none to be given to the places, yet I confess that (through the grace of God) the very places strucke me with a religious horrour, and filled my mind prepared to devotion, with holy motions. (1)

Moryson is typical of English Protestant travelers, who tended to write about Jerusalem with great ambivalence and anxiety. His visit to Jerusalem and its monuments, which were entirely under the care of non-Protestants, becomes a trial of Moryson's powers of unbelief. Moryson's exercise in skepticism is made easier by the contempt he holds for the citizens of the Holy City, whom he describes as "poore rascall people, mingled of the scumme of divers Nations, partly Arabians, partly Moores, partly the basest inhabitants of neighbor countries" (5). "[T]he Inhabitants of Jerusalem," writes Moryson, "at this day are as wicked as they were when they crucified our Lord . . ." (5). Stirred "by emulation and curiosity" (1) to see Jerusalem for himself, Moryson makes it clear nonetheless that he rejects the "superstitious inventions" of the "Sects of Christians there abiding" (2) and presents himself as a mere observer providing descriptive intelligence to his readers, and not as a religious pilgrim offering an account of his own religious practice or devotional response.

Like Fynes Moryson, Sir Henry Blount was a well-to-do English traveler to the Levant who claimed to make his journey more for self-edification than for commerce or politics. Blount's narrative, which first saw print in 1636, is titled *A Voyage into the Levant.* It describes a journey taken in 1634 from Venice to Constantinople to Egypt and then back to England. Referring to Ottoman rule in Cairo, Blount rejects the idea that "the Turkish domination there were nothing but sottish sensualitie, as most Christians conceive."[23] Blount is particularly impressed by the strength and discipline of the Ottoman army. He clearly admires and envies the political order and martial might of the Turks, declaring them to be "the only moderne people, great in action, and whose Empire hath so suddenly invaded the World, and fixt it selfe such firme foundations as no other ever did" (2). "I was of opinion," he goes on to say, "that he who would behold these times in their greatest glory, could not find a better scene then Turky. . . ." (2).

While travelling through the Danube River valley, Blount encounters the Turkish army on the march. He praises their discipline: "yet I wondered to see such a multitude so clear of confusion, violence, want, sicknesse, or any other disorder . . ." (13), and while drinking coffee with "Murath Basha," one of the Ottoman commanders, Blount is invited to serve in the Ottoman army "against the Polacke who is a Christian" (15). Murath asks Blount if it is permitted for an Englishman to fight with Muslims against Christians. Blount's response is striking in the way that it prefers Islam to Roman Catholicism:

> I humbly thanked him, for his favour, and told him that to an Englishman it was lawful to serve under any who were in League with our King, and that our King had not only a League with the Gran Signor, but continually held an embassadour at his Court, esteeming him the greatest Monarch in the World: so that my service there . . . would be exceedingly well received in England; and the Polacke, though in name a Christian, [was] yet of a Sect, which for Idolatry, and many other points, I much abhorred. . . ." (15)

Blount claims to have declined this invitation to serve with the Turks only because he "wanted language" (15) to command and communicate on the field of battle. He goes on to extol "the Turkes, whom we not only honored for their glorious actions in the world; but also loved, for the kinde Commerce of Trade which we find amongst them . . ." (15).

Blount was not the only traveler to be captivated by the Turks' commercial and military prowess. There were other English merchants who were receptive to Turkish culture and who traveled in order to turn a profit in Ottoman ports and markets. An example of the merchant-narrator is John Sanderson, who was involved in the Levant trade and whose autobiography is preserved in manuscript.[24] Sanderson and others, backed by investors like Osborne and Staper, exported woolen cloth, animal skins, wood, iron, and lead from England to the Levant and imported valuable commodities like oil, currants, wine, nuts, fine silk and cotton material, Turkish carpets, indigo, spices, and drugs (including "mummia," a powder made from mummies and sold by English apothecaries). Sanderson once oversaw the purchase for the Turkey Company of 600 pounds of mummy powder, "together with a whole bodie."[25]

Because there were no Protestant churches there, the growing community of English merchants in the Levant needed the services of Protestant ministers, and some of these clergymen, including William Biddulph, who served as chaplain to the English consulate in Aleppo, have left reports of their experiences in the region. Biddulph's letters describe travels that took place in the Levant between 1600 and 1608. As a Protestant divine, Biddulph is uncomfortable with both Islam and Roman Catholicism, and the preface to his letters claims that those who read them will "learne to love and reverence their Pastors . . . when they shall reade in what blindnesse and palpable ignorance other nations live, not knowing the right hand from the left in matters that concerne the kingdome of heaven . . . ."[26] Biddulph's account of his visit to Jerusalem is divided into three sections, one describing "Apparent Truths" about the Holy City, followed by a long section listing the "Manifest untruths" that were allegedly told by the Roman Catholic friars and monks who acted as tour guides for Christian pilgrims, and a short final section of "Doubtfull things." Biddulph takes pains to distinguish between those things that "have Scripture or reason for them, or both" and those that he declares to be "false and ridiculous" (130).

Protestant theology denied the spiritual effects of pilgrimage, and Lavender's "Preface" to Biddulph expresses this anti-Catholic view: "for Pilgrims goe with a superstitious devotion to worship Reliques at Jerusalem; but master Tymberley and his companions went thither onely as *travellers* to see the Holy Land."[27] As early modern Protestants, English visitors to Jerusalem did see themselves as performing an important religious function: they were there, not to perform good works but, rather, as "anti-pilgrims" who are present in order to express their skepticism and testify to the false "idolatry" or "superstition" of the other Christians who continued to uphold the importance of pilgrimage and the cult of saints. William Lithgow and George Sandys both play the part of the anxious, conflicted anti-pilgrim, sometimes moved to worship or reverence by the sacred spaces of Jerusalem, but just as often moved to contempt and scorn.

Perhaps the most unusual and fascinating example of the anti-pilgrimage is the case of the Quakers, both men and women, who traveled to the Mediterranean and to Jerusalem later in the seventeenth century. A narrative describing the travels of some of these Quakers was published in 1662 under the following title: *This is the short relation of some of the cruel sufferings (for the truth's sake) of Katherine Evans & Sarah Chevers.*[28] It was reprinted the next year with an additional text appended, "a short relation from George Robinson, of the sufferings that befel him in his journey to Jerusalem." This additional account describes Robinson's experience in Palestine, where he preaches publicly as a Quaker opposed to the "false religion" of the other Christians present in and around Jerusalem. His mission is not to visit the sacred sites, but to preach the Word and perhaps to save the souls of foreign misbelievers. Soon after his arrival in Jerusalem, Robinson is taken into custody by the friars overseeing the monuments. They bring him before the Turkish authorities who interrogate Robinson and then try to have him deported out of Palestine, but he manages to return to Ramallah. There, he is once again arrested by the friars, but taken away from them by some Muslims who escort him to a mosque:

> two Turks . . . brought me to one of their Houses of Worship or Mosco; and I being entered thereinto, many people gathered together therein, also the Priests of Mahomet, before whom I was called and caused to sit down; And then it was demanded of me, Whether I would turn unto the Turks religion? I answered, I could not turn unto them . . . but they pressed me very much, and said they would give me great things, and I should not need fear what the Christians might do unto me. Nevertheless I answered, I could not turn unto them for all the world: . . . Then some of the chief of them were displeased very much, and said, If I would not turn to their Religion, I should die. I answered, I should rather die than turn unto them. It was answered, I should then die. So they gave order to the Executioner . . . who haled me away to the place where it was expected I should have been burnt to death with camel's dung, and so sate me down upon the ground, where the Lord preserved me oer the fear of men, though I was as a Sheep prepared for the slaughter.[29]

After this "an ancient, tender man, a Turk, and of great reputation" (288) arrives in time to prevent the Quaker's execution. Robinson is then sent to Gaza, but he returns to Jerusalem one more time to confront the friars before sailing home to England at last.

The Quaker anti-pilgrims were unusual because they vented their hostility toward the Islamic and Roman Catholic Other face-to-face and in public. Their rash, confrontational fanaticism was risky behavior, but the very extremity of their public demonstrations seems to have puzzled officials who lived in a tolerant Levantine cultural environment that was multiethnic and religiously diverse. Taken aback by their histrionic protestations, and the uncompromising nature of their beliefs, Islamic and Roman Catholic officials preferred to send them home instead of making them into martyrs.

Unlike the Quakers, other English visitors in Jerusalem played it safe and kept their objections to themselves until they returned home. Among the other English travelers to Jerusalem during the seventeenth century were George Sandys and William Lithgow. In 1610, these two men, so different in social background and personal demeanor, set out on their separate journeys to the Levant and back. George Sandys (1578-1644) was the youngest son of the Archbishop of

York, and William Lithgow (1582-1645?) was the eldest son of James Lithgow, a burgess of Lanark, Scotland. Both visited the Holy City at Easter time and observed the celebrations there.

In 1609, Lithgow set out for Paris on the first of the three long journeys that are recounted in his *Totall Discourse of the Rare Adventures, and painefull Peregrinations of long nineteene Yeares Travayles, from Scotland, to the most Famous Kingdomes in Europe, Asia and Affrica* (1632). Lithgow claimed to have traveled, in total, more than 36,000 miles during the second decade of the seventeenth century, and most of this was accomplished on foot, without the privileges or comforts enjoyed by his contemporary, Sandys. Samuel Chew aptly describes Lithgow as "a hard, dour, truculent, and pugnacious Scot."[30] He is thrifty, too, and his observations and actions reveal an obsession with getting and spending. His most common complaint against the inhabitants of the Ottoman Empire is the continual need to bribe local officials with forced "tribute."

Lithgow is eager to impress readers with his humanist erudition. As he moves across the land, mostly on foot, Lithgow sometimes looks at the land through the lens of classical text and history, and at other times his perspective is Christian and scriptural. He fancies himself a poet, and his narrative is strewn with original poems inspired by the sights and situations that he encounters.

Lithgow refers with irony to his tour of Palestine as "my Pilgrimage," and like a traditional pilgrim, Lithgow arrived in Jerusalem on Palm Sunday, 1612.[31] There, and at the other Christian sites and shrines in the region, he observes early modern tourism at work. In *The Totall Discourse* he objects repeatedly to the collection of fees levied each day by various officials and tour guides in Palestine: "A journall tribute more fit for a Prince to pay, than a Pilgrime . . ." (213). In his description of the Holy Land, Lithgow alternates between genuine piety and awe (which support the seriousness and depth of his recorded experience), and, on the other hand, expressions of contempt for what he sees as ignorance, idolatry, and superstition. He rails against "all the illusions of their imaginary and false miracles, first invented partly by monasticall poverty, then confirmed by provincial bribery, and lastly they are faith-sold for consistoricall lucre" (218-19). These condemnations of Roman Catholics and of all non-Protestant Christians are delivered, not as a fellow Christian and pilgrim, but from the position of an intelligencer whose task it is to observe and report the religious practice of those Christians who control access to the sacred sites. Lithgow strongly condemns the other Christians as "sinfull, odious and damnable" heretics (217). The radical difference between Protestant dogma and what Lithgow observes other Christians to perform and believe moves him to say that these Christians are worse than Muslims. Lithgow goes so far as to claim that "if it were not for these Images, and superstitious Idolatries, . . . the Turkes had long agoe bene converted to the Christian Faith" (150). At the same time, though, he claims that the errors of Roman Catholic and Levantine Christianity are a corruption brought on by contact with the Turks: "Such is the villanie of these Orientall slaves under the Turkes; that not onely by conversing with them, learne some of their damnable Hethnicke customes, but also going beyond them in beastly sensualnesse, become worse then bruite beasts" (219).

Lithgow is eager to show that his Protestant faith has been tested and confronted by these other religions, and that his resistance to these temptations is the spiritual equivalent of his physical journey and its challenges: "Now thou bottomless Gulfe of Papistrie, here I forsake thee," he declares, "no Winter-blasting Furies of Satans subtile storms, can make ship-wracke of my Faith, on the stony shelfes of thy deceitful deepes" (218). Much of his narrative is thus a Protestant justification of faith, measured against the alleged folly, sensuality, and crookedness of all other religions. During the Easter festival, for example, Lithgow stands apart. At first he claims to offer the reader a "neutral" description of the Church of the Holy Sepulcher, though he does prostrate himself in prayer with the other pilgrims when they arrive at the shrine of Christ's tomb. Then, from the upper gallery of the church, during the three-day festival, Lithgow looks down on "all the Orientall people" (266) and their behavior. The "seven sorts of Nations, different in Religion, and language" who divide and occupy the church are for Lithgow a disorderly Babel, a manifestation of heresy and its capacity to divide and conquer the one truth. The ceremonies conducted on Holy Saturday by the six thousand pilgrims of various "Sects and sorts" are described as "an intollerable tumult, as if they had beene all mad or distracted of their wits," and Lithgow praises the efforts of the Turks to create order out of this chaos: "in the middest of all this hurley burley, [the Turks] runne amongst them with long Rods, correcting their misbehaviour with cruell stroakes: and so these slavish people, even at the height of their Ceremonious devotion are strangely abused" (269-70). At this point, Lithgow drops his descriptive neutrality and shifts into a polemical mode, condemning the "abhominable idolatry" that he witnesses when the pilgrims form a procession carrying an effigy of Christ to Calvary, where "they imbalmed the five timber holes" representing Christ's wounds. Lithgow concludes his account of the Easter rites with a vituperative expression of his Protestant iconoclasm, affiliating himself with the violent discipline of the Turks: "at some times, and in some parts I have torne a peeces those rich garments from their senseless images and blockes, thinking it a greater sinne not to do it than to stand staring on such prodigall prophannesse, with any superstitious respect, or with indifferent forbearance to winke at the wickednesse of Idolaters" (271). Here, the Protestant anti-pilgrim acts as a descecrator who strips the images of their rich coverings, revealing them to be mere dead things, without spirit. This action is symptomatic of Lithgow's general purpose in Jerusalem as an agent of Protestantism, anxiously refuting and violently reviling the rich and complex religious practice of local tradition. Islamic culture permitted diversity and tolerated difference within its imperial boundaries, but for Lithgow this religious tolerance is one of the heretical errors of the Turks' faith and as such it puts the Turkish order in question.

In Jerusalem, Lithgow is asked to join the chivalric order of the Knights of Jerusalem, but he refuses this offer. The crusading romance figure and the neo-medieval pilgrim are not acceptable roles for an English Protestant like Lithgow. For Lithgow, travel is a heroic endeavor because it is a dangerous test of body and soul—it is a grueling marathon of self-control, not an abandonment to irrational devotion or exotic sensuality. He advises all travelers to "constantly refraine from whoredome, drunkennesse, and too much familiarity with Strangers: for a Traveller that is not temperate, and circumspect in all his actions, although he were headed like that Herculean Serpent Hydra, yet it is impossible he can returne in safety from danger of Turkes, Arabs, Moores, wild beasts, & the deadly operative extremities of heat, hunger, thirst, and cold" (224-25).

Lithgow's contemporary, George Sandys, visited many of the same sites, but Sandys certainly faced fewer dangers from heat and hunger. Sandys had high connections at court, and though he is best-known today as a poet and translator of Ovid, he also pursued a successful career as a colonizer. He left England for the Levant in 1610 and returned in 1611. His travel narrative, *A Relation of a Journey . . . Containing a description of the Turkish Empire, of Aegypt, of the Holy Land, of the Remote Parts of Italy, and Ilands adjoyning. Begun A.D. 1610,* was printed in 1615 with further editions appearing in 1621, 1627, 1637, 1652, and 1673. Following his travels in the Mediterranean, Sandys became involved in various commercial and colonial enterprises: he was one of the undertakers named in the third Virginia charter of 1611 and he took shares in the Bermuda Company. In April 1621, Sandys was appointed by the Virginia Company as treasurer of the colony. He sailed to America, where he served on the colonial government's ruling council. Sandys left Virginia for home in 1631. Soon after his return to England, he became a gentleman of the privy chamber under his longtime patron, Charles I, to whom the *Relation* is dedicated.[32]

For Lithgow and Sandys, the three most attractive cities in the Ottoman territories were Jerusalem, Cairo, and Constantinople. Jerusalem was important as the site of Christ's crucifixion and resurrection, in spite of what these Protestants saw as the errors and decadence of later Christian civilization. In Egypt, Cairo and its pyramids evoked references to ancient civilization and to Hebrew scripture, especially the story of Moses in Egypt from Exodus. But Constantinople was of interest both for its past and its present, as the former seat of the Byzantine emperors, and as the site of Ottoman power and home for the sultan's court and harem. Sandys's narrative contains a long section describing life and customs in Constantinople and at the Ottoman palace there. The sultan's seraglio at Constantinople and the Church of the Holy Sepulcher in Jerusalem stand at opposite poles in the ideological construction of the Levant by English culture. Both are sites of disturbing contradiction: the holiest place in Christendom, traditionally thought of as the spiritual center of the world, was under the rule of the "infidel" Turk and dominated by what Protestants considered to be heretical sects. The sultan's harem was the hub of the world's greatest empire, an empire whose power, discipline, and masculine order were said to be enforced by extreme cruelty; but at the same time, the sultan's palace was the *locus classicus* for effeminate luxury, hidden sensuality, and sexual excess.

The sultan's seraglio was a particular obsession with Western describers of Turkish culture, and the early modern reports and translations that depict the Turks and their customs dwell voyeuristically in the forbidden space of the sultan's harem. In Robert Withers's 1625 translation of Ottaviano Bon, for example, the reader is tantalized by the secrets of the seraglio: "for there may none come near, nor be in sight of them [the sultan and his women], but himself and

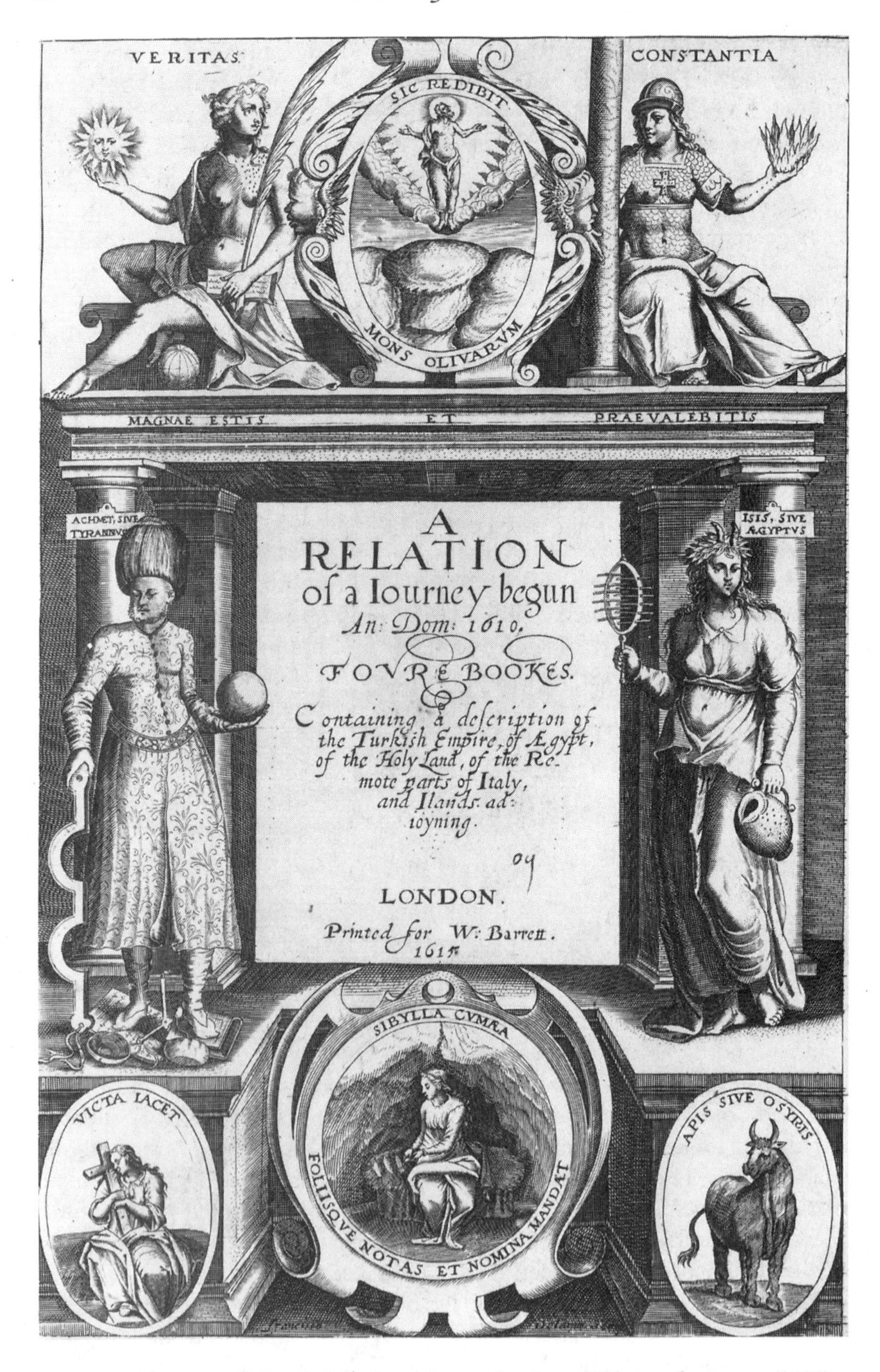

Figure 1. Title page and frontispiece from George Sandys, *A Relation of a Journey* (1615). Note the turbaned figure of the sultan, Ahmet I, on the left. He holds a sphere in his left hand, emblem of worldly power, and a yoke in his right, indicating tyranny. He tramples on the scales of justice and on a book, symbolizing his scorn for justice and learning.

his black eunuchs: nay if any other should but attempt, by some trick in creeping into some private corner, to see the women, and should be discovered, he should be put to death immediately."[33] The women never leave the precincts of the seraglio, and the eunuchs who return from the city are routinely searched. This is explained in a French description of the Ottoman court that was translated by Edward Grimestone in 1635:

> They doe not only search the women which enter, and the Eunuches at their returne from the Citie: But moreover they have a care of beasts: They will not allow the Sultanaes to keepe any Apes, nor Dogges of any stature. Fruits are sent unto them with circumspection: If their appetites demand any pompeons which are somewhat long, or cowcumbers, and such other fruits, they cut them at the Gate in slices, not suffering to passe among them any slight occasion of doing evill, so bad an opinion they have of their continencie. It is (without doubt) a signe of the Turks violent jealousie: for who can in the like case hinder a vicious women from doing evill? She is too industrious in her designs; and he which had his body covered with eyes always watching was deceived. In the meanetime if any woman in the Serrail be discovered in the effects of her lasciviousnesse, the Law long since established for them by the Sultan, condemnes her to die, the which is executed without remission: she is put into a sack, and in the night cast into the Sea, where she doth quench her flames with her life.[34]

This passage, with its emphasis on both sexual control and tyrannical cruelty, is typical of early modern representations of the sultan and his court. The libidinal regime at the center of Ottoman power is imagined as the core from which the masculine aggression of Turkish military power emanates. Some European authors (including, as we have seen, Blount) admired Ottoman power as a virile and masculine force capable of maintaining a strict discipline and order, but often they argued that such extreme discipline was only possible where relentless cruelty, confinement, and fear produce a slavish obedience. Spectacular punishment and assassination are often emphasized in early modern accounts of Turkish society and government: Blount, for example, describes the "horrid executions" such as "Empaling, Gaunching, Flaying alive, cutting off by the Waste with a red hot iron, [An]Ointing with honey in the Sunne, hanging by the Foot, planting in burning Lime, and the like" (52). Such extreme measures were said to be the key to the Ottomans' success as conquerors and rulers. At the same time, this Turkish cruelty was also linked to sensuality, as European authors sought to effeminate the masculine power they feared and envied. Again Blount: "for hee [the Prophet Mohammed] finding the Sword to be the foundation of Empires, and that to manage the Sword, the rude and sensuall are more vigorous, then wits softned in a mild rationall way of civilitie; did first frame his institutions to a rude insolent sensualitie . . ." (78). Islam itself was described by Christian writers as a religion based on sexual license, permitting polygamy and instant divorce. Furthermore, Turks or Muslims are represented as practitioners of "sodomy": Blount reports that beside their wives, "each Basha hath as many, or likely more, Catamites, which are their serious loves; for their wives are used (as the Turkes themselves told me) but to dresse their meat, to Launderesse, and for reputation . . ." (14).[35]

This construction of Turkish identity as both cruel and sensual is also present in Sandys's description of Ahmet I and his seraglio.[36] The sultan himself becomes an embodiment of the contradictions in the English construction of Islamic power: he is "strongly limd, & of a just stature, yet greatly inclining to be fat: insomuch as sometimes he is ready to choke as he feeds" (73). "His aspect is as hauty as his Empire is large," and he is "an unrelenting punisher of offences, even in his owne household: having caused eight of his Pages . . . to be throwne into the Sea for Sodomy (an ordinary crime, if esteemed a crime, in that nation) . . ." (73). Thus, the imperial household is represented as a microcosm of the empire, where extreme cruelty and sexual excess are both exercised. The sultan's sexual prodigiousness stands in contrast to the castrated eunuchs that surround him. Military order and hierarchy dominate Sandys's account of the male inhabitants of Topkapi palace, while the description of the sultan's women is a male fantasy, transgressive in its excess (Sandys's sultan possesses 500 beautiful virgin slaves, "the choisest beauties of the Empire"). Sexual supply and demand comprise a paradoxical economy by which the excess of bodies provided for the sultan's pleasure also becomes his procreative weakness or lack: "But for all his multitude of women, he hath yet begotten but two sonnes and three daughters, though he be that way unsatiatably given, (perhaps the cause he hath so few) and vieth all sorts of foods that may inable performance" (75). Though the harem is secret and off limits to all men but the sultan himself and his imperial eunuchs, Sandys's narrative intrudes voyeuristically in the seraglio by offering a physical description of the imperial concubines, and an engraved image of a Turkish woman is included with Sandys's text.

In this picture of imperial sexuality, and throughout Sandys's description of the Turkish nation, there is a condescending and superior tone that thinly veils a deeper sense of imperial envy, sometimes exacerbated by an anxious feeling of competition with the other Europeans in the Levant. Although he was a privileged representative of the English nation, Sandys's confrontation with the wealth and power of the Turk did not allow him to articulate a discourse of domination. Rather, an inferiority complex marks his text. This is indicated in Sandys's formal dedication to Prince Charles. Here, we can detect nuances of the future colonizer, who would seek power and wealth in a new empire where no Turk or Muslim would be found:

> The parts I speak of are the most renowned countries and kingdomes: once the seats of most glorious and triumphant Empires; the theaters of valour and heroicall actions; the soiles enriched with all earthly felicities . . . where wisedome, vertue, policie, and civility have been planted, have flourished: and lastly where God himselfe did place his owne Commonwealth. . . .

According to Sandys's dedication, though, this noble Mediterranean world has been taken over and ruined by Turkish barbarism and tyranny:

> Which countries once so glorious, and famous for their happy estate, are now through vice and ingratitude, become the most deplored spectacles of extreme miserie: the wild beasts of mankind having broken in upon them, and rooted out all civilitie; and the pride of a stern and barbarous Tyrant possessing the thrones of ancient and just dominion.

The text that follows the dedication, however, belies the claim made in the above quotation: the impression we get from reading the whole of Sandys's *Relation* is of a sophisticated and prosperous Ottoman Empire, not a wasteland. In fact, for Sandys, Blount, and others, one way that the construction of an Islamic Other helped in the formation of English national and religious identity was by pointing to the Turks as a model for emulation. In the above passage from his dedication, however, Sandys dismisses the Old Mediterranean World like sour grapes—though he praises the ancient vintage of classical and Christian empires. Later, in the body of his *Relation,* Sandys moves away from the rhetoric that denounces and belittles Ottoman power, and writes instead of Constantinople as a place of "marvellous splendor" (31). The palace grounds of the sultan are praised: "Luxury being the steward, and the treasure unexhaustable" (32), and Sandys commends the palace buildings for their "costly curiousnesse, matter, and amplitude" (33). Though often envious in tone, and sometimes scornful of "the Mahometan superstition" (43) and of Turkish customs, Sandys's description of the Ottoman territories, from Greece to Egypt, acknowledges a fertile, populous land and an elaborately ordered, sophisticated civilization.

While expressing both admiration and envy for the Turks, Sandys and his English contemporaries could not help but pay tribute to the Ottoman imperial achievement. English writers acknowledged Turkish supremacy, and English merchants sought profits, not conflict. The Islamic conquests were explained away as punishment for Christian (especially non-Protestant) sin and corruption, teaching the lesson "of the frailty of man, and mutability of what so ever is worldly . . ."[37] The notion that papist superstition and sultanic tyranny had brought ruin to the Old Imperial World encouraged some English merchants to turn toward the New World and new empires beyond the reach of Ottoman power. The "rich lands" that Sandys describes so longingly in his dedicatory epistle, areas that supposedly remained "waste and overgrowne" because of Turkish violence and oppression, with "large territories dispeopled, or thinly inhabited," prefigure the colonialist fantasy of a fruitful but vacant land that awaits conquest at the hands of a noble, Christian people. This fantasy would be revived and redirected toward the project of the New World plantations, including the colony at Virginia, where Sandys himself would serve for nearly ten years as a colonial administrator and profiteer.

During the course of the seventeenth century, the development of English trade in the Mediterranean and the colonizing voyages to the Americas were both propelled by the same proto-imperialist energy. England's cultural and textual trafficking with the Turk helped to develop an ideological basis for the colonizing process that would continue and intensify during the coming centuries, and the representation of Turks and Muslims in early modern travel narratives played a small but significant role in this process. The early modern encounter with Islam and the Turks encouraged English merchants and ship captains to maintain an aggressive, outward-looking posture. Their knowledge of the Ottoman empire made it easier to negotiate with its power and to make profitable exchange as partners-in-trade with the Turks. This built confidence and filled the coffers of the merchant-investors who then looked to express their developing imperial identity in the New World. There, the English themselves would be the cruel and powerful masters of a new order, and the profits and commodities produced or obtained in America would be taken from peoples and places that the English could subdue and dominate. English merchant venturers imagined a fresh and uncontaminated "commonwealth," waiting to be built by the Elect Nation in Virginia, far from Constantinople and Jerusalem.

Notes

1. Nabil Matar discusses the presence of "Britons among the Muslims," including pirates and captives, in chapter two of *Turks, Moors and Englishmen in the Age of Discovery* (New York: Columbia University Press, 1999).
2. For one historian's account of the degree to which a sense of anti-Islamic unity was preserved among the Christians of Western Europe, see Franklin L. Baumer, "England, the Turk, and the Common Core of Christendom," *The American Historical Review* (1945): 26-48.
3. See Dorothy M. Vaughan, *Europe and the Turk: A Pattern of Alliances, 1350-1700* (Liverpool: Liverpool University Press, 1954) for a survey of European relations with the Ottoman sultanate during this period. On the competition between England and other Christian nations, see Niels Steensgard, "Consuls and Nations in the Levant from 1570 to 1650" in *Merchant Networks in the Early Modern World,* ed. Sanjay Subrahmanyam (Aldershot: Variorum, 1996), 13-53.
4. This connection between trade in the Levant and commercial expansion elsewhere is made by Robert Brenner in *Merchants and Revolution: Commercial Exchange, Political Conflict and London's Overseas Traders, 1550-1653* (Princeton: Princeton University Press, 1993) and in his article, "The Social Basis of English Commercial Expansion," in *Merchant Networks in the Early Modern World,* 361-84. A recent study by Gigliola Pagano de Divitis, *English Merchants in Seventeenth-Century Italy,* trans. S. Parkin (Cambridge: Cambridge University Press, 1997) also sheds some light on this issue.
5. Theodore K. Rabb, *Enterprise and Empire: Merchant and Gentry Investment in the Expansion of England, 1575-1630* (Cambridge, MA: Harvard University Press, 1967), 2-3. See also T. S. Willan, "Some Aspects of English Trade with the Levant in the Sixteenth Century," *English Historical Review* 70 (1955): 399-410; Ralph Davis, "England and the Mediterranean, 1570-1670" in *Essays in the Social and Economic History of Tudor and Stuart England,* ed. F. J. Fisher (Cambridge: Cambridge University Press, 1961); and Kenneth R. Andrews, *Trade, Plunder and Settlement: Maritime Enterprise and the Genesis of the British Empire, 1480-1630* (Cambridge: Cambridge University Press, 1984), 87-100.
6. Continuous trade with autonomous Morocco had begun around 1550. When relations with Spain deteriorated in the 1570s, the English agreed to trade bullets to the Moroccans in exchange for saltpeter. During the early 1580s, the Barbary Company was established to control official trade between England and Morocco. Relations between these two countries continued to flourish, culminating in the arrival of a large Moroccan embassy in London in 1600. On the Barbary trade of this period, see Henry de Castries, *Les Sources Inedites de L'Histoire du Maroc,* 1st series, Dynastie Saadienne (1530-1660), Archives et Biblioteques D'Angleterre, vol. 1 of 2 (Paris: E. Leroux Luzac, 1918), 445-54; and T. S. Willan, *Studies in Elizabethan Foreign Trade* (Manchester: Manchester University Press, 1959), 184ff.
7. For a careful reconstruction of these events, accompanied by relevant documents, consult S. A. Skilliter, *William Harborne and the Trade with Turkey, 1578-1582: A documentary study of the first Anglo-Ottoman relations* (Oxford: Oxford University Press, 1977).
8. Cited in Willan, "Some Aspects of English Trade," 400.
9. For a more detailed history of the Levant company's foundation and later development, consult Mortimer Epstein, *The Early History of the Levant Company* (London: G. Routledge & Sons, 1908) and Alfred C. Wood, *A History of the Levant Company* (London: Oxford University Press, 1935).

10. Richard Hakluyt, *The Principal Navigations, Voyages, Traffiques and Discoveries of the English Nation made by sea or over-land to the remote and farthest distant quarters of the earth at any time within the compass of these 1600 yeares,* 3 vols. (London, 1598-1600); reprinted in 12 vols. (Glasgow: J. MacLehose and Sons, 1903-1905), 1 of 12 vols:lxx.
11. For a survey of English travel writings from this period, see Samuel Chew, *The Crescent and the Rose: Islam and England during the Renaissance* (Oxford: Oxford University Press, 1937), 33-53. Chew provides the fullest and most useful survey, but two other studies are also worth consulting: Brandon H. Beck, *From the Rising of the Sun: English Images of the Ottoman Empire to 1715* (New York: P. Lang, 1987) and Orhan Burian, "Interest of the English in Turkey as Reflected in English Literature of the Renaissance," *Oriens* 5 (1952): 208-29.
12. See William H. Sherman, *John Dee: The Politics of Reading and Writing in the English Renaissance* (Amherst: University of Massachusetts Press, 1995).
13. See James P. Helfers, "The Explorer or the Pilgrim? Modern Critical Opinion and the Editorial Methods of Richard Hakluyt and Samuel Purchas," *Studies in Philology* 94:2 (Spring 1997): 160-86.
14. For a list of such texts, consult the bibliography by Edward G. Cox, *A Reference Guide to the Literature of Travel, Including Voyages, Geographical Descriptions, Adventures, Shipwrecks, and Expeditions* (Seattle: University of Washington Publications in Language and Literature), vols. IX (November 1935), X (May 1938), and XII (May 1949). Reprinted New York: Greenwood Press, 1969. Three vols.
15. In chapter four, "The Voyages of a Nation," of Richard Helgerson's *Forms of Nationhood: The Elizabethan Writing of England* (Chicago: University of Chicago Press, 1992). For a more recent analysis of Hakluyt's texts as propaganda for English overseas expansion, see Pamela Neville-Sington, "'A Very Good Trumpet': Richard Hakluyt and the Politics of Overseas Expansion" in Cedric C. Brown and Arthur F. Marotti, *Texts and Cultural Change in Early Modern England* (New York: Macmillan, 1997): 66-79. Also worth consulting is Andrew Hadfield, *Literature, Travel, and Colonial Writing in the English Renaissance, 1524–1625* (Oxford: Clarendon Press, 1998).
16. David B. Quinn, ed., *The Hakluyt Handbook* (London: Hakluyt Society, 1974), xii.
17. Quinn, *The Hakluyt Handbook,* xiii.
18. Hakluyt, *Principall Navigations,* 1: 2v.
19. For examples of this kind of information, see the sections entitled "Notes concerning the trade of Argier" (190-91) and "Notes concerning the trade in Alexandria" (191-92) in the first volume of Richard Hakluyt, *The Principall Navigations* (1598-1600). These texts are discussed in P. Wittek, "The Turkish Documents in Hakluyt's 'Voyages,'" *Bulletin of the Institute of Historical Research* 19: 57 (Nov. 1942): 121-39.
20. C. F. Beckingham, "The Near East: North and Northeast Africa," ch. 14 in vol. 1 of Quinn, *The Hakluyt Handbook,* 184.
21. See G. B. Parks on "Tudor Travel Literature" in chapter 9 of Quinn, *The Hakluyt Handbook;* and E. G. R. Taylor, *Late Tudor and Early Stuart Geography* (1934; reprint New York: Octagon Books, 1968). Anthony Parr's excellent introduction to *Three Renaissance Travel Plays* (Manchester: Manchester University Press, 1995) is also helpful.
22. Fynes Moryson, *An Itinerary Containing His Ten Yeeres Travell,* vol. 1 (Glasgow: James MacLehose & Sons, 1907).
23. Sir Henry Blount, *A Voyage into the Levant: A Breife Relation of a Journey, Lately Performed by Master H[enry] B[lount], Gentleman, from England by the way of Venice, into Dalmatia, Sclavonia, Bosnia, Hungary, Macedonia, Thessaly, Thrace, Rhodes and Egypt unto Gran Cairo: With Particular Observations Concerning the moderne condition of the Turkes and other people under that Empire* (London, 1636), 3.
24. Sanderson's collected papers were discovered and edited by Sir William Foster as *The Travels of John Sanderson in the Levant, 1584–1602* (London: Hakluyt Society, 1931).
25. Foster, *The Travels of John Sanderson,* 45. This was part of a cargo aboard the *Hercules,* a ship that returned to London from the Levant in 1588 with a cargo reported to be worth 70,000 pounds! See Willan, "Some Aspects of English Trade," 407-8.
26. [William and Peter Biddulph], *The Travels of certaine Englishmen into Africa, Asia, Troy, Bythnia, Thracia, and to the Black Sea. And into Syria, Cilicia, Pisidia, Mesopotamia, Damascus, Canaan, Galile, Samaria, Judea, Palestina, Jerusalem, Jericho, and to the Red Sea and to sundry other places. Begunne in. . .1600, and by some of them finished this year 1608. The others not yet returned* (London, 1609), A2r. This text begins with a "Preface" by (pseud.) Theophilus Lavender, who claims that the text is a collation of more than 20 letters sent by William Biddulph ("Preacher to the Company of Eng-

lish Merchants resident in Aleppo") and his brother, Peter Biddulph ("Lapidarie and Diamond cutter in those Countries" [A1r]) to his relation, one Bezaliel Biddulph, in England.

27. Biddulph, *The Travels of certaine Englishmen,* A4v-B1r. My emphasis. While in Jerusalem, Biddulph and his group met with two other Englishmen, Henry Timberlake (or Tymberley) and John Burrell. Timberlake also wrote a letter describing his experiences in the Holy Land, and it was printed as *A true and strange discourse of the late travailes of two English Pilgrimes* (London, 1603).
28. The writings of Evans and Cheevers are discussed and contextualized in Elaine Hobby, *Virtue of Necessity: English Women's Writing, 1649–88* (Ann Arbor: University of Michigan Press, 1987), 36-41.
29. *A true account of the great tryals and cruel sufferings undergone by those faithful servants of God, Katherine Evans and Sarah Cheevers . . . To which is added, a short relation from George Robinson, of the sufferings that befel him in his journey to Jerusalem; and how God saved him from the hands of cruelty when the sentence of death was passed against him* (London, 1663), 286-87.
30. Chew, *The Crescent and the Rose,* 39.
31. Lithgow, *Totall Discourse,* 213.
32. See Richard Beale Davis, *George Sandys, Poet Adventurer: A Study in Anglo-American Culture in the Seventeenth Century* (London: Bodley Head, 1955).
33. Ottaviano Bon, trans. Robert Withers, *A Description of the Grand Signor's Seraglio, or Turkish Emperours Court* (London, 1625), 65. A version of this text was originally published in Purchas. It was reprinted in 1650 and 1653. See the editor's introduction to *The Sultan's Seraglio,* trans. Robert Withers, ed. Godfrey Goodwin (London: Saqi Books, 1996).
34. Seigneur Michael de Baudier de Languedoc. *Histoire Generalle du Serrail, et de la Cour du Grand Seigneur Empereur des Turcs* (Paris, 1624) trans. Edward Grimestone, *The History of the Imperiall Estate of the Grand Seigneurs* (London, 1635), 65.
35. See Nabil Matar, chapter 4, "Sodomy and Conquest" in *Turks, Moors and Englishmen in the Age of Discovery.*
36. For a description of what the Ottoman seraglio was really like, based on primary archival work, consult Leslie Penn Peirce, *The Imperial Harem: Women and Sovereignty in the Ottoman Empire* (New York: Oxford University Press, 1993).
37. Sandys, dedicatory epistle.

SECTION TWO

THOMAS DALLAM

Thomas Dallam was born in Dallam, Lancashire, sometime around 1570. He joined the Blacksmith's Company to learn the trade of organmaking, and was admitted as a liveryman. After building the mechanical organ described in the text, along with his journey accompanying it to Constantinople, Dallam returned to England, where his son Robert was born in 1602 (the date of his marriage is not known). In 1605 he was commissioned to build an organ for King's College Chapel, Cambridge; either Dallam or one of his sons continued to tune this organ until 1641. In 1613 he was commissioned to build double organs for Worcester Cathedral and in 1617 for the Chapel Royal at Holyrood House, Edinburgh. From 1624 to 1627 he collaborated with his son Robert on an organ for Durham Cathedral. Dallam had a second son, Ralph, as well as a daughter, whose name is unknown. Both Ralph and Robert, as well his daughter's son, René Harris, followed Thomas Dallam in his trade; the family were perhaps the most important organmakers in seventeenth-century England.

The excerpts from his diary printed here are taken from J. Theodore Bent, ed., *Early Voyages and Travels to the Levant,* by the Hakluyt Society, no. 87 (London, 1893).

—Mary C. Fuller

"In this Book is the Account of an Organ Carryed to the Grand Seignor and Other Curious Matter"

[Thomas Dallam, the narrator, has come to Constantinople to service an elaborate mechanical organ he has made, which the Levant Company is to present to the Turkish ruler as a gift.]

SEPTEMBER

The 11th Daye, beinge Tusdaye, we Carried our instramente over the water to the Grand Sinyors Courte, Called the surralya [*seraglio*], and thare in his moste statlyeste house I began to sett it up. . . . At everie gate of the surralia thare alwayes sitethe a stoute Turke, abute the calinge or degre of a justis of the peace, who is caled a chia; not withstandinge, the gates ar faste shut, for thare pasethe none in or oute at ther owne pleasures. . . . These gates ar made all of massie iron; tow men, whom they do Cale jemeglans,[1] did open them.

The 15th, I finished my worke in the Surraliao, and I wente once everie daye to se it, and dinede Thare almoste everie Daye for the space of a monthe; which no Christian ever did in there memorie that wente awaye a Christian.

The 18 daye (stayinge somthinge longe before I wente), the Coppagawe[2] who is the Grand Sinyor's secritarie, sente for me that one of his frendes myghte heare the instramente. Before I wente awaye, the tow jemaglanes, who is keepers of that house, touke me in theire armes and Kised me, and used many perswations to have me staye with the Grand Sinyor, and sarve him.

[*Dallam recounts the presentation of the organ.*]

The Grand Sinyor, beinge seated in his Chaire of estate, commanded silence. All being quiett, and no noyes at all, the presente began to salute the Grand Sinyor; for when I lefte it I did alow a quarter of an houre for his cominge thether. Firste the clocke strouke 22; than The chime of 16 bels went of, and played a songe of 4 partes. That beinge done, tow personagis which stood upon to corners of the seconde storie, houldinge tow silver trumpetes in there handes, did lifte them to theire heades, and sounded a tantarra.[3] Than the muzicke went of, and the orgon played a song of 5 partes twyse over. In the tope of the orgon, being 16 foute hie, did stande a holly bushe full of blacke birds and thrushis, which at the end of the musick did singe and shake theire wynges. Divers other motions thare was which the Grand Sinyor wondered at. Than the Grand Sinyor asked the Coppagawe[4] yf it would ever doo the lyke againe. . . . Cothe he: I will se that. In the meane time, the Coppagaw, being a wyse man, and doubted whether I hade so appoynted it or no, for he knew that it would goo of it selfe but 4 times in 24 houres, so he cam unto me, for I did stand under the house sid, wheare I myghte heare the orgon goo, and he asked me yf it would goo againe at the end of the nexte houre; but I tould him . . . yf it would please him, that when the clocke had strouk he would tuche a litle pin with his finger, which before I had shewed him, it would goo at any time. Than he sayde that he would be as good as his worde to the Grand Sinyor. When the clocke began to strick againe, the Coppagaw went and stood by it; and when the clocke had strouke 23, he tuched that pinn, and it did the lyke as it did before. Than the Grand Sinyor sayed it was good. He satt verrie neare vnto it, ryghte before the Keaes [keys], wheare a man should playe on it by hande. He asked whye those keaes did move when the orgon wente and nothinge did tuche them. He Tould him that by those thinges it myghte be played on at any time. Than the Grand Sinyor asked him yf he did know any man that could playe on it. He sayd no, but he that came with it coulde, and he is heare without the dore. Fetche him hether, cothe the Grand Sinyor, and lett me se how he dothe it. Than the Coppagaw opemed that Dore which I wente out at, for I stoode neare unto it. He came and touke me by the hande, smylinge upon me; but I bid my drugaman[5] aske him what I should dow, or whither I shoulde goo. He answered that it was the Grand Sinyore's pleasur that I should lett him se me playe on the orgon. So I wente with him. When I came within the Dore, That which I did se was verrie wonderfull unto me. I cam in direcktly upon the Grand Sinyore's ryghte hande, som 16 of my passis (paces) from him, but he would not turne his head to louke upon me. He satt in greate state, yeat the sighte of him was nothinge in Comparrison of the traine that stood behinde him, the sighte whearof did make me almoste to thinke that I was in another worlde. The Grand Sinyor satt stille, behouldinge the presente which was befor him, and I stood daslinge my eyes with loukinge upon his people that stood behinde him, the which was four hundrethe persons in number. Two hundrethe of them weare his princepall padgis [*pages*], the yongest of them 16 yeares of age, som 20, and som 30. They weare apparled in ritche clothe of goulde made in gowns to the mydlegge; upon theire heades litle caps of clothe of goulde, and some clothe of Tissue;[6] great peecis of silke abowte theire wastes instead of girdls, upon their leges Cordivan buskins,[7] reede. Theire heades wear all shaven, savinge that behinde Their ears did hange a locke of hare like a squirel's taile; theire beardes shaven, all savinge theire uper lips. Those 200 weare all verrie proper men, and Christians borne.

When I had stode almost one quarter of an houre behouldinge this wonder full sighte, I harde the Grande Sinyore speake unto the Coppagaw, who stood near unto him. Than the Coppagaw cam unto me, and touke my cloake from aboute me, and laye it Doune upon the Carpites, and bid me go and playe on the organ; but I refused to do so, because the Grand Sinyor satt so neare the place wheare I should playe that I could not com at it, but I muste needes turne my backe Towardes him and touche his Kne with my britchis, which no man, in paine of deathe, myghte dow, savinge only the Coppagaw. So he smyled, and lett me stande a litle. Than the Grand Sinyor spoake againe, and the Coppagaw, with a merrie countenance, bid me go with a good curridge, and thruste me on. When I cam verrie neare the Grand Sinyor, I bowed my heade as low as my kne, not movinge my cape, and turned my backe righte towardes him, and touched his kne with my britchis.

He satt so righte behinde me that he could not se what I did; tharfore he stood up, and his Coppagaw removed his Chaire to one side, wher he myghte my handes; but in his risinge from his chaire, he gave me a thruste forwardes, which he could not otherwyse dow, he satt so neare me; but I thought he had bene drawinge his sorde to cut of my heade.

I stood thar playinge suche thinge as I coulde untill the cloke stroucke, and than I boued my heade as low as I coulde, and wente from him with my backe towardes him. As I was taking of my cloake, the Coppagaw came unto me and bid me stand still and lett my cloake lye; when I had stood a litle whyle, the Coppagaw bid me goo and cover the Keaes of the organ; then I wente Close to the Grand Sinyor againe, and bowed myselfe, and then I wente backewardes to my Cloake. When the Company saw me do so theye semed to be glad, and laughed. Than I saw the Grand Sinyor put his hande behind him full of goulde, which the Coppagaw Receved, and broughte unto me fortie and five peecis of gould called chickers[8] and than was I put out againe wheare I came in, beinge not a little joyfull of my good suckses.

The laste of September I was sente for againe to the surralia to sett som thinges in good order againe, which they had altered, and those tow jemoglans which kepte that house made me verrie kindly welcom, and asked me that I would be contented to stay with them always, and I should not wante anythinge, but have all the contentt that I could desire. I answered them that I had a wyfe and Childrin in Inglande, who did expecte my returne. Than they asked me how long I had been married, and how many children I hade. Thoughe in deede I had nether wyfe nor childrin, yeat to excuse my selfe I made them that Answeare.

Than they toulde me that yf I would staye the Grand Sinyor would give tow wyfes, ether tow of his Concubines or els tow virgins of the beste I Could Chuse my selfe, in Cittie or contrie.

The same nyghte, as my Lorde was at supper, I tould him what talke we had in the surralya, and whate they did offer me to staye thare, and he bid me that by no mearies I should flatly denie them anythinge, but be as merrie with them as I could, and tell them that yf it did please my Lorde that I should stay, I should be the better contented to staye; by that meanes they will not go about to staye you by force, and yow may finde a time the better to goo awaye when you please.

OCTOBER

The 12, beinge Fridaye, I was sente for to the Courte, and also the Sondaye and Monday folloinge, to no other end but to show me the Grand Sinyors privie Chamberes, his gould and silver, his chairs or estate; and he that showed me them would have me to sitt downe in one of them, and than to draw that sord out of the sheathe with the which the Grand Sinyor doth croune his kinge.

When he had showed me many other thinges which I wondered at, than crossinge throughe a litle squar courte paved with marble, he poynted me to goo to a graite in a wale, but made me a sine that he myghte not goo thether him selfe. When I came to the grait the wale was verrie thicke, and graited on bothe the sides with iron verrie strongly; but through that graite I did se thirtie of the Grand Sinyors' Concobines that weare playinge with a bale in another courte. At the firste sighte of them I thoughte they had bene yonge men, but when I saw the hare of their heades hange doone on their backes, platted together with a tasle of smale pearle hanginge in the lower end of it, and by other plaine tokens, I did know them to be women, and verrie prettie ones in deede.

Theie wore upon theire heades nothinge bute a litle capp of clothe of goulde, which did but cover the crowne of her heade; no bandes a boute their neckes, nor anythinge but faire cheans of pearle and a juell hanginge on their breste, and juels in their ears; their coats weare like a souldier's mandilyon,[9] som of reed sattan and som of blew, and som of other collors, and grded like a lace of contraire collor; they wore britchis of scamatie,[10] a fine clothe made of coton woll, as whyte as snow and as fine as lane[11]; for I could desarne the skin of their thies throughe it. These britchis cam doone to their mydlege; som of them did weare fine cordevan buskins, and som had their leges naked, with a goulde ringe on the smale of her legg; on her foute a velvett panttoble[12] 4 or 5 inches hie. I stood so longe loukinge upon them that he which had showed me all this kindnes, began to be verrie angrie with me. He made a wrye mouthe, and stamped with his foute to make me give over looking; the which I was verrie lothe to dow, for that sighte did please me wondrous well.

Than I wente awaye with this Jemoglane to the place wheare we lefte my drugaman or intarpreter, and I tould my intarpreter that I had sene 30 of the Grand Sinyores Concobines; but my intarpreter advised me that by no meanes I should speake of it, whearby any Turke myghte hear of it; for if it weare knowne to som Turks, it would presente deathe to him that showed me them. He durste not louke upon them him selfe. Although I louked so longe upon them, theie saw not me, nether all that whyle louked towards that place. Yf they had sene me they would all have come presently thether to louke upon me, and have wonddred as moche at me, or how I cam thether, as I did to se them.

The nexte daye our shipp caled the Heckter, beinge reddie to departe, I wente to carrie my beed and my Chiste aborde the shipp. Whyleste I was aborde the shipp, thar came a jemoglane or a messenger from the surralia to my lord imbassador, with an express comand that the shipp should not departd, but muste stay the Grand Sinyores pleasur. When my lord hard this messidge, with suche a comande, he begane to wonder, what the Cause should be. . . .

The messenger tould him that he did not know the cause whre, nether whearfore, but he did hearde the chia say that yf the workman that sett up the presente in the surralia would not be perswaded to stay be hind the shipe, the ship muste staye untill he had removed the presente unto another place.

Than my Lorde inquiered for me and sente one to the ship whear I was, who tould me that I muste com presently to my Lorde; so when I came to my lorde I found with him another messinger, who broughte the sartaintie of the matter that it was for no other cause but for my stainge to remove the organ; but when my lord tould me that I muste be contented to staye and Lette the ship goo, than was I in a wonderfull perplixatie, and in my furie I tould my lorde that that was now com to pass which I ever feared, and that was that he in the end would betray me, and turne me over into the Turkes hands, whear I should Live a slavish Life, and never companie againe with Christians, with many other suche-like words.

My Lord verrie patiently gave me leve to speake my mynde. Than he lay his hand on my shoulder and tould that as he was a Christian him selfe, and hooped tharby to be saved, it was no plote of his, nether did he know of any suche matter as this till the messinger came. In the end cothe he: Be yow contented to staye, and let the ship goo; and it shall cost me 5 hundrethe pound rether than yow shalbe Compeled to stay a day Longer than yow are willinge your selfe after yow have removed the presente. . . .

My Lorde did speake this so frindly and nobly unto me, that upon a sodon he had altered my mynde, and I tould him that I would yeld my selfe into Godes hand and his.

Than said my Lorde: I thanke yow, I will send to the shipe for suche thinges as yow desier to have lefte behinde, for yow muste goo presently to the surralia to se the place wheare yow muste sett up the presente, or els they thinke that yow mean not to com at all; so away wente I with my drugaman or interpreter my ould way to the surralia gates, the which they willingly opened, and bid me welcome when I came to that house wheare the present did stande. Those jenoglanes, my ould acquitance which kept that house, and had bene appointed by the Grand Sinyor to perswade me to staye thare allwayes, as indeed theie had done diveres times and diveres wayes, now they thoughte that I would staye in deed, theye imbraced me verrie kindly, and kiste me many times. What my dragaman said to them I know not, but I thinke he told them that I would not staye, tharfore, when I was gone oute of the house doune som 4 or 5 steps into a courte, as I was putting on my pantabls, one of these jemoglanes cam behinde me and touke me in his armes and Carried me up againe into the house, and sett me doune at that dore wheare all the Grand Sinyore's brothers weare strangeled that daye he was made Emprore. My intrpreter folloed apase. When he that carried me had sette me doune, I bid my drugaman aske him why he did so, and he, seinge me louk merrely, he him selfe laughed hartaly, and saide that he did so but to see how I would tak it yf they should staye me by force. Than I bid my dragaman tell him that they should not need to go aboute to staye me by force, for I did staye willingly to doo the Grand Sinyor all the sarvis that I could.

Than these 2 jemoglanes wente with me to show me the house wheare unto the present should be removed.

The 24 my worke was finished.

The 25 I went to that place againe with the Coppagaw, to show him somethinges in the presente, and to se that I had lefte nothinge amise.

And that those jemoglanes was verrie earneste with me in perswation to stay and live thare.

NOVEMBER

The 12th of November. . . . This daye, in the morninge, I put on a pare of new shoues, and wore them quite oute before nyghte; but this daye I touke a great could with a surfett, by means whear-of I was sore trubled with a burninge fever, and in great dainger of my Life. When I was somthinge recovered, by the helpe of God and a good fisition, it hapemed that thar was good Company reddie to com for Inglande, suche as in 2 or 3 years I could not have had the lik, if I had stayed behinde them, and they weare all desierus to have my company. My Lord was verrie unwillinge that I should goo at that time, because I was verrie wayke, not able to goo on foute one myle in a daye. But I desiered my lord to give me leve, for I had rether die by the way in doinge my good will to goo hom, than staye to die thare, wheare I was perswaded I could not live if I did staye behinde them.

Notes

* Notes in italics are the author's; all others cited from Bent's *Early Voyages*.

1. Jemeglans=adjemoglans=sons of strangers (*adjemi*). The adjemoglans were either captives in war, or sons of Christian parents taken when young, and designed for the more servile offices of the seraglio, which a Turk would not do. The Bostangee-basha, or head-gardener, rose from their ranks and often obtained great power.
2. The Qapu Agha, or Chief Eunuch.
3. Spanish tantarara, the redoubled beating of a drum.
4. Gatekeeper.
5. *drugaman: interpreter (Dallam's interpreter was "a Cornishman born")* .
6. Tissue = interwoven or variegated. "The chariot was covered with cloth of gold Tissued upon blue." (Bacon)
7. Made of Spanish leather. "I will send you the Cordovan pockets and gloves." (Howell, Familiar Letters, 1650.)
8. Sequins.
9. Mandilion = a soldier's cloak. "A mandilion that did with button meet." (Chapman: Hom., *Il.*, x.)
10. *Scamatie*, deriv. Italian *scamatareh*, to beat off the dust of wool.
11. Muslin or lawn.
12. *A slipper or sandal.*

JOHN RAWLINS

John Rawlins was born in Rochester, and lived for 23 years in the port of Plymouth, on the southwest coast of England. Employed by two Plymouth merchants as the master of their ship, the Nicholas, Rawlins set out late in 1621 on a trading voyage to Gibraltar, on which his ship was taken by Turkish corsairs. The English prisoners were taken to Algiers: two younger men were "by force and torment . . . compelled . . . to turn Turks," and the rest were sold as slaves. Rawlins himself, by reason of a lame hand, was last to be sold, finally being purchased by his own captor for the equivalent of seven pounds ten shillings on account of his experience as a pilot and master. Yet when he proved slow at shipboard work, his owner threatened to sell him upcountry if he could not furnish a ransom of twice his purchase price. At this moment, a converted Englishman named John Goodale and his associates had bought and begun fitting out another English prize, the *Exchange of Bristol.* As both Goodale, the master, and Henry Chandler, the captain, were English converts, they "concluded to have all English slaves to go in her"; among those purchased were two of Rawlins's men, James Roe and John Davies. Being asked where a skilled pilot and navigator might be found, Davies informed them that Rawlins was for sale, and accordingly he was purchased. The text that follows records what happened on the voyage, as told by an unidentified captive from the *Nicholas*: Rawlins's discontentment with his lot, his recruitment of other slaves, free workers, and converts among the crew, and finally his successful mutiny and return to England along with his fellow captives and those officers and crew who had joined them, some willingly and some under coercion. The text that appears here is taken from the abbreviated version printed by Samuel Purchas in 1625 (*Purchas his Pilgrimes,* vol. 2, part 1, book 6 [London, 1625], 889-96); the original, printed in 1622—the year of the *Exchange's* return—survives in a single copy in the Bodleian Library, Oxford.

—Mary C. Fuller

The wonderfull recouery of the Exchange of Bristow, from the Turkish Pirats of Argier, published by IOHN RAWLINS, heere abbreuiated.

[Rawlins was master on one of two English merchant ships taken near Gibraltar by the Turks, and brought into Algiers along with their crews.]

The 26. of the same moneth, Iohn Rawlins his Barke, with his other three men and a boy, came safe into the Mould,[1] and so were put all together to be carried before the Bashaw, but that they tooke the Owners seruant, and Rawlings Boy, and by force and torment compelled them to turne Turkes:[2] then were they in all seuen English, besides Iohn Rawlins, of whom the Bashaw tooke one, and sent the rest to their Captaines, who set a valuation vpon them, and so the Souldiers hurried vs like dogs into the Market, whereas men sell Hacknies[3] in England, we were tossed vp and downe to see who would giue most for vs; and although we had heauy hearts, and looked wth sad countenances, yet many came to behold vs, sometimes taking vs by the hand, sometime turning vs round about, sometimes feeling our brawnes and naked armes, and so beholding our prices written in our breasts, they bargained for vs accordingly, and at last we were all sold, and the Souldiers returned with the money to their Captaines.

[T]he *Exchange of Bristow,* a ship formerly surprised by the Pirats, lay all vnrigged in the Harbour, till at last one Iohn Goodale, an English Turke, with his confederates, vnderstanding shee was a good sailer, and might be made a proper Man of Warre, bought her from the Turkes that tooke her, and prepared her for their owne purpose: now the Captaine that set them on worke, was also an English Renegado,[4] by the name of Rammetham Rise, but by his Christen name Henrie Chandler, who resolued to make Goodale Master ouer her; and because they were both English Turkes, hauing the command notwithstanding of many Turkes and Moores, they concluded to have all English slaves to goe in her, and for their Gunners, English and Dutch Renegadoes, and so they agreed with the Patrons of nine English, and one French Slaue for their ransoms, who were presently imployed to rig and furnish the ship for a Man of Warre, and while they were thus busied, two of Iohn Rawlins men, who were taken with him, were also taken vp to serue in this Man of Warre, their names, Iames Roe, and Iohn Dauies, the one dwelling in Plimmoth, and the other in Foy, where the Commander of this ship was also borne, by which occasion they came acquainted, so that both the Captaine, and the Master promised them good vsage, vpon the good seruice they should performe in the voyage, and withall demanded of him, if he knew of any Englishman, to be bought, that could serue them as a Pilot, both to direct them out of Harbour, and conduct them in their voyage. For in truth neither was the Captaine a Mariner, nor any Turke in her of sufficiency to dispose of her through the Straites in securitie, nor oppose any enemie, that should hold it out brauely against them. Dauies quickly replied, that as farre as he vnderstood, Villa Rise[5] would sell Iohn Rawlins his Master, and Commander of the Barke which was taken, a man euery way sufficient for Sea affaires, being of great resolution and good experience; and for all he had a lame hand, yet had he a sound heart and noble courage for any attempt or aduenture.

[T]he Turks a ship-boord conferred about the matter, and the Master whose Christen name was Iohn Goodale ioyned with two Turkes, who were consorted with him, and disbursed one hundred Dooblets a piece, and so bought him of Villa Rise, sending him into the said ship, called the *Exchange of Bristow,* as well to supervise what had been done, as to order what was left vndone, but especially to fit the sailes, and to accommodate the ship, all which Rawlins was very carefull and indulgent in, not yet thinking of any particular plot of deliuerance, more then a generall desire to be freed from this Turkish slauerie, and inhumane abuses.

By the seuenth of Ianuarie, the ship was prepared with twelue good cast Pieces,[6] and all manner of munition and prouision, which belonged to such a purpose, and the same day haled out of the Mould of Argier, with this company, and in this manner.

There were in her sixtie three Turkes and Moores, nine English Slaues, and one French, foure Hollanders that were free men, to whome the Turkes promised one prise or other and so to returne to Holland; or if they were disposed to goe backe againe for Argier, they should have great reward and no enforcement offered, but continue as they would, both their religion and their customes: and for their Gunners they had two of our Souldiers, one English and one Dutch Renegado; and thus much for the companie. For the manner of setting out, it was as vsual as in other ships, but that the Turkes delighted in the ostentous[7] brauerie of their Streamers, Banners, and Top-sayles; the ship being a handsome ship, and well built for any purpose: the Slaues and English were imployed vnder Hatches about the Ordnance, and other workes of order, and accommodating themselues: al which Iohn Rawlins marked, as supposing it an intolerable slauerie to take such paines, and be subiect to such dangers, and still to enrich other men and maintaine their voluptous filthinesse and liues, returning themselues as Slaues, and liuing worse then their Dogs amongst them. Whereupon hee burst out into these, or such like abrupt speeches: Oh Hellish slauerie to be thus subiect to Dogs! Oh, God strengthen my heart and hand, and something shall be done to ease vs of these mischiefes, and deliuer vs from these cruell Mahumetan Dogs. The other Slaues pittying his distraction (as they thought) bad him speake softly, lest they should all fare the worse for his distemperature. The worse (quoth Rawlins) what can be worse? I will either attempt my deliuerance at one time, or another, or perish in the enterprise: but if you would be content to hearken after a release, and ioyne with me in the action, I would not doubt of facilitating the same, and shew you a way to make your credits thriue by some worke of amazement, and augment your glorie in purchasing your libertie, I prethee be quiet (said they againe) and thinke not of impossibilities: yet if you can but open such a doore of reason and probabilitie, that we be not condemned for desperate and distracted persons, in pulling the Sun as it were out of the Firmament: wee can but sacrifice our liues, and you may be sure of secrecie and taciturnitie.

All this while our slauery continued, and the Turkes with insulting tyrannie set vs still on worke in all base and seruile actions, adding stripes and inhumane reuilings, euen in our greatest labour, whereupon Iohn Rawlins resolued to obtayne his libertie, and surprize the ship; prouiding Ropes with broad speckes of Iron, and all the Iron Crowes,[8] with which hee knew a way, vpon consent of the rest, to rame vp or tye fast their Scuttels, Gratings, and Cabbins, yea to shut vp the Captaine himselfe with all his consorts, and so to handle the matter, that vpon the watch-word giuen, the English being Masters of the Gunner roome, Ordnance, and Powder, they would eyther blow them into the Ayre, or kill them as they aduentured to come downe one by one, if they should by any chance open their Cabbins. But because hee would proceed the better in his enterprise, as he had somewhat abruptly discouered himselfe to the nine English slaues, so he kept the same distance with the foure Hollanders, that were free men, till finding them comming somewhat toward them, he acquainted them with the whole Conspiracie, and they affecting the Plot, offered the aduenture of their liues in the businesse. Then very warily he vndermined the English Renegado, which was the Gunner, and three more his Associats, who at first seemed to retract. Last of al were brought in the Dutch Renegadoes, who were also in the Gunner roome, for alwayes there lay twelue there, fiue Christians, and seuen English, and Dutch Turkes: so that when another motion had settled their resolutions, and Iohn Rawlins his constancie had put new life as it were in the matter, the foure Hollanders very honestly, according to their promise, sounded the Dutch Renegadoes, who with easie perswasion gaue their consent to so braue an Enterprize.

For we sayled still more North-ward, and Rawlins had more time to tamper with his Gunners, and the rest of the English Renegadoes, who very willingly, when they considered the matter, and perpended[9] the reasons, gaue way vnto the Proiect, and with a kind of ioy seemed to entertayne the motiues: only they made a stop at the first on-set, whoe should begin the enterprize, which was no way fit for them to doe, because they were no slaues, but Renegadoes, and so had alwayes beneficiall entertaynment amongst them. But when it is once put in practice, they would be sure not to faile them, but venture their liues for God and their Countrey. But once againe he is disappointed, and a suspitious accident brought him to recollect his spirits anew, and studie on the danger of the enterprize, and thus it was. After the Renegado Gunner, had protested secrecie by all that might induce a man to bestow some beliefe vpon him, he presently went vp the Scottle, but stayed not aloft a quarter of an houre, nay he came sooner down, & in the Gunner roome sate by Rawlins, who tarryed for him where he left him: he was no sooner placed, and entred into some conference, but there entered into the place a furious Turke, with his Knife drawne, and presented it to Rawlins his body, who verily supposed, he intended to kill him, as suspitious that the Gunner had discouered something, whereat Rawlins was much moued, and so hastily asked what the matter meant, or whether he would kill him or no, obseruing his countenance, and according to the nature of iealousie, conceiting that his colour had a passage of change, whereby his suspitious heart, condemned him for a Traytor: but that at more leisure he sware the contrary, and afterward proued faithfull and industrious in the enterprize. And for the present, he answered Rawlins in this manner, no Master, be not afraid, I thinke hee doth but iest. With that Iohn Rawlins gaue backe a little and drew out his Knife, stepping also to the Gunners sheath and taking out his, whereby he had two Kniues to one, which when the Turke perceiued, he threw downe his Knife, saying, hee did but iest with him. But (as I said) when the Gunner perceiued, Rawlins tooke it so ill, hee whispered something in his eare, that at last satisfied him, calling Heauen to witnesse, that he neuer spake word of the Enterprize, nor euer would, either to the preiudice of the businesse, or danger of his person: Notwithstanding, Rawlins kept the Kniues in his sleeue all night, and was somewhat troubled, for that hee had made so many acquainted with an action of such importance; but the next day, when hee perceiued the Coast cleere, and that there was no cause of further feare, hee somewhat comforted himselfe.

All this while, Rawlins, drew the Captaine to lye for the Northerne Cape, assuring him, that thereby he should not misse purchase, which accordingly fell out, as a wish would haue it.

So in the name of God, the Turkes and Moores being placed as you haue heard, and fiue and forty in number, and Rawlins having proined[10] the touch-holes, Iames Roe gaue fire to one of the peeces, about two of the clocke in the afternoone, and the confederates vpon the warning, shouted most cheerefully: the report of the peece did teare and breake downe all the Bitickell,[11] and compasses, and the noise of the slaues made all the Souldiers amased at the matter, till seeing the quarter of the ship rent, and feeling the whole body to shake vnder them: vnderstanding the ship was surprised, and the attempt tended to their vtter destruction, neuer Beare robbed of her whelpes was so fell and mad: for they not onely cald vs dogs, and cried out, Vsance de Lamair, which is as much as to say, the Fortune of the wars: but attempted to teare vp the planckes, setting a worke hammers, hatchets, kniues, the oares of the boate, the boat-hooke, their curtle-axes, and what else came to hand, besides stones and brickes in the Cooke-roome; all which they threw amongst vs, attempting still and still to breake and rip vp the hatches, and boordes of the steering, not desisting from their former execration, and horrible blasphemies and reuilings.

When Iohn Rawlins perceiued them so violent, and vnderstood how the slaues had cleared the deckes of all the Turkes and Moores beneath, he set a guard vpon the Powder, and charged their owne Muskets against them, killing them from diuers scout-holes, both before and behind, and so lessened their number, to the ioy of all our hearts, whereupon they cried out, and called for the Pilot, and so Rawlins, with some to guard him, went to them, and vnderstood them by their kneeling, that they cried for mercy, and to haue their liues saued, and they would come downe, which he bad them doe, and so they were taken one by one, and bound, yea killed with their owne Curtleaxes; which when the rest perceiued, they called vs English dogs, and reuiled vs with many opprobrious tearmes, some leaping ouer-boord crying, it was the chance of war, some were manacled, and so throwne ouer-boord, and some wer slaine and mangled with the Curtleaxes, till the ship was well cleared, and ourselues assured of the victory.

At the first report of our Peece, and hurliburly in the decks, the Captaine was a writing in his Cabbin, and hearing the noyse, thought it some strange accident, and so came out with his Curtleaxe in his hand, presuming by his authority to pacifie the mischiefe: But when hee cast his eyes vpon vs, and saw that we were like to surprise the ship, he threw down his Curtleaxe, and begged vs to saue his life, intimating vnto Rawlins, how he had redeemed him from Villa-Rise, and euer since admitted him to place of command in the ship, besides honest vsage in the whole course of the Voyage. All which Rawlins confessed, and at last condescended to mercy, and brought the Captaine and fiue more into England. The Captaine was called Ramtham-Rise, but his Christen name, Henry Chandler, and as they say, a Chandlers sonne in Southwarke. Iohn Good-ale, was also an English Turke. Richard Clarke, in Turkish, Iasar; George Cocke, Ramdam; Iohn Browne, Mamme; William Winter, Mustapha; besides all the slaues and Hollanders, with other Renegadoes, who were willing to be reconciled to their true Sauiour, as being formerly seduced with the hopes of riches, honour, preferment, and such like deuillish baits, to catch the soules of mortall men, and entangle frailty in the tarriers[12] of horrible abuses, and imposturing deceit.

Notes

1. Mould: mole or breakwater, i.e., harbor.
2. Turn Turks: convert to Islam. "Turk," in this document, indicates religious and cultural affiliation rather than ethnic origin.
3. Hacknies: riding horses.
4. Renegado: renegade or apostate, in this case a Christian convert to Islam.
5. Villa Rise: Rawlins's Turkish captor. "Reis," or "leader," was a title given corsair captains.
6. Pieces: cannons or guns.
7. Ostentous: ostentatious.
8. Crowes: crowbars or grappling hooks.
9. Perpended: pondered, considered.
10. Proined: primed.
11. Bitickell: binnacle, the box near the helm of a ship that holds the compass.
12. Tarriers: obstructions.

ENGLISH TURKS AND RESISTANT TRAVELERS: CONVERSION TO ISLAM AND HOMOSOCIAL COURTSHIP

Mary C. Fuller

These two autobiographical documents—an unpublished diary kept by the organmaker Thomas Dallam, and a twice-printed account by a skilled mariner named John Rawlins—narrate the trajectory of two Englishmen's experiences as they move into and away from engagement with the Islamic world. In many respects, Dallam's and Rawlins's experiences differ sharply. Dallam, in 1599, spent several months at the court of Mahomet III in Istanbul, where he oversaw the assembly and working of the elaborate mechanical organ sent as a present to the sultan by Queen Elizabeth. He had what he describes as exceptional access to this elite and alien milieu: both to the presence of the sultan, on one occasion the diary describes at length, and also to the seraglio, "which no Christian ever did in there memorie that wente awaye a Christian."[1] Some 20 years later, Rawlins found himself in a very different corner of the Islamic Mediterranean, and on different terms: captured by corsairs, he was taken to the corsair capital of Algiers and sold as a slave, his eventual owners proving to be Turks in name (literally) but Englishmen by birth.[2] As a slave to these Turks, he soon found himself again on shipboard, practicing his craft in known waters. Dallam was a privileged visitor to the metropolitan center of Ottoman culture and power, Rawlins a coerced laborer at the fringes of that empire. Yet both men possessed skills that made them desirable subjects for recruitment and assimilation, whether voluntary or forced. Both narratives, written from the metropolitan center of the Ottoman empire and from its margins, are notably populated with converts—English Turks, Dutch Turks, dragomen, and adjemoglans. Along with the experience of Ottoman desire for Europeans to join them by converting, these narratives register the lived possibility of conversion, a possibility that both narrators in the end reject.

On September 18th, Dallam writes, the "two jemaglanes . . . touke me in their armes and Kised me, and used many perswations to have me staye. . . ."[3] These two men were guards at the inner gates of the seraglio where Dallam had spent the preceding week setting up and checking on his instrument, which he refers to as "the present." Dallam describes his visits to the seraglio in terms of privilege and danger. The house where the organ is installed Dallam characterizes as "a house of pleasur, and lyke wyse a house of slaughter"; on his accession to the throne, the current ruler had had his brothers strangled there, a practice the English believed to be customary for the Ottoman sultans. Dallam makes two allusions to the privilege he enjoyed of dining in the seraglio daily during this time: once to note that there one ate grapes after every meal, and the second time to note that this unconditional access has never been granted before. When other Christians had been allowed in, that access formed part of a process of recruitment and conversion. Dallam's own admission to the seraglio and to its secret places (whether the secrets are those of male power, or of women and sex) form part of an effort at recruitment, albeit an unsuccessful one. On the day when these persuasions begin, Dallam had demonstrated the present to the Grand Signior's secretary; his expertise leads immediately to efforts at winning him to the Grand Signior's service. The tenor of these efforts is one of allurement, persuasion, and what manifests as affection in the guards' repeated embraces and kisses.

Dallam comments later in the text that the two gatekeepers "had bene appointed by the Grand Sinyor to perswade me to staye there allwayes." On September 30, they tell him he will want for nothing if he stays, and may have his choice of two wives among the sultan's concubines or "tow virgins of the best I Could Chuse"; on October 12, he is given a tour of the sultan's privy chambers, invited to sit in his state chair and to draw his sword, and given a privy look at some of the concubines playing ball; on October 13 he is prevented from sailing for home, and the jemoglanes "merrily" pretend (on learning that he still hopes eventually to leave) to kidnap him, carrying him bodily into the house where the sultan's brothers are strangled "to see how I would tak it yf they should staye me by force." On the 25, when Dallam revisits his organ, they are again "verrie earneste" in their invitation, but on November 12 he succeeds in boarding a ship bound for home.

Dallam seldom responds explicitly to these offers, though the text leaves little ambiguity in his intention to reject them. We can explore Dallam's responses in more detail by considering two scenes in which he literally stands and observes Ottoman culture: the first, when he enters the sultan's presence to demonstrate the mechanical organ; the second, when he is led to spy on the sultan's concubines.

The sultan proves—at least on Dallam's account—to be fascinated by the ingenious English machine, and his questions quickly exhaust the information of his attendants; thus, it becomes necessary to summon Dallam himself, unprepared and very surprised, into the sultan's presence. When Dallam enters, he perceives the sultan to be gazing in wonder at this western machine; he, in turn, finds the scene "verrie wonderfull," but his wonder is directed less at the sultan himself than at his attendants, the sight of which makes him feel he is in "another worlde." This moment of silent wonder is not a short one, either in the text or in its actual duration, which Dallam estimates at almost a quarter of an hour. Dallam numbers the sultan's attendants at 400: 200 pages, 100 dwarves, and 100 mutes.[4] Dallam says the mutes caused him the most wonder for their fluency in communicating to him by signs, though they and the dwarves are described only briefly. The pages, who come first, are described in far more detail: their rich dress, unusual hairstyle, their ages, that all were "verrie proper men," and finally that they were "Christians borne." These Christian courtiers generate for Dallam a wonder or fascination that mirrors the Grand Signior's wonder at his mechanical organ.

Dallam's fascinated gaze at the sultan's 200 pages is repeated and intensified later in the diary, when he is directed to watch the sultan's concubines at play; the manner of his description echoes and amplifies the description of the pages, and he stands staring until his companion angrily pulls him away. At first, indeed, he believes the concubines to be handsome young men, and only close scrutiny of their bodies reveals them to be women—"and verrie prettie ones in deede."

This unusual access to the sultan's forbidden women presumably serves as a kind of teaser for the offer to choose wives for himself from among them—yet this heterosexual pleasure or liberty takes place alongside or even within a homosocial or homoerotic context. As Dallam admires the pages, or the concubines thought to be pages, he is himself courted by the adjemoglans—and the physical gestures of that relationship fluctuate between pleasure and danger, affection and overpowering force, as if in their kisses and embraces is always the potential for sexual coercion. In other words, this courtship, which is primarily social or economic, is nonetheless enacted largely in the register of the body and the language of affection or passion.

Dallam's anxious perception or projection of an occulted, coercive, and indeed eroticized force shapes the description of his musical demonstration for the sultan, which is marked by a keen awareness of his dangerous proximity to the ruler's body. Dallam initially refuses to play the organ, until the Qapu Agha "thruste me on"—he describes in some detail the exact movements of his body, the way he is forced by the sultan's nearness to the organ to turn his back to him, and to touch his breeches to the sultan's knee. As Dallam turns his back, in the narrative, to sit at the keyboard, he gives the only visual description of the sultán, who sits behind him but is projected vividly in his mind's eye. Then, as the sultan stands to look more closely over Dallam's shoulder, "in his risinge from his chaire, he gave me a thruste forwardes, which he could not otherwyse dow, he satt so neare me; but I thought he had bene drawinge his sorde to cut of my heade." The English ambassador, Henry Lello, had told Dallam earlier that when he himself was presented to the sultan there would be an elaborate formality of permitted and proscribed movements, and in particular that "in payne of my heade I muste not turne my backe upon him."

In the context of Dallam's narrative, however, the ceremonial associated with entering the sultan's presence has additional resonance. While Dallam is himself drawn in as the spectator of the fascinating bodies of others—mutes, concubines in trousers, Christian youths as Turkish pages—at the same time his own body becomes the object of affectionate, dangerous, or exciting contacts. When the sultan commands that he be prevented from leaving, Dallam's sense of personal objectification colors the outburst in which this generally tolerant and interested account reveals an underlying judgment of a more definite kind. Early in the diary, Dallam had consulted Lello on how to respond to the advances of the adjemoglanes. Lello counselled him to be "merrie" with them, hint at an interest and never directly say no; by this flirtatious delay of a definite commitment, he could hope to avoid being retained in Istanbul by force. When, on October 13, Dallam is taken off the ship on which he had hoped to sail for England, he finally in a fury speaks his mind: that serving the sultan would be a "slavish Life," that it would be intolerable not to be among Christians, and that all along he has feared being betrayed, handed over by the ambassador as another present to accompany the organ he made. In the end, describing a return voyage undertaken despite a dangerous illness, he writes that rather than stay in Turkey, and die of unhappiness, he would rather die trying to go home.

Dallam remains in the Ottoman court under the ambassador's diplomatic protection, thus as a kind of encapsulated observer of culturally different surroundings. While he sometimes feels himself to be in jeopardy, the boundary separating him from the culture around him can't be breached without his assent, or at least without abrogating larger understandings about sovereignty and immunity. Perhaps it is not so much that these understandings were inviolable, as that Dallam wasn't quite worth the trouble of violating them. Rawlins's relations to the culture that tries to absorb him are marked by at least two crucial differences: first, as a captive and later a slave, he is immersed in it by main force, without immunity or consent; second, many of the Turks around him are by origin (and presumably upbringing) European. Not only his fellow slaves, but also some of the Turks who tyrannize over them, are English—and the latter are still English enough to ally with Rawlins or at least to return home, reconcile themselves, and resume being Christians and Englishmen again. In a sense, the second condition holds for Dallam as well, though the narrative registers it only occasionally—despite their places and names, the jemoglanes, the pages, and indeed Dallam's dragoman are Turks only by confession, and not by birth. Dallam describes his dragoman as "an Inglishe man, borne in Chorlaye in Lancashier . . . in religion a perfit Turke, but he was our trustie frende."[5] The presence of foreign converts signifies differently in the Rawlins text because of a further difference in setting. The bulk of the narrative takes place on shipboard, a place with its own, inevitably somewhat polyglot and cosmopolitan, requirements and norms. Isolated at least provisionally from larger structures of custom and force, a ship can end up just about anywhere. This mobility is at several levels crucial to the narrative.

When the English sailors of the Rawlins narrative arrive as prisoners in Algiers, they are depicted as initially faced with a brutally limited set of possibilities—slavery or conversion—without necessarily even the ability to choose between them.[6] The narrator describes the experience of being sold as literally dehumanizing, likening their treatment to that given dogs or horses; the text sets against the subjective experience of "heauy hearts, and . . . sad countenances," the objective interest of purchasers not in hearts and faces, but in "brawnes and naked armes . . . and . . . our prices writen in our breasts." Meanwhile, two younger men of lower status—"the Owners seruant, and Rawlings Boy"—are "compelled . . . to turne Turkes." From other countrymen, the prisoners hear that besides 500 Englishmen recently brought to the market as slaves, "aboue a hundred hansome youths [were] compelled to turne Turkes, or made subiect to more vilder prostitution, and all English."[7] The handsome youths who convert are if anything coerced more violently than the men simply sold as slaves, according to the details of "Execrable tortures by Hellish Pirates inflicted on the English, to make them Renegadoes and Apostates." One pauses over "more vilder prostitution." The phrasing leaves ambiguous whether that prostitution is literal—the handsome youths were forced into sexual service—or whether conversion itself is being made, along with an unnamed but "more vilder" second term, to signify as a form of prostitution, a possibility suggested by a later comment that some converts caring only for "sensuall lusts and pleasures . . . for preferment or wealth very voluntarily renounced their faith." Or perhaps the youths face the linked alternatives of being circumcised (conversion) or being made eunuchs. On any of these readings, the passage coheres around a perceived linking of confession and transgressive (male) sexuality, whether it is the prostitution of boys, genital surgery and mutilation, or selling one's faith for profit under the sway of "sensuall lusts." Here, the narrative suggests that forced conversion is a form of sexual predation on males whose age or status makes them vulnerable.

Unlike Dallam's diary, the "Wonderfull Recouery" was written for publication in a deliberately anonymous voice (either that of an unidentified English captive, or Rawlins himself in the Caesarian third person). Its polemic condemnation of Turks is made more extreme and explicit throughout than is the case in Dallam's diary—who, though he may have found assimilation to Ottoman life and belief unthinkable, didn't need to say so constantly to himself. Yet the actual narrative of Rawlins's experience as a slave in the *Exchange* gives a more complicated picture both of slavery and of conversion than one would suspect from its polemic opening. Rawlins finds the terms of the English slaves' service to be, in an oddly literal phrasing, "an *intolerable* slaverie" (italics mine). The intolerable conditions are particularized in terms of the kind of work they are made to do ("all base and seruile actions"); their treatment by those who set them to work ("stripes and inhumane reuilings, euen in our greatest labour"); their inability to profit from their own labor ("to take such paines . . . and still to enrich other men"); and a condemnation of those who are supported by that labor ("still to maintaine their voluptuous filthinesse and liues"). The language of dehumanization which serves initially to characterize the experience of slaves also, reciprocally, describes those they serve: "Oh Hellish slauerie to be thus *subiect* to Dogs" (italics mine), and in particular, to "Mahumetan dogs." Rawlins asserts an absolute cultural difference and superiority in relation to those who have in practice subjected him. Nor is this rhetorical abjecting of the cultural other merely rhetorical. When Rawlins gains control of the ship, some of the crew "by their kneeling . . . cried for mercy, and to have their lives saved"; he induces them to come out voluntarily, then takes, binds, and kills them one by one, hacking them with their own axes or throwing them overboard in chains. To his concerns with status and reward, labor and power, we might add what seems like rage against difference, a rage that allows him to kill without mercy and to embody the inhumanity he decries. Yet in this text the terms of nationality cross against the dichotomous rhetoric that makes slaves and masters dogs to each other. Although the terms of Rawlins's objections to his captors (that they are voluptuous, tyrannous) echo popular stereotypes of Turkish behavior, here the tyrannous Turks are English.

What difference does their nationality of origin make in the text? In some obvious ways, none at all. To Rawlins, his masters *are*—behaviorally, relationally—Turks. Goodale buys Rawlins to secure his services as a slave, not to free him as an Englishman. In part, however, he buys Rawlins as a slave because Rawlins, like Goodale, is English. Chandler and Goodale "because they were both English Turkes . . . concluded to have all English slaves to goe in her, and for their Gunners, English and Dutch Renegadoes." English sailors were valued on the Barbary Coast for their skills[8]; yet the text also suggests simply a national preference, that English slaves were preferred because they were of the same nationality as the owners themselves. The ship's company consists initially of 63 "Turkes and Moores," one French and nine English slaves, four free Dutchmen, and a few Dutch and English gunners who are converts. After the manning of two prizes and taking on board the English crew of one, the total ship's company falls to sixty-four, of whom fifteen are English slaves and prisoners, and six are converts who join with Rawlins—three more sympathizing converts have gone with a prize crew, suggesting that Rawlins won over nine in all. At the narrative's end, Chandler successfully pleads with Rawlins and is spared to return to England along with five more converts. The narrator gives these figures to show how many fight on each side; I cite them to make an additional point, which is that the English, Dutch, and French converts are identified *as* converts only when they are ready to switch allegiances again. In other words, the English and other converts are indistinguishably "Turks" until they change sides and return to the fold, and thus one is never quite certain how many there are, or that there are no additional Englishmen subsumed in the number of "Turkes and Moores" who fail to receive Rawlins' mercy.

Who are these converts? The narrative tells us only of those who will renege on their new religion and allegiance; and among them, those who join Rawlins exhibit both a willingness to risk their lives in his project and a real resistance to it. The English gunner is particularly reluctant, and Rawlins has to "undermine" and "tamper with" him over the course of time; even then, his real intentions remain suspect. The converts who agree to join the rebellion balk at beginning it, as they "had alwayes had beneficiall entertaynment" among the Turks; if they had been "seduced" by hopes of gain, these hopes must have been to some extent gratified. Yet the loyalty the converts exhibit doesn't suffice to hold them to their choice, nor does anyone Rawlins approaches betray his plans. The English and Dutch converts would seem to be missing the motives that animate Rawlins: bad treatment, lack of reward, a disgust with confessional or cultural difference. Yet they follow him, whatever their initial resistance, with what the narrator describes as "a kind of joy."

In the end, we don't know and the narrative doesn't tell us why the converts do what they do, what mixture of homesickness, patriotism, glory, or religious remorse animates their choices. In one case the process seems clear: the captain, who we recall is an English Turk, begs Rawlins for mercy and is reconciled to Christianity on the ship's return to England. Though the captain is not actually subject to violence, the coerced nature of this (re)conversion requires little comment. What does seem worth noting is that Rawlins spares him *not* as an act of unearned mercy, displaying the qualities of Englishman and Christian, but as an acknowledgement of "beneficiall entertaynment," one that contravenes the previous claim that he was motivated by bad treatment. Rawlins agrees that the captain had "redeemed him from Villa-Rise, and euer since admitted him to place of command in the ship, besides honest vsage in the whole course of the Voyage." Moreover, at the moment when Chandler/Rammetham Rise is about to reemerge as Henry Chandler, reconciled to Christianity, the earlier purchase of Rawlins in order to secure his services is retrospectively characterized as *redemption*.

Perhaps it is not surprising that the narrator is willing to rehabilitate the returned convert, whose status makes him something of a special case; yet to understand Chandler's treatment of Rawlins as fair and even benevolent removes what might have seemed to be a premise of his violence: namely, that in treating him with inhumanity Rawlins's captors have merited inhuman treatment in return. The captain appeals to Rawlins in terms of his privileged status as having been "admitted to position of command": as if in exchange for this recognition Rawlins is willing to give up as provisional or positional the account of brutality that seemed to motivate the rebellion, exchanging for a bond with the captain his identification with other English slaves *not* admitted to a position of command. The quasi-voluntary surrender, both material and moral, of the ship's highest officer doesn't quite cover over the slaughter of his subordinates, and in this narrative of mobility the Christianity of Rawlins's Christian heroism seems itself to be positional, and nominal.

To conclude, Dallam exhibits a passive, persistent resistance to the seductions of Islamic courtship; Rawlin's, whose very persuasions are couched in the vocabulary of war, mounts a violent response aimed not only at resisting absorption but at extracting what has been converted and destroying the residue. These narratives should, however, be read against a larger background: as texts like those by Dallam and Rawlins suggest, the voluntary conversion of Christian Englishmen to Islam was not uncommon.[9] Yet as those converts did not write their own stories, they survive most prominently as figures in the margins of narratives where resistance, passive or active, is at the center.

Notes

1. "Seraglio" here means something closer to "palace" than "harem." Mahomet or Mehmet III is referred to in Dallam's text—thus here as well—as "the Grand Signior."
2. On Algiers as "corsair capital," see Fernand Braudel, *The Mediterranean and the Mediterranean World in the Age of Philip II* (Berkeley, CA: University of California Press, 1995), trans. Siân Reynolds, vol. II: 870; see the rest of this chapter for a discussion of Mediterranean piracy in general, and Algiers in particular. On early seventeenth-century England and Algiers, see also Sir Godfrey Fisher, *Barbary Legend: War, Trade, and Piracy in North Africa, 1415-1830* (Oxford: Clarendon Press, 1957).
3. While both texts have been significantly abridged, I'll try to indicate omissions that might alter the sense of the text. The part of Dallam's diary that appears here has been edited to focus on his reiterated encounters with two of the sultan's gatekeepers, his responses to them, and his one recorded encounter with the sultan.
4. Here, I'm referring to material abridged in the text presented here; see J. Theodore Bent, *Early Voyages and Travels to the Levant,* works issued by the Hakluyt Society, no. 87 (London, 1893), 69-70.
5. Bent, *Early Voyages and Travels to the Levant,* 84.
6. As Nabil Matar points out, the alternatives of conversion or slavery were not always real alternatives; some captives managed to gain their liberty without apostasizing, and converts were not always enfranchised. Matar's chapter, "'Turning Turke': Conversion to Islam in English writings," gives a broader picture of the motives that might have induced Englishmen of Rawlins's time to convert (*Islam in Britain, 1558-1685* [Cambridge, UK: Cambridge University Press, 1998], 21-49).
7. Again, I refer to material abridged in this edition; see Samuel Purchas, *Purchas his Pilgrimes,* vol. II, part I, book VI (London, 1625), 889.

8. Several contemporary narratives turn on the superior skill of the English as mariners. See, for instance, the narrative printed by Purchas of a captured English ship simply sailed home by its remaining crew with no hindrance from their captors on board: "for they made the Turkes beleeue, the wind was come faire, and that they were sayling to Argier, till they came within sight of England" (Purchas, *Pilgrimes* II, part I, book VI, 895).
9. Matar, "Turning Turke," comments that "no account has survived by an early modern Briton who chose freely to convert to Islam and to remain in the Muslim Empire because such a convert never returned to his home in order to publish his work" (40).

SECTION THREE

BISHOP HENRY KING

Bishop Henry King (1592-1669), son of John King, Bishop of London, was educated at Westminster and Christ Church, Oxford (M.A. 1614), where, in tune with the times, he published Latin verses in occasional collections on public events such as the death of Prince Henry (1612) and the marriage of Princess Elizabeth (1613). After Oxford, church appointments came quickly: by 1617 he was prebend of St. Pancras, rector of Chigwell, Essex, archdeacon of Colchester, rector of Fulham, and a royal chaplain. In 1642 King was appointed Bishop of Chichester, but dispossessed the following year; he was reappointed at the Restoration. Throughout his clerical career, King remained interested in secular writing, stimulated no doubt by friendships with John Donne, James Howell, Ben Jonson, George Sandys, and Izaak Walton. In 1657 his *Poems* were published, and in 1664 reissued with additional elegies. King himself probably had no hand in these productions, which contain several false attributions. In 1700 the original selection—without the additional elegies—was reissued under the title *Ben Jonson's Poems, Paradoxes, and Sonnets.* J. Hannah collected and edited *King's Poems and Psalms* (Oxford, 1843).

A manuscript of King's poem "To my Noble and Judicious Friend Sir Henry Blount upon his Voyage" is in the Bodleian Library, Oxford (shelfmark: Ms. *Eng.poet.e.30, fol. 71). The poem, first published in *Poems* (1657), is given here from George Saintsbury, ed., *Minor Poets of the Caroline Period,* 3 vols. (1905; reprint, Oxford: Clarendon Press, 1968), vol. 3, pp. 223-26.

—Gerald MacLean

To my Noble and Judicious Friend Sir Henry Blount upon his Voyage

Sir, I must ever own myself to be
Possess'd with human curiosity
Of seeing all that might the sense invite

By those two baits of profit and delight:
And since I had the wit to understand
The terms of native or of foreign land;
I have had strong and oft desires to tread
Some of those voyages which I have read.
Yet still so fruitless have my wishes prov'd,
That from my Country's smoke I never mov'd: 10
Nor ever had the fortune (though design'd)
To satisfy the wand'rings of my mind.
Therefore at last I did with some content,
Beguile myself in time, which others spent:
Whose art to provinces small lines allots,
And represents large kingdoms but in spots.
Thus by Ortelius and Mercator's aid[1]
Through most of the discover'd world I stray'd.
I could with ease double the Southern Cape,
And in my passage Afric's wonders take: 20
Then with a speed proportion'd to the scale
Northward again, as high as Zembla sail.
Oft hath the travel of my eye outrun
(Though I sat still) the journey of the Sun:
Yet made an end, ere his declining beams
Did nightly quench themselves in Thetis' streams.
Oft have I gone through Egypt in a day,
Not hinder'd by the droughts of Lybia;
In which, for lack of water, tides of sand
By a dry deluge overflow the land. 30
There I the Pyramids and Cairo see,
Still famous for the wars of Tomombee,
And its own greatness; whose immured sense
Takes forty miles in the circumference.
Then without guide, or stronger caravan
Which might secure the wild Arabian,
Back through the scorched deserts pass, to seek
Once the world's Lord, now the beslaved Greek,
Made by a Turkish yoke and fortune's hate
In language as in mind degenerate. 40
And here, all wrapp'd in pity and amaze
I stand, whilst I upon the Sultan gaze;
To think how he with pride and rapine fir'd
So vast a territory hath acquir'd;
And by what daring steps he did become
The Asian fear, and scourge of Christendom:
How he achiev'd, and kept, and by what arts
He did concentre those divided parts;
And how he holds that monstrous bulk in awe,

By settled rules of tyranny, not Law: 50
So rivers large and rapid streams began,
Swelling from drops into an Ocean.

Sure, who e'er shall the just extraction bring
Of this gigantic power from the spring;
Must there confess a Higher Ordinance
Did it for terror to the earth advance.
For mark how 'mongst a lawless straggling crew,
Made up of Arab, Saracen, and Jew,
The world's disturber, faithless Mahomet
Did by impostures an opinion get: 60
O'er whom he first usurps as Prince, and than[2]
As prophet does obtrude his Alcoran.[3]
Next, how fierce Ottoman[4] his claim made good
From that unblest religion, by blood;
Whilst he the Eastern kingdoms did deface,
To make their ruin his proud Empire's base.
Then like a comet blazing in the skies,
How death-portending Amurath[5]did rise,
When he his horned crescents did display
Upon the fatal plains of Servia; 70
And farther still his sanguine tresses spread,
Till Croya life and conquests limited.[6]
Lastly, how Mahomet thence styl'd the Great,[7]
Made Constantine's his own Imperial seat;
After that he in one victorious bond
Two Empires grasp'd, of Greece and Trebizond.

This, and much more than this, I gladly read,
Where my relators it had storyed;
Besides that people's manners and their rites,
Their warlike discipline and order'd fights; 80
Their desp'rate valour, hard'ned by the sense
Of unavoided Fate and Providence:
Their habit, and their houses, who confer
Less cost on them than on their sepulchre:
Their frequent washings, and the several bath
Each Meschit[8] to itself annexed hath:
What honour they unto the Mufty[9] give,
What to the Sovereign under whom they live:
What quarter Christians have; how just and free
To inoffensive travellers they be: 90
Though I confess, like stomachs fed with news,
I took them in for wonder, not for use,
Till your experienc'd and authentic pen

Taught me to know the places and the men;
And made all those suspected truths become
Undoubted now, and clear as axiom.

Sir, for this work more than my thanks is due;
I am at once inform'd and cur'd by you.
So that, were I assur'd I should live o'er
My periods of time run out before; 100
Ne'er needed my erratic wish transport
Me from my native lists to that resort,
Where many at outlandish marts unlade
Ingenuous manners, and do only trade
For vices and the language. By your eyes
I here have made my full discoveries;
And all your countries so exactly seen,
As in the voyage I had sharer been.
By this you make me so; and the whole land
Your debtor: which can only understand 110
How much she owes you, when her sons shall try
The solid depths of your rare history,
Which looks above our gadders' trivial reach,
The commonplace of travellers, who teach
But table-talk; and seldomly aspire
Beyond the country's diet or attire:
Whereas your piercing judgement does relate
The policy and manage of each State.
And since she must here without envy grant
That you have further journey'd the Levant 120
Than any noble spirit by her bred
Hath in your way as yet adventured;
I cannot less in justice from her look,
Than that she henceforth canonize your book
A rule to all her travellers, and you
The brave example; from whose equal view
Each knowing reader may himself direct,
How he may go abroad to some effect,
And not for form: what distance and what trust
In those remoter parts observe he must: 130
How he with jealous people may converse,
Yet take no hurt himself by that commerce.
So when he shall embark'd in dangers be,
Which wit and wary caution not foresee;
If he partake your valour and your brain,
He may perhaps come safely off again,
As you have done; though not so richly fraught
As this return hath to our staple brought.

I know your modesty shuns vulgar praise,
And I have none to bring; but only raise 140
This monument of Honour and of Love,
Which your long known deserts so far improve,
They leave me doubtful in what style to end,
Whether more your admirer or your friend.

HENRY BLOUNT

Before his fourteenth birthday, Henry Blount (1602-1682) entered Trinity College, Oxford, where despite his youth he established a reputation for learning and wit. In 1619 he took his B.A. and left Oxford to study law at Grey's Inn, where he presumably pursued an interest in foreign trade, traveling to France, Italy, and Spain. In 1634 he set off for the Levant, sailing from Venice down the Adriatic coast to Dalmatia, thence inland to Belgrade, Nis, Sophia, Edirne (Adrianople), and Constantinople, where he stayed only a few days before setting off with a Turkish fleet for Rhodes, Alexandria, and Cairo. In Egypt he visited the pyramids at Giza and the "labyrinth" or mortuary temple at Hawarah in the Fayyum, which Herodotus had considered surpassed the pyramids. Blount was knighted by Charles I on March 21, 1640. He accompanied the king during the early stages of the Civil War, saw action at the battle of Edgehill, and later joined the court in Oxford. After the king's execution in 1649, he accepted commissions by Parliament to investigate legal and trade abuses, a role in which he continued after the Restoration.

Blount's account of his 11-month journey, *A Voyage Into The Levant,* was first published in 1636 and reprinted seven times before 1671; a German translation appeared in 1687, and a Dutch translation in 1707. Several signed copies of the first and second editions in the British Library are, presumably, presentation copies. The extracts below are from the first edition: *A Voyage Into The Levant. A Breife Relation of a Journey, lately performed by Master H. B. Gentleman, from England by way of Venice, into Dalmatia, Sclavonia, Bosnah, Hungary, Macedonia, Thessaly, Thrace, Rhodes, and Egypt, unto Gran Cairo: With particular observations concerning the moderne condition of the Turkes, and other people under that Empire,* London, Printed by J. L. [John Leggatt] for Andrew Crooke, 1636. I have silently modernized as follows: long "s" has been shortened; "u" and "v", "i" and "j" are distinguished according to modern usage.

—Gerald MacLean

Blount Gets Ready

Intellectuall Complexions have no desire so strong, as that of *knowledge;* nor is any knowledge unto man so certaine, and pertinent, as that of humane affaires: This *experience* advances best, in observing of people, whose *institutions* much differ from ours; for customes conformable to our owne, or to such wherewith we are already acquainted, doe but repeat our old observations, with little acquist of new. So my former time spent in viewing *Italy, France,* and some little of *Spaine,* being countries of Christian institution, did but represent in a severall dresse, the effect of what I knew before.

Then seeing the customes of men are much swayed by their naturall dispositions, which are originally inspired and composed by the Climate whose ayre, and influence they receive, it seemes naturall, that to our Northwest parts, of the World, no people should be more averse, and strange of behaviour, then those of the *South-East:* Moreover, those parts being now possesst by the *Turkes,* who are the only moderne people, great in action, and whose Empire hath so suddenly invaded the World, and fixt it selfe such firme foundations as no other ever did; I was of opinion, that hee who would behold these times in their greatest glory, could not find a better *scene* then *Turky:* these considerations sent me thither; where my generall purpose gave me foure particular cares: First, to observe the Religion, Manners, and policie of the *Turkes,* not perfectly, (which were a taske for an inhabitant rather then a passenger,) but so farre forth, as might satsifie this scruple, (to wit) whether to an unpartiall conceit, the *Turkish* way appeare absolutely barbarous, as we are given to understand, or rather another kind of civilitie, different from ours, but no lesse pretending: Secondly, in some measure, to acquaint my selfe with those other sects which live under the *Turkes,* as *Greekes, Armenians, Freinks,*[10] and *Zinganaes,*[11] but especially the *Jewes;* a race from all others so averse both in nature and institution, as glorying to single it selfe out of the rest of mankinde, remaines obstinate, contemptible, and famous: Thirdly, to see the *Turkish* Army, then going against *Poland,* and therein to note, whether their discipline *Military* encline to ours, or else be of a new mould, though not without some touch, from the countries they have subdued; and whether it be of a frame apt to confront the Christians, or not: The last and choice peece of my intent, was to view *Gran Cayro,* and that for two causes; first, it being clearly the greatest concourse of Mankinde in these times, and perhaps that ever was; there must needs be some proportionable spirit in the Government: for such vast multitudes, and those of wits so deeply malicious, would soone breed confusion, famine, and utter desolation, if in the *Turkish* domination there were nothing but sottish sensuality, as most *Christians* conceive: Lastly, because *Egypt* is held to have beene the fountaine of all *Science,* and *Arts civill,* therefore I did hope to find some sparke of those cinders not yet put out; or else in the extreme contrarietie, I should receive an impression as important, from the ocular view of so great a revolution; for above all other senses, the eye having the most immediate, and quicke commerce with the soule, gives it a more smart touch then the rest, leaving in the *fancy* somewhat unutterable; so that an eye witnesse of things conceives them with an *imagination* more compleat, strong, and intuitive, then he can either apprehend, or deliver by way of relation; for relations are not only in great part false, out of the relaters mis-information, vanitie, or interest; but which is unavoidable, their choice, and frame agrees more naturally with his judgement, whose issue they are, then with his readers; so as the reader is like one feasted with dishes fitter for another mans stomacke, then his owne: but a traveller takes with his eye, and eare, only such *occurrents* into observation, as his owne apprehension affects, and through that *sympathy,* can digest them into an experience more naturall for himselfe, then he could have done the notes of another: Wherefore I desiring somewhat to informe my selfe of the *Turkish* Nation, would not sit downe with a booke knowledge thereof, but rather (through all the hazard and edurance of travell,) receive it from mine owne eye not dazled with any affection, prejudicacy, or mist of education, which preoccupate the minde, and delude it with partiall *ideas,* as with a false glasse, representing the *object* in colours, and proportions untrue: for the just censure of things is to be drawn from their *end* whereto they are aymed, without requiring them to our customes, and ordinances, or other impertinent respects, which they acknowledge not for their touch-stone: wherefore he who passes through the severall educations of men, must not try them by his owne, but weyning his minde from all former habite of *opinion,* should as it were putting off the old man, come fresh and sincere to consider them: This preparation was the cause, why the superstition, policie, entertainments, diet, lodging, and other manners of the Turkes, never provoked me so far, as usually they doe those who catechize the world by their owne home; and this also barres these observations from appearing beyond my owne closet, for to a minde possest with any set doctrine, their unconformitie must needs make them seeme unsound, and extravagant, nor can they comply to a rule, by which they were not made. Neverthelesse considering that *experience* forgotten is as if it never had beene, and knowing how much I ventured for it, as little as it is, I could not but esteeme it worth retaining in my owne memory, though not transferring to others: hereupon I have in these lines registred to my selfe, whatsoever most tooke me in my journey from *Venice* into *Turky.*

First, I agreed with a *Janizary*[12] at *Venice,* to find me Diet, Horse, Coach, passage, and all other usuall charges, as farre as *Constantinople:* Then upon the seventh of *May,* 1634. I embarq'd on a *Venetian Galley* with a *Caravan* of *Turkes,* and *Jewes* bound for the *Levant,* not having any *Christian* with them besides my selfe: this occasion was right to my purpose; for the familiaritie of bed, board, and passage together, is more opportune to disclose the customes of men, then a much longer habitation in Cities, where societie is not so linkt, and behaviour more personate, then in travell, whose common sufferings endeare men, laying them open, and obnoxious to one another: The not having any other *Christian* in the *Caravan,* gave mee two notable advantages: First, that no other mans errors could draw either hatred, or engagement upon mee; then I had a freedome of complying upon occasion of questions by them made; whereby I became all things to all men, which let me into the breasts of many.

[Blount encounters Ottoman hospitality in a passage worth comparing with the accounts by Dallam and Rawlins discussed by Mary Fuller in this volume. Having crossed the Adriatic from Venice, Blount's Caravan joins up and travels with Ottoman forces being summoned against Poland. Blount reports on the condition of the troops and the splendid traveling court of Murad Basha, the Ottoman general, who is not to be confused with Sultan Murad IV (1623-1640), the current Grand Seignior.[13] *Blount's arguments about the shared Anglo-Ottoman hatred of Catholics for their idolatry recall Elizabeth's diplomatic letters to Murad III (1574-1595), printed by Richard Hakluyt in* The Principall Navigations *(1589).*[14] *Blount may also have known that Edward Barton, the second representative of the Levant Company resident in Istanbul, had accompanied Sultan Mehmed*

III against the Hungarians in 1595/96, gaining the respect of the Ottomans without compromising his reputation among his own countrymen.[15]*]*

Blount Makes Friends

That which secured and emboldned my enquiry and passage these twelve dayes March, was an accident the first night; which was thus: the *Campe* being pitch'd on the Shoare of *Danubius,* I went, (but timorously) to view the Service about *Murath Bashaes* Court, where one of his favorite Boyes espying me to bee a Stranger, gave me a Cup of *Sherbet;* I in thanks, and to make friends in Court, presented him with a Pocket *Looking Glasse,* in a little *Ivory* Case, with a *Combe;* such as are sold at *Westminster-hall* for foure or five shillings a peece: The youth much taken therewith, ran, and shewed it to the *Bashaw,* who presently sent for me, and making me sit, and drinke *Cauphe* in his presence, called for one that spake *Italian;* then demanding of my *condition, purpose, countrey,* and many other particulars, it was my fortune to hit his humour so right, as at last, he asked if my Law did permit me to serve under them going against the *Polacke* who is a *Christian;* promising with his hand upon his breast, that if I would, I should be inrolled of his Companies, furnished with a good Horse, and of other necessaries be provided with the rest of his Household; I humbly thanked him, for his favour, and told him that to an *Englishman* it was lawfull to serve under any who were in League with our *King,* and that our *King* had not only a League with the *Gran Signior,* but continually held an *Embassadour* at his Court, esteeming him the greatest *Monarch* in the *World:* so that my Service there, especially if I behaved my selfe not unworthy of my *Nation,* would be exceedingly well received in *England;* and the *Polacke,* though in name a *Christian,* yet of a *Sect,* which for *Idolatry,* and many other points, we much abhorred; wherefore the *English* had of late, helpt the *Muscovite* against him, and would be forwarder under the *Turkes,* whom we not only honoured for their glorious actions in the world; but also loved, for the kinde *Commerce* of *Trade* which we finde amongst them: But as for my present engagement to the warre, with much sorrow, I acknowledged my incapacitie, by reason I wanted *language,* which would not only render me uncapable of *Commands,* and so *unserviceable,* but also endanger me in *tumults,* where I appearing a Stranger, and not able to expresse my affection, might be mistaken, and used accordingly; wherefore I humbly entreated his Highnesse leave to follow my poore affaires, with an eternall *oblige* to *Blazon* this honourable favour wheresoever I came: He forthwith bade me doe as liked me best; wherewith I tooke my leave, but had much confidence in his favour, and went often to observe his Court.

[On the Island of Rhodes Blount investigates a graveyard and finds himself suspected of being an enemy spy. He has been reporting on the strength of the fortifications set up around the harbor when he stops and writes about the sun.]

Blount Becomes Scottish

In this *Iland* the Sunne is so *powerfull,* and *constant,* as it was anciently Dedicate to *Phoebus:* they have a kinde of Grape as big as a Damsen, and of that colour, the *Vines* if *watered* beare all the yeare, both ripe Grapes, half ripe, and knots; all together upon the same Vine: yet that they may not weare out too soone, they use to forbeare watering of them in *December,* and *January,* during which time they beare not, till after a while that they are watred again: Upon my first landing I had espyed among divers very *honourable Sepultures,* one more brave then the rest, and new; I enquired whose it was; a *Turke* not knowing whence I was, told me it was the *Captaine Basha,* slaine the yeare before by two *English Ships;* and therewith gave such a Language of our *Nation,* and threatning to all whom they should light upon, as made me upon all demands professe my selfe a *Scotchman,* which being a name unknowne to them, saved me, nor did I suppose it any quitting of my Countrey, but rather a *retreat* from one corner to the other; and when they enquired more in particular, I intending my owne *safetie* more then their *instruction,* answered the truth both of my *King,* and *Country,* but in the old obsolete *Greeke,* and *Latine* titles, which was as dark to them as a discourse of *Isis* and *Osyris;* yet the third day, in the morning, I prying up, and downe alone, met a *Turke* who in *Italian* told me ah! you are an *Englishman,* and with a kinde of malicious posture, laying his *fore-finger* under his eye, methought he had the lookes of a designe, he presently departed, I got to my *Galleon,* and durst goe to land no more: The next morne we departed for *Alexandria* . . .

Notes

1. *Ortelius his Epitome of the Theatre of the World* was issued in an augmented English edition in 1603; a complete *Theatrum* was reissued in England in 1606, and an amplified edition in 1610. Mercator's *Atlas* was not issued in English until 1635. See Samuel Chew, *The Crescent and the Rose: Islam and England during the Renaissance* (New York: Oxford University Press, 1937), p. 23.
2. Than: i.e., "then"
3. King may have seen a small tract entitled *Here begynneth a lytell treatyse of the turkes called Alcaron. And also it spekeIth of Machamet the Nygromancer,* printed in 1515.
4. The Ottoman dynasty takes its name from one Osman who, according to legend, was the first of the Turkish warlords in Anatolia to declare himself an independent sovereign, ca. 1302. Osman's name variously appears as Othman, Othoman, and Ottoman.
5. Murad II (1421-1451).
6. Murad II's reign was marked by near continuous fighting throughout Serbia, Bosnia, Hungary, Poland, Wallachia, and Albania. Murad's defeat of the Hungarian army at the Battle of Varna on November 10, 1444 effectively subjugated Serbia and Bosnia. However, the Albanian George Kastriotis, popularly known as "Scanderbeg," held out against Ottoman control from the mountain city of Krüje, or "Croya." Here King repeats the legend that Murad died during the unsuccessful seige of Krüje, to be found in Richard Knolles's *Generall Historie of the Turkes,* p. 331. Murad abandoned the seige in October 1450, and died in February 1451 (Colin Imber, *The Ottoman Empire 1300-1481* [Istanbul: Isis Press, 1990], pp. 142-43).
7. Mehmet II (1444-1446, 1451-1481), who successfully captured Constantinople in 1453, was called "Great" by Knolles but "Fatih"—the lawgiver—in Turkish.
8. *Meschit*: not uncommon form for Mosque.
9. *Mufty*, Mufti: a senior Muslim theologian who gives legal counsel; often head of the *Ulema* or religious authorities.
10. *Freinks*, Franks: Europeans, speakers of the form of Italian called "Lingua Franca."
11. *Zinganaes*: gypsies.
12. *Janizary*: member of the military elite formally trained in the seraglio; by the time of Blount's voyage, however, they had evidently started taking second jobs as travel agents.

13. See Thomas Nabbes's "Continuation" of Richard Knolles's *Generall Historie, of the Turkes* (1638) (p. 24), which evidently relies for some of its details on Blount's account.
14. But see Susan A. Skilleter (*William Harborne and the Trade with Turkey, 1578-1582* [London: British Academy, 1977], for an accurate transcription of the original correspondence.
15. The pious William Biddulph termed Barton "the mirror of all Ambassadours that ever came to *Constantinople*," continuing: "who for his wisedome, good government, policie, and Christian cariage hath left an immortal fame behind him in those Countries, to this present day, and lieth buried at an Iland of the Greeks, within twelve miles of *Constantinople* called *Bartons* Iland to this day" (*The Travels of certaine Englishmen into Africa, Asia, Troy, Bithnia, Thracia, and to the Blacke Sea* [1609], p. 40). Barton died in 1597, possibly of the dysentery, and was buried on Heybeli (Halki), one of the Prince's Islands in the sea of Marmara. In September 1999 I was unable to find Barton's gravestone, reported as still *in situ* by John Freely (*Blue Guide: Instanbul* [London: Black, 1991], pp. 347-48). Photographs of the gravestone and site, however, appear in Ahmet Refik, *Türkler ve Kraliçe Elizabet (1200–1255)* (Instanbul: Matbaacilik ve Nesriyat, 1932), pp. 11, 14.

OTTOMANISM BEFORE ORIENTALISM? BISHOP KING PRAISES HENRY BLOUNT, PASSENGER IN THE LEVANT

Gerald MacLean

I. BEFORE LADY MARY: OTTOMANIST DISCOURSE AND THE MAKING OF THE BRITISH IMPERIAL SUBJECT, 1580-1720.

When Lady Mary Wortley Montagu writes home, a peculiarly British form of neocolonial occupation becomes evident: the one being practiced whenever visitors cannot help but imagine themselves agents while in Turkey, eager to find parts to play in the great game of European diplomacy. Having "past a Journey that has not been undertaken by any Christian since the Time of the Greek Emperours," as she informed the Princess of Wales, how could Lady Mary not feel that simply being here and now in Adrianople was to be an agent, part of history in the making?[1] Lady Mary knew that others had traveled into Ottoman territory before, often going out of her way to name and to correct the accounts of Paul Rycaut and Aaron Hill. Yet she insists that her readers notice how she is also an original, not simply a woman subverting class and gender norms by traveling, but something rather more than this, an informed adventurer going somewhere for the first time and with a mission; seeing inside the *haremlik* and bringing back a secret from the East—inoculation against smallpox. It may be that the fulfillment of the Oriental stereotype can only finally be achieved by a European—preferably a woman—cross-dressed in Oriental disguise.[2] Anxious to claim that she was a first, Lady Mary nevertheless knew that others had preceded her, most of them men. Who were those who went before?[3]

By their own reports, the earliest British visitors to the Ottoman Empire often were, or found themselves acting as if they were, national agents, but perhaps not with the fully developed sense of empowerment that can be mapped in Lady Mary's epistles. How did this sense of personal entitlement and agency develop among visitors for whom the Islamic Empire continued to provide a threatening but also enviable model of successful imperial might? How, in most general terms, did these early visitors to the Ottoman Empire come to imagine and represent themselves? What conventions of self-construction and self-representation did they employ when writing about themselves and those they met? Who were these specters of Lady Mary? Those British visitors who had gone before, especially those whom she appears to ignore?

I will turn to one of my favorites among these specters—Henry Blount, whose *Voyage Into The Levant* (1636) recorded a journey made in 1634—but first wish to offer some further general and historical contexts that will help clarify the terms of my discussion of Blount's text. I will show that in its own day, Blount's text was perceived to mark a generic breakthrough in travel writing, and suggest that this can be explained by how British national identities were coming into focus at the time in terms of global imaginaries, most specifically a discourse about the Islamic Ottoman Empire that was quite explicitly linked to emergent imperial ambitions. In other words, Blount might be said to have been among the first to imagine a British Empire into being; and he did so by setting out to learn from the Turks.

By the term "Ottomanism," I want to describe the tropes, structures, and fantasies by means of which Europeans sought to make knowable the imperial Ottoman other: both the imperial dynasty and the vast maritime and territorial areas that they governed. Ottomanism will be found to be both strategic and interested. Like all systems of knowledge production, Ottomanism arises from both lack and desire, and in this sense tells us perhaps rather more about the desiring subject than about the object of knowledge. European desire for the "worldly goods" of the East saturates Renaissance art, literature, political thought, and commercial practice—as Lisa Jardine has recently shown in *Worldly Goods.*[4] Most often, what stood in the way of that desire was the Islamic Ottoman Empire. From the late sixteenth to the early eighteenth centuries, a period roughly corresponding to that empire's greatest territorial reach into Europe, knowledge about the imperial dynasty and the vast maritime and territorial areas that they governed was obviously crucial for competing European trading interests, keen to capture Eastern markets.

Yet early British interests in the Ottoman Empire—the terrible Turk, his religion, society, and empire—demonstrate the inseparability of commercial from cultural concerns. In the Renaissance and early modern periods, personal and national identities were busily remaking themselves in accordance with what was newly becoming known of the world through a wide variety of different kinds of exchange and encounter. For the insular British, personal and national desires and identities were no longer constructed only from within the local, the familiar, and the traditional, but increasingly became inseparably connected to the global, the strange, and the alien. When, in *The Way of the World* (1700), William Congreve's Mirabel announces he will permit Millimant to serve only "native" drinks once they are married, and then specifies "tea, chocolate, and coffee" (act 3, scene 5), something very strange has clearly occurred in the concept of what is native to England.[5]

II. Ottomanism and Imperial Envy: British Representations of the Ottoman Empire, 1580-1720

The term "Ottomanism" directly and in very obvious ways looks back to 1978 when Edward Said's *Orientalism* appeared, but not without some sense of what has happened since then, most specially the crucial theoretical and political development of feminist engagements within colonial discourse studies that have linked the gender politics of travel with forms of cultural imperialism and thereby complicated Said's initial Orientalism thesis.[6] When they represented Ottoman civilization, early modern British writers framed a discursive Anglo-Ottoman relationship in ways that complicate our understanding of both Orientalism and the cultural history of British imperialism. Calling this discursive relationship "imperial envy" helps to describe, analyze, and even explain, some of the structures and features that are most distinctive about the ways that attitudes toward the empire of the Ottoman Turks come into focus during the first great age of British visitors to the Islamic East. How did knowledge of, attitudes toward, commercial interests in, and fantasies about the Turks and the Ottoman Empire contribute to the making of British imperial subjects? How might Ottomanist discourse be said to have produced an imperial subject such as Lady Mary Wortley Montagu?[7]

What is most distinctive if not peculiar about the desires of British travelers to the Ottoman Empire from the 1580s, when the Turkey and Levant Companies began trading, until the arrival of Lady Mary Wortley Montagu, often becomes most clear in reference to and distinction from Said's initial thesis in *Orientalism*. Unlike "the Orient," the Ottoman Empire really existed. The Ottomans were well-aware of the fact, and were quite capable of representing themselves. Said himself is perfectly clear that although "there were—and are—cultures and nations whose location is in the East,"[8] Orientalism describes the ways that "Westerners" set about understanding and eventually controlling those cultures and nations by designating them the "Orient." When early European visitors set about understanding, misunderstanding, overlooking, or ignoring the self-representations of Ottoman culture, the reports they left behind may often owe rather more to their own imperial fantasies and personal ambitions than to really existing conditions.[9] But they never doubted that the Ottoman Empire existed. *Orientalism* is also almost entirely concerned with the period following the defeat of Islamic imperialism, the post-Napoleonic period of Enlightenment. For Said, the previous era is dominated by "Christian supernaturalism,"[10] but as we will see, this generalization is only partly useful.

In sixteenth- and seventeenth-century Britain, thinking about Turks meant thinking about religious differences but it also meant thinking about empires in the East. The Ottomans had one, a great one. The British wanted an empire and devoted the seventeenth century, investing a great deal of energy and resources, to sorting out how to get theirs. Their imperial ambitions involved dealing not only with the Ottomans and their great territorial empire, but with a previous history of British discourse about the Turks. Behind the ambiguous representation of the Ottoman Empire as both a realm of tyrannous slavery and a space where Britons might do rather well for themselves, lurked those twin developments of trade and knowledge called the Renaissance, an era that may be said to have begun with the Christian loss of Constantinople to Islam in 1453.

If it is correct to think that the capture of Constantinople in 1453 by Fatih Mehmet in significant ways signaled the start of the European Renaissance, the consequences have yet to be fully explored and detailed. To a divided Christian Europe, this crucial moment of territorial loss created not an absence but a lack; and in consequence of that lack, a desire that, among the British at least, most often took shape as what I call imperial envy. To European Protestants and Catholics alike, the loss of Constantinople to Islam not only ended a historical era initiated by the capture of sacred lands by Islam, it also finally sundered eastern and western Christianities. Clearly, if we are to view the Renaissance as, in part, a systematic reconstruction of Graeco-Roman foundations that helped Europeans explain the increasing loss of territory within the expanding borders of the Ottoman Empire, we need also to notice how the Renaissance involved enormous investments of material goods and energies finding ways to cross those borders in order to exchange other material goods and skills. For the British, at least, the desire to trade is where the story of Ottomanism and imperial envy begins, during the reign of Elizabeth, little more than a century after Mehmet seized Constantinople.

III. Bishop King Praises Henry Blount's Voyage

Henry Blount's strenuous self-representations, in the opening pages of *A Voyage Into The Levant,* reveal his rationale and methods for preparing himself before starting on his journey from Christian Europe into the Ottoman Empire. A precocious learner, much celebrated for his wit while at Oxford, and trained in law, Blount presents himself to us as, in many ways, he no doubt was: a well-educated man of his times, a well-informed and yet skeptical observer who is seeking to contribute to knowledge currently unavailable from previous accounts. Casting himself right from the start as a man of the new sciences, Blount proposes a comparative and rationalist inquiry into the Islamic world, thereby indicating how Christian supernaturalism may not have dominated discourse about the East until the European Enlightenment. Blount claims, at least, that his desires to travel to Ottoman lands are both skeptical and rational; he wants to test tradition and authority, to find out if the "*Turkish* way" isn't at all the way it has been represented, but "rather another kind of civility, different from ours, but no lesse pretending."[11] Perhaps the Turks weren't so terrible as they had often been made out to be.

Blount rather portentously explains how preparing to travel to Ottoman lands requires an extended process of unlearning that turns into nothing less than a "putting off the old man," if one is to test received wisdom free "from all former habit of opinion."[12] Yet this remarkable moment of spiritual rebirth in Blount describes something quite unlike the "innocence" of vision that Mary Louise Pratt finds in the figure she calls the "seeing-man," the European male traveler who imagines his own "innocence in the same moment" that he asserts "European hegemony . . . he whose imperial eyes passively look out and possess."[13] Pratt has in mind here the activities of those she calls the "bourgeois" naturalists of the late eighteenth century who used scientific classification to "naturalize" their own "global presence and authority."[14] But there is nothing passive or possessive about Blount's gaze. Pratt's "seeing-man" does not describe the situation of an apparently independent member of the landed squirearchy traveling into the Ottoman Empire with—so he declares—a carefully prepared program of skeptical inquiry and analysis necessitating a complete personal, intellectual, cultural, and spiritual makeover on the part of the inquiring visitor. As we will see, Blount calls this figure a "passenger."[15]

In its own day, Blount's *Voyage* was praised by a learned bishop for transforming travel writing by refiguring the agency of the British visitor to Ottoman lands. In panegyrical verses, "To my Noble and Judicious Friend Sir Henry Blount upon his Voyage," Bishop Henry King explains how Blount's example teaches others how best to manage their business in the lands of the Turks.[16] Apparently, reading Blount's *Voyage* did not so much add to King's prior knowledge of Ottoman imperial history as alter his attitude by informing him of useful things that British travelers to the East clearly needed to know. According to King, Blount's various reports of encountering and contending with cultural differences transformed the familiar registers of both travel writing and the knowledge they offered about the Ottomans into a new form of practical knowledge, a guide to the kinds of agency available to future British travelers to Ottoman lands.

The first half of King's verses summarizes what he claims to have known and thought about the Ottomans before reading Blount. The Anglican prelate admits that "those two baits of profit and delight" (line 4) had often tempted him, while reading about the origins of the Ottoman Empire and looking over the maps of Ortelius and Mercator, to feel "strong and oft desires to tread / Some of those voyages which I have read" (lines 7-8). He describes an imaginary journey north through Africa, pauses in Cairo, then sets off to Constantinople: "Once the world's Lord, now the beslaved Greek, / Made by a Turkish yoke and fortune's hate / In language as in mind degenerate" (lines 37-40). King offers the standard European Renaissance view of how the expansion of the ungodly Turkish Empire has brought slavery and degeneration to the once glorious Greeks. Here too, in this imaginary moment of gazing on Constantinople, King describes himself "all wrapp'd in pity and amaze," a state that I would call imperial envy. What he ponders is the way God has allowed the ungodly to use terrible means to achieve and maintain control over so vast an empire (see lines 41-50). To the Anglican divine, the Ottoman empire is both a moral and a theological problem: how can God allow pride, rapine, and tyranny, the emperor's base motives and base means, to continue enabling him to hold such a monstrous bulk of an empire together? The answer, of course, has already been provided in the terms of the question: the Ottoman Empire is the scourge of Christendom.

> Sure, who e'er shall the just extraction bring
> Of this gigantic power from the spring;
> Must there confess a Higher Ordinance
> Did it for terror to the earth advance. (lines 53-56)

After this reassurance of Christian providentialism, King offers a summary of the rise of Islam up to the capture of Constantinople. King emphasizes Turkish cruelty while listing all of the features of life in that empire to be gleaned from reading previous travel books. He repeats one of Blount's key observations regarding the hospitality of Turks toward Christian travelers, but even here King insists that he already knew about "What quarter Christians have; how just and free / To inoffensive travellers they [the Turks] be" (lines 89-90).

What King does claim is that Blount has transformed travel writing from "the commonplace of travellers, who teach / But table talk" (lines 114-115) into a systematic program of knowledge that will directly serve Britain's national interests in the East. King confesses that his own earlier reading among travel writers was "for wonder, not for use," so an immediate effect of reading Blount's

Voyage is that the Bishop now feels himself "cured" of his own desire ever to travel from his "native lists" (see lines 92-112). But King expands his personal obligation to Blount into a national debt. King insists that Britain owes Blount something for providing future "sons" with necessary knowledge for them to travel and deal successfully in Ottoman lands. The nation should, he argues, "canonize" Blount's book as a "rule to all her travellers," because of the greater profits reading it will encourage: Blount's mercantile metaphors are surely most apt (see lines 124-138). Blount's book is itself the rich freight, the "return" to the nation's "staple," the knowledge of how to travel and "come safely off." Dismissive of those who travel "for form" not "effect," King praises Blount's "example" by which future "sons" will learn how they might most profitably pursue their nation's commercial enterprise, going abroad "to some effect." Blount, the exemplary guide to "dangers . . . / Which wit and wary caution [may] not forsee" (lines 133-134) exemplifies strategies needed for British travelers encountering the dangers and diplomacies of everyday life inside a supremely powerful but nervous empire, notoriously hostile to Christians. According to King, Blount deserves national recognition and reward for extending British agency within Ottoman territories by providing such a detailed account of how Britons might visit, avoid dangers, and deal profitably.

IV. Blount Gets Ready

How well does Blount's *Voyage* live up to the learned bishop's praise? From the start, Blount's account is couched in secular logic that owes rather more to Neoplatonism, to the Florentine humanists, and to Bacon, than to Christian supernaturalism. He does not defend traveling on religious grounds, but in order to advance knowledge by means of rational inquiry. "Intellectuall Complexions," he starts out, desire "*knowledge,*" especially "of humane affaires." Proposing that "*experience* advances" knowledge of human institutions "best," Blount argues that knowledge will increase proportionally with the degree of novelty and difference experienced. "So my former time spent in viewing *Italy, France,* and some little of *Spain,*" he concludes, "being countries of Christian institution, did but represent in a severall dresse, the effect of what I knew before."[17] Cultural difference is crucial to knowledge, and will best advance free from religion.

Intellect desires knowledge; experience of difference increases knowledge as difference increases; therefore, intellect desires experience of radical difference. Within this logic, Blount's designation of "Christian" countries is clearly a cultural and political category that suspends those religious differences between Protestants and Catholics that were currently devastating Europe. Nowhere in Blount's *Voyage* do we find the trace of Christian providentialism so evident in King's poem. For Blount, national institutions are the historical result of geography, nature, and climate. Neither a product or evidence of a theogenetic or providential design, national and cultural differences appear to knowledge as matters of space. Since no culture could appear more different to an inquiring observer from the "Climate"[18] of the "*North-west* parts of the World" than that practiced by "those of the *South-East,*" Blount proposes that knowledge of the geographical space over which Islam currently rules will necessarily prove specially desirable. Here "the *Turkes*" have established an empire "and fixt it selfe such firme foundations as no other ever did; I was of opinion, that hee who would behold these times in their greatest glory, could not find a better *scene* then *Turky:* these considerations sent me thither."[19] In 1634, when Blount set out, the lands northwest of the Danube were continuing to be destroyed by rival Christian armies in a devastating war that, begun in 1618, had reached its peak of intensity with the intervention of Gustavus Adolphus four years earlier. Southeast of Vienna, on the other hand, the Turks—"the only moderne people, great in action"[20]—were successfully ruling an actually existing empire that, while Blount was there, constituted the most extensive and most firmly established empire in that region for some time. How could anyone of an intellectual complexion not want "to behold these times in their greatest glory" by going to see for themselves just what was going on inside the Ottoman Empire?

Even as he appears to conclude when writing "these considerations sent me thither," Blount evidently felt the need to explain further his purposes and preparations before setting out, describing four "particular cares": observing religion, viewing the various "sects," examining the army, and visiting Cairo. Expecting to find "another kind of civility, different from ours, but no lesse pretending," Blount distinguishes himself from expatriate "inhabitants," and modestly assumes for himself the guise of what he terms a "passenger"; someone with an "unpartiall" mind, a comparative project, and a healthy disrespect for the common prejudices of Christians about Turks.[21] Passengers might never achieve the perfect knowledge available to "inhabitants," but neither do they simply pass by without seeing what is about them. Blount, the exemplary passenger, sets forth with an agenda that requires stopping long enough to test received opinion by means of experience. Intellectual passengers travel to seek critical distance from their own cultural specificity, already expecting to find that things may not be the way they are commonly said to be. And to do this, Blount points out, passengers need to be capable of imagining and even attempting to live through the possibility that other cultures might be "different from ours, but no lesse pretending."[22] Litotes—the rhetorical figure of negative understatement being used here—is never, of course, a trope of equality but one of relative negation; "no lesse pretending" points to the absent foundations of European normativity.[23]

Each of Blount's stated reasons for traveling to the Levant exemplifies a general interest in comparative imperial identities, an interest that is characteristic of imperial envy. His second declared aim—to investigate the seemingly peaceful coexistence of distinct religious, racial, and cultural communities within *millets*—is an aspect of Ottoman society that proved constantly fascinating to British visitors and readers. Blount provocatively calls these groups "sects"—a term he will use throughout the *Voyage* in order to distance himself from any religious persuasion, Christian or otherwise. Blount writes that he wanted to "acquaint my selfe with those other sects which live under the *Turkes,* as *Greekes, Armenians, Freinks,* and *Zinganaes,* but especially the *Jewes.*"[24] His third professed reason for traveling is "to see the *Turkish* Army, then going against *Poland,*"[25] since doing so will enable him to compare Turkish and Christian military discipline. Blount states that his fourth reason for traveling was a personal desire to visit Cairo, "and that for two causes,"[26] the first of which continues his agenda of skeptically interrogating received Christian opinion about Ottoman administration; the second: "because *Egypt* is held to have been the fountaine of all *Science,* and *Arts civill,* therefore I did hope to find some spark of those cinders not yet put out."[27] Blount's suspicion that Christian images of Ottoman "sottish sensuality" might greatly misrepresent actually existing conditions is indeed "shrewd," as the historian Samuel Chew once remarked.[28] But it also draws attention to how Blount's interests are persistently imperial, arising from the desire to assess the nature, range, and authority of Ottoman power on the one hand, while seeing if there is anything left of the great African imperial civilization on the other.

In describing his own interests in the workings of Ottoman imperialism, Blount notably does not impose the kind of Christian moral-providential frame we noticed in King's poem, where the ungodly empire of the Turks was a contributing cause of Greek degeneracy. Although Blount presumes that the vast multitudes living in Cairo are "deeply malicious," he does not hold the Turks responsible for current conditions in the "greatest concourse of Mankind in these times";[29] rather he wonders at their ability to rule over it. Whatever may have been the cause, our Oxford rhetorician and wit boldly mixes metaphors when acknowledging Egypt as the "fountaine of all *Science,* and *Arts civill,*" and expressing his hope to find "some spark of those cinders not yet put out."[30] By the 1630s, evidently, the European Renaissance had not entirely erased knowledge of the African roots of civility, a matter about which Blount is strangely lacking in his usual skepticism.

V. Blount Puts Off the Old Man

Throughout his opening pages, Blount distinguishes himself from writers on Turkey who never went there, for whom "Turk" was synonymous with Islam and therefore principally an enemy to be feared and righteously destroyed. But he also, as we have seen, considers himself a "passenger" in distinction from inhabitants, insisting nevertheless that the experience of the passenger allows knowledge of a sort unavailable from reading the frequently false reports of others. Having found himself "desiring somewhat to informe my selfe of the *Turkish* Nation, [he] would not sit down with a booke knowledge therof," but had to go and see for himself in order to avoid any "affection, prejudicacy, or mist of education," all of which "delude" the mind with " partial *ideas.*"[31] Passengers may know less than inhabitants, who can provide visitors with reliable, local information—such as the "Family" who had been resident in Cairo for 25 years, who "informed me of many things, with much certaintie."[32] But passengers know more than those who stay at home and read books full of their author's prejudices and opinions. It was in this self-styled role of passenger that Blount performed most of his traveling from the moment he left Venice on May 7, 1634, a passenger on a Venetian galley amidst "a *Caravan* of *Turkes,* and *Jewes,*"[33] since it is their trading interests that determine the route and schedule he finds himself following. Passengers may have to rely upon others, going where and when they are ready, but the well-prepared passenger is not simply passive and dependent.

Blount reserves one of his more startling revelations of what being a passenger can mean for the moment just before his journey actually begins. About to set forth, Blount retrospectively assures us that his "preparation" served him well, bringing about nothing less than a form of personal reconstruction that would seem to be essential for all who would travel "through the severall educations of men."[34] To do so, Blount insists, one "should as it were putting off the old man come fresh and sincere to consider them."[35] Evoking the Pauline injunction to the Ephesians to put aside the "old man," Blount secularizes the trope: not so much a figure of spiritual rebirth, more a proposal to unlearn received prejudice. Passengers of an intellectual complexion need to put aside their ingrained domestic pieties and provincial prejudgments, since only by doing so will they prepare themselves—in mind and body—for the change of diet and lodging.

Only thus prepared can Blount allow his travels to begin. But his narrative has barely begun carrying him forth on his journey before circumstances once again demand that he engages in yet another remarkable moment of self-reconstruction by stepping outside the burden of Christianity. Delighted to find himself sailing among Turks and Jews, "not having any Christian with them besides my selfe," Blount explains that traveling without other Christians offered two advantages: he wouldn't get into trouble because of association with the views of other Christians and secondly: "I had a freedome of complying upon occasion of questions by them made; whereby I became all things to all men, which let me into the breasts of many."[36]

Once he starts leaving Europe, Blount doesn't simply imagine himself prepared and ready to step outside his own culture, he joyfully shakes it off and declares himself better able to perform without it. He certainly finds any association with Christianity burdensome, later describing himself not "loving company of *Christians* in Turky,"[37] and avoids them whenever possible. Later, he will note that the Tartars are specially keen on seizing known Christians and selling them into slavery.[38] Consequently, acting free from any customary cultural identification as a Christian, Blount can become compliant, "all things to all men," the prototypical agent who, keeping himself unobtrusive and secretive, finds a way into "the breasts of many." Blount's "passenger" may, after all, be none other than the ideal spy.

VI. Blount Boldly Goes

The opening pages of Blount's *Voyage* offer a detailed account of his reasons for traveling, but by the time one has finished reading his report of the journey, it is hard not to suspect that he has been rather cagey from the start; certainly with respect to his material means, but also about his motives for adventuring into Ottoman lands. One cannot resist speculating, looking for indications and hints in his elaborately stated intentions, and elsewhere that crucial aspects of his "purpose" would have been too obvious to contemporary readers to need mentioning. Although he provides constant circumstantial reminders and asides about traveling conditions, not once does Blount feel any need to tell us directly how it was that he undertook this journey in the first place. Who paid? After offering his detailed rationale, Blount simply and suddenly projects himself to Venice, where he finds himself employing a Janissary to serve as travel agent to organize his trip as far as Constantinople.

Here, the terms of King's verses help us to grasp the contemporary sense in which it may have been the very act of writing the *Voyage,* of publishing the book, of describing in detail and illustrating with exemplary episodes how to be a passenger, that constituted both Blount's most obvious and therefore unspoken goal and, according to King, his great achievement. By writing to instruct future sons of empire how they might most profitably travel among and, more importantly, deal with the Turks, Blount perhaps fulfilled a mission that did not need explaining at the time. As Sir Thomas Palmer had observed in 1606, since princes no longer traveled abroad themselves, they needed others to do it for them.[39] In this sense, Bishop King's verses urge to be read as a professional letter of reference, a commendation for the knighthood that Blount received from Charles I in 1640, four years after initial publication of *Voyage.* For Blount, it would seem, the Ottoman Empire was, as Disraeli said of the East, a career: but that is another story for which we would need to examine the whole of Blount's *Voyage.*

I have suggested ways in which we need to complicate and historicize Said's account of how the Christian west approached and perceived the Islamic Orient during the pre-Napoleonic period. Bishop King does, it is true, offer the mysteries of divine providence—Said's "Christian supernaturalism"—to explain the rise and success of the Ottoman Empire. But there is nothing supernatural about his concern regarding the power of the sultan or, for that matter, the commercial metaphors that he uses to praise Blount for advancing the kinds of knowledge and skill that will promote the nation's trading interests in the East. King was doubtless correct in drawing attention to the usefulness of the knowledge Blount gained for encouraging future trade. Yet it was advancing knowledge in the face of traditional hostility and religious prejudice toward the Turks that animated Blount's project, according to his avowed purpose. By shifting knowledge of the Ottoman Empire to an empirical rather than religious frame of reference, Blount's achievement was also to demystify the ingrained certainties of national identity. Blount conceives of traveling to the lands of the Turks as an exercise in self-reconstruction that must begin, at least, with a suspension of normative domestic perceptions and attitudes, a stepping outside native religion and national identity. Far from perceiving the Ottoman Empire to be an orientalized space awaiting western penetration and dominance, Blount realized that understanding the great eastern empire required that European visitors put aside received opinion before going and seeing for themselves. Before there was any question of Europeans setting out to control the East, it was necessary that they should become capable of recognizing and understanding what was already there, firmly in place, rather than simply denouncing it for being barbarous. Before empire, there was imperial envy. Before Orientalism, there was Ottomanism.

Notes

1. Lady Mary Wortley Montagu, *The Complete Letters of Lady Mary Wortley Montagu,* ed. Robert Halsband, 3 vols. (Oxford: Clarendon Press, 1965), vol.1, p. 310.

2. See important recent studies of Montagu's letters by Srinivas Aravamudan, "Lady Mary Wortley Montagu in the Hamman: Masquerade, Womanliness, and Levantinization," *ELR* 62.1 (1995): 69-104; Isobel Grundy, *Lady Mary Wortley Montagu: Comet of the Enlightenment* (Oxford: Oxford University Press, 1999); Robert Halsband, *The Life of Mary Wortley Montagu* (New York: Oxford University Press, 1960); Felicity Nussbaum, *Torrid Zones: Maternity, Sexuality, and Empire in Eighteenth-Century English Narratives* (Baltimore: Johns Hopkins University Press, 1995); Mary Louise Pratt, *Imperial Eyes: Travel Writing and Trans-culturation* (London: Routledge, 1992); and Meyda Yegenoglu, *Colonial Fantasies: Towards a Feminist Reading of Orientalism* (Cambridge: Cambridge University Press, 1998).
3. See notable studies of writings by early travelers by Brandon Beck, *From the Rising of the Sun: English Images of the Ottoman Empire to 1715* (New York: Lang, 1987); Orhan Burian, "Interest of the English in Turkey as Reflected in English Literature of the Renaissance," *Oriens* 5 (1952): 209-29; Samuel Chew, *The Crescent and the Rose: Islam and England during the Renaissance* (New York: Oxford University Press, 1937); Hamid Dereli, *Kiraliçe Elizabeth Devrinde Türkler ve Ingilizler—Bir Arastirma* (Istanbul: Anil Matbassi, 1951); and Nabil Matar, *Turks, Moors, and Englishmen in the Age of Discovery* (New York: Columbia University Press, 1999).
4. Lisa Jardine, *Worldly Goods: A New History of the Renaissance* (London: Macmillan, 1996).
5. As Fernand Braudel points out, "the abundance of literature on coffee defies description" (*The Mediterranean and the Mediterranean World in the Age of Philip II,* trans. Sian Reynolds, 2 vols. [London: Collins, 1972], vol. 1, p. 762 n. 35), but see K. N. Chaudhuri, *Asia before Europe: Economy and Civilization of the Indian Ocean from the Rise of Islam to 1750* (Cambridge: Cambridge University Press, 1990), and, on the arrival of coffee-drinking to early modern Britain, see Nabil Matar, *Islam in Britain, 1558-1685* (Cambridge: Cambridge University Press, 1998), pp. 110-17.
6. See important studies by Nandini Bhattacharya, *Reading the Splendid Body: Gender and Consumerism in Eighteenth-Century British Writing on India* (Newark: University of Delaware, 1997); Inderpal Grewal, *Home and Harem: Nation, Gender, Empire, and the Cultures of Travel* (London: Leicester University Press, 1996); Reina Lewis, *Gendering Orientalism: Race, Feminity and Representation* (London: Routledge, 1996); Lisa Lowe, *Critical Terrains: French and British Orientalism* (Ithaca: Cornell University Press, 1991); Jyotsna Singh, *Colonial Narratives/Cultural Dialogues* (New York: Routledge, 1996); Gayatri Spivak, *Imaginary Maps: Three Stories by Mahasweta Devi* (New York: Routledge, 1995); and Yegenoglu, *Colonial Fantasies.*
7. For Anglo-Ottoman relations during this period, see invaluable studies by Palmira Brummett, *Ottoman Seapower and Levantine Diplomacy in the Age of Discovery* (Albany: State University of New York Press, 1994); Daniel Goffman, *Britons in the Ottoman Empire, 1642-1660* (Seattle: University of Washington Press, 1998); Halil Inalcik, *The Ottoman Empire: The Classical Age 1300-1600* (1973; reprint, London: Phoenix, 1994); Cemal Kafadar, *Between Two Worlds: The Construction of the Ottoman State* (Berkeley: University of California Press, 1995); Leslie Peirce, *The Imperial Harem: Women and Sovereignty in the Ottoman Empire* (Ithaca: Cornell University Press, 1993); Susan A. Skilleter, *William Harborne and the Trade with Turkey, 1578-1582* (London: British Academy, 1977); and Christine Woodhead, "'The Present Terrour of the World'? Contemporary Views of the Ottoman Empire, c. 1600," *History* 72 (1987): 20-37.
8. Edward Said, *Orientalism: Western Conceptions of the Orient* (1978; reprint, Harmondsworth: Penguin, 1995), p. 5.
9. See Ezal Kural Shaw's "The Double Veil: Travelers' Views of the Ottoman Empire, Sixteenth through Eighteenth Centuries," in *English and Continental Views of the Ottoman Empire, 1500-1800* (Los Angeles: William Andrews Clark Memorial Library, 1972), pp. 1-29, on the problem of linguistic interference, and Alain Grosrichard (*The Sultan's Court: European Fantasies of the East,* trans. Liz Heron [1979; reprint, London: Verso, 1998]) on the fantasmatic elements of European fears of "despotism."
10. Said, *Orientalism,* p. 122.
11. Henry Blount, *A Voyage Into The Levant* (London, 1636), p. 2.
12. Blount, *Voyage,* p. 4.
13. Pratt, *Imperial Eyes,* p. 7.
14. Pratt, *Imperial Eyes* , p. 28.
15. Blount, *Voyage,* p. 2.
16. Henry King, "To My Noble and Judicious Friend Sir Henry Blount upon his Voyage," in *Minor Poets of the Caroline Period,* ed. George Saintsbury, 3 vols. (1905; reprint, Oxford: Clarendon Press, 1968). See Chew's account of King's poem (*The Crescent and the Rose,* pp. 24-25).

17. Blount, *Voyage,* p. 1.
18. Blount, *Voyage,* p. 1.
19. Blount, *Voyage,* p. 2.
20. Blount, *Voyage,* p. 2.
21. Blount, *Voyage,* p. 2
22. Blount, *Voyage,* p. 2.
23. Blount's discussion is worth considering in terms of Gayatri Spivak's notions of "worlding" and "the ethical embrace." See Shankar Raman's essay in this volume; and Spivak, "Echo," in *The Spivak Reader,* ed. Donna Landry and Gerald MacLean (New York: Routledge, 1996) and the editors' introduction in the same volume, pp. 5-6.
24. Blount, *Voyage,* p. 2.
25. Blount, *Voyage,* p. 2.
26. Blount, *Voyage,* p. 3.
27. Blount, *Voyage,* p. 3.
28. Chew, *The Crescent and the Rose,* p. 43.
29. Blount, *Voyage,* p. 3.
30. Blount, *Voyage,* p. 3.
31. Blount, *Voyage,* p. 4.
32. Blount, *Voyage,* p. 38.
33. Blount, *Voyage,* p. 5.
34. Blount, *Voyage,* p. 4.
35. Blount, *Voyage,* p. 4.
36. Blount, *Voyage,* p. 5.
37. Blount, *Voyage,* p. 38.
38. Blount, *Voyage,* p. 69.
39. Thomas Palmer, *An Essay of the Meanes how to make our Travailes, into forraine Countries, the more profitable and honorable* (London, 1606), sigs. A2-A2v.

SECTION FOUR

LADY MARY WORTLEY MONTAGU

Published in 1763, Lady Mary Wortley Montagu's *Travels* are the only known eighteenth-century travelogue written and printed by an Englishwoman who had seen Asia, Africa, or the Ottoman Empire. Montagu, born Mary Pierrepont, was the firstborn child of Evelyn and Mary Pierrepont (Fielding). In 1690 her father became Earl of Kingston, and she became Lady Mary. Her mother died in 1692. By the time Montagu was 12, she was writing poetry and calling herself an author; she soon taught herself Latin. In 1712 her father arranged her marriage with an Irish peer; in response, she eloped with Edward Wortley Montagu. He was connected to London literary and political circles, and the marriage supported Montagu's writerly and civic aspirations.

In 1716 Montagu's husband became ambassador to the Ottoman Porte. The Wortley Montagus intended to stay five years abroad, but in 1717 Wortley was recalled without cause. They traveled to Constantinople overland, through Holland, Germany, Austria, Hungary, Serbia, and Bulgaria. The couple returned home by ship over the Mediterranean Sea, visiting Greece, Sicily, Tunisia, and Italy, and then traveling overland through France. In Constantinople, Montagu observed live smallpox vaccination, inoculated her son, and in 1721 introduced the technique to England by inoculating her daughter, while the College of Physicians witnessed the procedure. Inoculation generated public controversy, yet in 1789, Henrietta Inge commemorated Montagu's action with a wall plaque in Lichfield Cathedral.

While traveling, Montagu wrote letters and kept journals. Almost all this writing is unrecovered except through the *Travels* text; Montagu used letters and journals as source material to write a nonfiction prose work that uses epistolary conventions as organizational or thematic devices. In 1724 Montagu's friend, feminist philosopher Mary Astell, called the travelogue *Travels of an English Lady,* which is the title I use. Between 1763 and 1815, over 38 British, Continental, and U.S. editions were published.

Note On Text

I edit the *Travels* as a literary work of nonfiction prose intended for posthumous print publication. My copytext is not Montagu's holograph but the first edition, which I follow for spelling, capitalization, and regularized page layout. But I take word meaning and punctuation as substantive, and so I correct the first edition's alterations to both. I have also emended into the copytext a few holograph spellings, if a copytext spelling alters sound or rhythm too much. I also follow Montagu's alterations to holograph punctuation; she frequently scratched out comma tails to make periods, or scratched out one mark and replaced it with another. The full text of the *Travels* is available in *The Complete Letters of Lady Mary Wortley Montagu* (Oxford University Press, 1965/66), edited by Robert Halsband. Halsband edited the *Travels* differently, as the surviving witness to Montagu's (mostly unrecovered) letters from her travels; please check his textual note.

Of the various source documents surviving from the *Letters,* those significant to choosing the copytext of a critical edition are the first edition and two manuscripts of the *Letters:* the holograph and the printer's copy. Montagu completed the holograph by 1724. In December 1761 she gave it to the Reverend Benjamin Sowden in Rotterdam, on her way home from Venice to England. Montagu died in August 1762; soon afterwards, Sowden sold the holograph to Montagu's son-in-law, the third Earl of Bute.[1] No positive evidence for Sowden's role in publication has yet been recovered, but comparison of the surviving printer's copy[2] with Sowden's surviving letters make a strong case for the printer's copy being written out in Sowden's hand.

Five accounts of the publication transaction are known, all of which contradict each other to some degree: Horace Walpole's, James Dallaway's, Hannah Sowden's (Sowden's daughter), the unknown author of the first edition's "Advertisement of the Editor," and Lady Louisa Stuart's (Montagu's granddaughter). The consistencies among the accounts center around the people named by or associated with them: Lady Mary Wortley Montagu, Benjamin Sowden, Archibald Maclaine, and the Bute family. Walpole's account fits most closely the surviving evidence: "Lady Mary Wortley has left twenty-one large volumes in prose and verse in manuscript—nineteen are fallen to Lady Bute, and will not see the light in haste. The other two Lady Mary in her passage gave to somebody in Holland, and at her death expressed great anxiety to have them published. Her family are in terrors lest they should be, and have tried to get them: hitherto the man is inflexible."[3]

My gratitude to: Lord Harrowby, M.A. Bosson, and the Harrowby Manuscripts Trust for permission to cite for publication; the Holborn Public Library; Bernard Crystal and the Columbia University Library; the Bodleian Library; the British Library; and the University of Chicago Library; and the Lichfield Cathedral Chapter Office.

—Rebecca Chung

LETTER XXVI.

To the Lady—.

Adrianople, April. 1, O.S. 1717.

I AM now got into a new world, where every thing I see appears to me a change of scene, and I write to your ladyship with some content of mind, hoping at least that you will find the charm of novelty in my letters, and no longer reproach me, that I tell you nothing extraordinary. I won't trouble you with a relation of our tedious journey, but I must not omit what I saw remarkable at *Sophia,* one of the most beautiful towns in the Turkish Empire, and famous for its hot baths, that are resorted to both for diversion and health.[4] I stop'd here one day, on purpose to see them, designing to go *incognito,* I hired a Turkish coach. These voitures are not at all like ours, but much more convenient for the country, the heat being so great that glasses would be very troublesome. They are made a good deal in the manner of the Dutch coaches having wooden lattices painted and gilded, the inside being painted with baskets and nosegays of flowers, intermixed commonly with little poetical motto's, they are covered all over with scarlet cloth, lined with silk, and very often richly embroidered and fringed. This covering entirely hides the persons in them, but may be thrown back at pleasure, and the ladies peep through the lattices. They hold four people very conveniently, seated on cushions, but not raised.

In one of these covered waggons I went to the *Bagnio* about ten a clock, it was already full of women, it is built of stone, in the shape of a dome, with no windows but in the roof, which gives light enough, there was five of these domes joined together, the outmost being less than the rest and serving only as a hall, where the *Portress* stood at the door. Ladies of quality generally give this woman the value of a crown or ten shillings and I did not forget that ceremony.[5] The next room is a very large one, paved with marble, and all round it, raised two Sofas of marble one above another. There were four fountains of cold water in this room, falling first into marble basons, and then running on the floor in little channels made for that purpose, which carried the streams into the next room, something less than this, with the same sort of marble Sofas, but so hot with steams of sulphur proceeding from the baths joining to it, 'twas impossible to stay there with one's cloaths on. The two other domes were the hot baths, one of which had cocks of cold water turning into it, to temper it to what degree of warmth, the bathers have a mind to.[6]

I was in my travelling habit, which is a riding dress, and certainly appeared very extraordinary to them, yet there was not one of 'em that shewed the least surprize or impertinent curiosity, but received me with all the obliging civility possible. I know no European court where the ladies would have behaved themselves in so polite a manner to a stranger. I believe in the whole there were two hundred women, and yet none of those disdainful smiles, or satyric whispers, that never fail in our assemblies, when any body appears that is not dressed exactly in fashion. They repeated over and over to me, UZELLE, PEK UZELLE, which is nothing but *charming, very charming.* The first Sofas were covered with cushions and rich carpets, on which sat the ladies, and on the second their slaves behind 'em, but without any distinction of rank by their dress, all being in the state of nature, that is, in plain English, stark naked, without any beauty or defect concealed. Yet there was not the least wanton smile, or immodest gesture amongst 'em. They walked, and moved with the same majestic grace which Milton describes of our General Mother.[7] There were many amongst them as exactly proportioned as ever any goddess was drawn by the pencil of Guido, or Titian,[8] and most of their skins shiningly white, only adorned by their beautiful hair divided into many tresses hanging on their shoulders braided either with pearl or ribbon, perfectly representing the figures of the graces.[9]

I was here convinced of the truth of a reflection that I had often made, *that if it twas the fashion to go naked, the face would be hardly observed*. I perceived that the ladies with the finest skins, and most delicate shapes, had the greatest share of my admiration, though their faces were sometimes less beautiful than those of their companions. To tell you the truth, I had wickedness enough to wish secretly, that Mr. *Gervais* could have been there invisible,[10] I fancy it would have very much improved his art, to see so many fine women naked, in different postures, some in conversation, some working, others drinking coffee or sherbet, and many negligently lying on their cushions while their slaves (generally pretty girls of seventeen or eighteen) were employ'd in braiding their hair in several pretty manners. In short 'tis the women's coffee-house, where all the news of the town is told, scandal invented, &c. They generally take this diversion once a week, and stay there at least four or five hours, without getting cold by immediate [sic] coming out of the hot-bath into the cool room, which was very surprizing to me. The lady that seemed the most considerable amongst them, entreated me to sit by her, and would fain have undressed me for the bath. I excused myself with some difficulty, they being all so earnest in persuading me, I was at last forced to open my shirt[11] and shew them my stays, which satisfied them very well, for I saw they believed I was so locked up in that machine that it was not in my own power to open it, which contrivance they attributed to my husband. I was charmed with their civility and beauty, and should have been very glad to pass more time with them, but Mr. W—resolving to pursue his journey the next morning early, I was in haste to see the ruins of Justinian's church, which did not afford me so agreeable a prospect as I had left, being little more than a heap of stones.

Adieu Madam I am sure I have now entertained you with an account of such a sight as you never saw in your life, and what no book of travels could inform you of, 'tis no less than death for a man to be found in one of these places.

LETTER XXIX.

To the Countess of—.

Adrianople, April 1, O.S. 1717.

I WISH to God (dear sister)[12] that you was as regular in letting me have the pleasure of knowing what passes on your side of the globe, as I am careful in endeavouring to amuse you by the account of all I see, that I think you care to hear of. You content yourself with telling me over and over that the town, is very dull, it may possibly be dull to you, when every day does not present you with something new, but for me that am in arrear at least two months news, all that seems very stale with you, would be fresh and sweet here. Pray let me into more particulars, I will try to awaken your gratitude, by giving you a full and true relation of the novelties of this place, none of which would surprize you more, than a sight of my person as I am now in my Turkish habit, though I believe you would be of my opinion that 'tis admirably becoming. I intend to send you my picture, in the mean time accept of it here.

The first piece of my dress is a pair of drawers very full that reach to my shoes, and conceal the legs more modestly than your petticoats. They are of a thin rose-colour damask brocaded with silver flowers, my shoes of white kid leather embroidered with gold. Over this hangs my smock. Of a fine white silk gauze edged with embroidery, this smock has wide sleeves hanging half-way down the arm, and is closed at the neck with a diamond button, but the shape and colour of the bosom very well to be distinguished through it. The *Antery* is a waist-coat made close to the shape, of white and gold damask, with very long sleeves falling back, fringed with deep gold fringe, and should have diamond or pearl buttons. My *Caftan* of the same stuff with my drawers, is a robe exactly fitted to my shape, and reaching to my feet, with very long strait falling sleeves. Over this is the girdle of about four fingers broad, which all that can afford have entirely of diamonds or other precious stones, those that will not be at that expence, have it of exquisite embroidery on sattin, but it must be fastened before with a clasp of diamonds. The *Curdée* is a loose robe they throw off, or put on, according to the weather, being of a rich brocade (mine is green and gold) either lined with ermine, or sables, the sleeves reach very little below the shoulders. The headdress is composed of a cap, called *Talpock* which is in winter of fine velvet embroidered with pearls or diamonds, and in summer of a light shining silver stuff, this is fixed on one side of the head, hanging a little way down with a gold tassel, and bound on, either with a circle of diamonds (as I have seen several) or a rich embroidered handkerchief. On the other side of the head, the hair is laid flat, and here the ladies are at liberty to show their fancies, some putting flowers, others a plume of heron's feathers, and in short what they please. But the most general fashion is a large *Bouquet* of jewels, made like natural flowers, that is, the *buds* of pearl, the *roses* of different coloured rubies, the *jessamines* of diamonds, *jonquils* of topazes &c., so well set and enameled, 'tis hard to imagine any thing of that kind so beautiful. The hair hangs at its full length behind, divided into tresses braided with pearl, or ribbon, which is always in great quantity. I never saw in my life so many fine heads of hair, I have counted one hundred ten of these tresses, of one lady, all natural, but it must be owned that every beauty is more common here than with us. 'Tis surprising to see a young woman that is not very handsome, they have naturally the most beautiful complexions in the world, and generally large black eyes. I can assure you with great truth that the court of England (though I believe it the fairest in Christendom) cannot shew so many beauties, as are under our protection here. They generally shape their eye-brows, and the Greeks and Turks have a custom of putting round their eyes, on the inside a black tincture, that at a distance, or by candle-light adds very much to the blackness of them. I fancy many of our ladies would be overjoyed to know this secret, but 'tis too visible by day. They dye their nails rose-colour, I own I cannot enough accustom myself to this fashion, to find any beauty in it.

As to their morality or good conduct, I can say like Harlequin, 'tis just as 'tis with you,[13] and the Turkish ladies don't commit one sin the less for not being Christians.[14] Now I am a little acquainted with their ways, I cannot forbear admiring either the exemplary discretion, or extreme stupidity, of all the writers that have given accounts of 'em.[15] 'Tis very easy to see, they have more liberty than we have, no woman of what rank soever, being permitted to go in the streets, without two *Muslins,* one that covers her face all but her eyes, and another that hides the whole dress of her head, and hangs half way down her back, and their shapes are wholly concealed by a thing they call a *Ferigee,* which no woman of any sort appears without. This has strait sleeves that reaches to their fingers ends, and it laps all round 'em not unlike a riding-hood, in winter 'tis of cloth, and in summer plain stuff or silk. You may guess how effectually this disguises them, that there is no distinguishing the great lady from her slave, and tis impossible for the most jealous husband to know his wife when he meets her, and no man dare touch or follow a woman in the street.

This perpetual masquerade gives them entire liberty of following their inclinations without danger of discovery. The most usual method of intrigue, is to send an appointment to the lover to meet the lady at a Jews shop, which are as notoriously convenient as our Indian-houses. And yet even those who don't make use of 'em, do not scruple to go, to buy pennorths and tumble over rich goods, which are chiefly to be found amongst that sort of people. The great ladies seldom let their gallants know who they are, and 'tis so difficult to find it out, that they can very seldom guess at her name, they have corresponded with, above half a year together. You may easily imagine the number of faithful wives very small in a country where they have nothing to fear from their lover's indiscretion, since we see so many have the courage to expose themselves to that in this world, and all the threatned punishment of the next, which is never preached to the Turkish damsels. Neither have they much to apprehend from the resentment of their husbands, those ladies that are rich having all their money in their own hands, which they take with 'em, upon a divorce, with an addition which he is obliged to give 'em. Upon the whole, I look upon the Turkish women as the only free people in the Empire, the very Divan pays a respect to 'em, and the Grand Signior himself[16] when a Bassa[17] is executed, never violates the privileges of the *Haram* (or womens apartment) which remains unsearched entire to the widow. They are Queens of their slaves, which the husband has no permission so much as to look upon, except it be an old woman or two that his lady chuses. 'Tis true their law permits them four wives, but there is no instance of a man of quality that makes use of this liberty, or of a woman of rank that would suffer it.[18] When a husband happens to be inconstant (as those things will happen) he keeps his mistress in a house apart, and visits her as privately as he can. Just as 'tis with you. Amongst all the great men here I only know the *Testerdar* (*i.e.* Treasurer)[19] that keeps a number of she slaves for his own use, (that is, on his own side of the house, for a slave once given to serve a lady, is entirely at her disposal) and he is spoke of as a libertine, or what we should call a rake, and his wife won't see him, though she continues to live in his house. Thus you see dear sister the manners of mankind do not differ so widely, as our voyage writers would make us believe. Perhaps it would be more entertaining to add a few surprizing customs of my own invention, but nothing seems to me so agreeable as truth, and I believe nothing so acceptable to you, I conclude with repeating the great truth of my being

Dear Sister

&c.

LETTER XXXIII.

To the Countess of—.

Adrianople, April 18, O.S.

I WRIT to you (dear sister) and to all my other English correspondents by the last ship, and only Heaven can tell when I shall have another opportunity of sending to you. But I cannot forbear writing, though perhaps my letter may lye upon my hands this two months, to confess the truth my head is so full of my entertainment yesterday, that 'tis absolutely necessary for my own repose to give it some vent, without farther preface I will then begin my story.

I was invited to dine with the Grand *Vizier*'s lady,[20] and 'twas with a great deal of pleasure I prepared myself for an entertainment which was never before given to any Christian. I thought I should very little satisfy her curiosity (which I did not doubt was a considerable motive to the invitation) by going in a dress she was used to see, and therefore dressed myself in the court habit of Vienna, which is much more magnificent than ours. However I chose to go *incognito,* to avoid any disputes about ceremony, and went in a Turkish coach only attended by my woman that held up my train, and the Greek lady who was my interpretress. I was met at the court-door by her black Eunuch[21] who helped me out of the coach with great respect, and conducted me through several rooms, where her she slaves finely dressed were ranged on each side. In the innermost, I found the lady sitting on her sofa, in a sable vest, she advanced to meet me, and presented me half a dozen of her friends with great civility. She seemed a very good woman, near fifty year old, I was surprized to observe so little magnificence in her house, the furniture being all very moderate, and except the habits and number of her slaves, nothing about her appeared expensive.[22] She guessed at my thoughts, and told me she was no longer of an age to spend either her time or money in superfluities, that her whole expence was in charity, and her employment praying to God. There was no affectation in this speech, both she and her husband are entirely given up to devotion. He never looks upon any other woman, and what is much more extraordinary touches no bribes, notwithstanding the example of all his predecessors. He is so scrupulous in this point, he would not accept Mr. W—'s present till he had been assured over and over, 'twas a settled perquisite of his place at the entrance of every Ambassador. She entertained me with all kind of civility till dinner came in, which was served one dish at a time, to a vast number, all finely dressed after their manner, which I do not think so bad, as you have perhaps heard it represented. I am a very good judge of their eating, having lived three weeks in the house of an *Effendi* at Belgrade,[23] who gave us very magnificent dinners dressed by his own cooks, which the first week pleased me extremely, but I own, I then begun to grow weary of it, and desired our own cook might add a dish or two after our manner, but I attribute this to custom. I am very much inclined to believe, an Indian that had never tasted of either would prefer their cookery to ours. Their sauces are very high, all the roast very much done, they use a great deal of rich spice, the soop is served for the last dish, and they have at least as great a variety of ragouts[24] as we have. I was very sorry I could not eat of as many as the good lady would have had me, who was very earnest in serving me of every thing. The treat concluded with coffee, and perfumes which is a high mark of respect. Two slaves kneeling *censed* my hair, cloaths, and handkerchief. After this ceremony, she commanded her slaves to play and dance which they did with their guitars in their hands, and she excused to me, their want of skill, saying she took no care to accomplish them in that art.

I returned her thanks and soon after took my leave, I was conducted back in the same manner I entered, and would have gone strait to my own house, but the Greek lady with me earnestly solicited me to visit the *Kahya*'s lady, saying he was the second officer in the Empire, and ought indeed to be looked upon as the first, the Grand Vizier having only the name while he exercised the authority.[25] I had found so little diversion in this *Haram,* that I had no mind to go into another, but her importunity prevailed with me, and I am extreme glad that I was so complaisant. All things here were with quite another air than at the Grand Vizier's, and the very house confessed the difference between an old devotee and a young beauty. It was nicely clean and magnificent, I was met at the door by two black Eunuchs who led me through a long gallery between two ranks of beautiful young girls, with their hair finely plaited almost hanging to their feet, all dressed in fine light damasks brocaded with silver. I was sorry that decency did not permit me to stop to consider them nearer, but that thought was lost upon my entrance into a large room, or rather pavilion, built round with gilded sashes, which were most of 'em thrown up, and the trees planted near them gave an agreeable shade, which hindered the Sun from being troublesome. The jessamines and honeysuckles that twisted round their trunks shedding a soft perfume, increased by a white marble fountain playing sweet water, in the lower part of the room, which fell into three or four basons, with a pleasing sound. The roof was painted with all sort of flowers falling out of gilded baskets, that seemed tumbling down. On a sofa raised three steps and covered with fine Persian carpets, sat the Kayha's lady, leaning on cushions of white sattin embroidered, and at her feet sat two young girls, the eldest about twelve year old, lovely as angels, dressed perfectly rich, and almost covered with jewels. But they were hardly seen near the fair *Fatima* (for that is her name) so much her beauty effaced every thing. I have seen all that has been called lovely either in England or Germany, and I must own that I never saw any thing so gloriously beautiful, nor can

I recollect a face that would have been taken notice of near her's. She stood up to receive me, saluting me after their fashion, putting her hand upon her heart, with a sweetness full of majesty that no court breeding could ever give. She ordered cushions to be given me, and took care to place me in the corner which is the place of honour, I confess though the Greek lady had before given me a great opinion of her beauty, I was so struck with admiration, that I could not for some time speak to her, being wholly taken up in gazing. That surprizing harmony of features! that charming result of the whole! that exact proportion of body! that lovely bloom of complexion unsullied by art! the unutterable enchantment of her smile! but her eyes! large and black with all the soft languishment of the blue! every turn of her face discovering some new charm.

After my first surprize was over, I endeavoured by nicely examining her face to find out some imperfection, without any fruit of my search, but my being clearly convinced of the error of that vulgar notion, that a face perfectly regular would not be agreeable. Nature having done for her with more success, what *Apelles* is said to have essayed, by a collection of the most exact features to form a perfect face,[26] and to that a behaviour so full of grace and sweetness, such easy motions, with an air so majestic yet free from stiffness or affectation, that I am persuaded could she be suddenly transported upon the most polite throne of Europe, no body would think her other than born and bred to be a Queen, though educated in a country we call barbarous. To say all in a word, our most celebrated English beauties would vanish near her.

She was dressed in a *Caftan* of gold brocade flowered with silver, very well fitted to her shape and shewing to advantage the beauty of her bosom, only shaded by the thin gauze of her shift, her drawers were pale pink, green, and silver, her slippers white finely embroidered, her lovely arms adorned with bracelets of diamonds, and her broad girdle set round with diamonds. Upon her head, a rich Turkish handkerchief of pink and silver, her own fine black hair hanging a great length in various tresses, and on one side of her head some bodkins of jewels. I am afraid you will accuse me of extravagance in this description, I think I have read some where that women always speak in rapture when they speak of beauty, but I can't imagine why they should not be allowed to do so. I rather think it virtue to be able to admire without any mixture of desire or envy. The gravest writers have spoke with great warmth of some celebrated pictures and statues; the workmanship of Heaven certainly excells all our weak imitations, and I think has a much better claim to our praise, for me, I am not ashamed to own, I took more pleasure in looking on the beauteous *Fatima,* than the finest piece of sculpture could have given me. She told me, the two girls at her feet were her daughters, though she appeared too young to be their mother. Her fair maids were ranged below the Sofa to the number of twenty, and put me in mind of the pictures of the antient nymphs, I did not think all nature could have furnished such a scene of beauty. She made them a sign to play and dance, four of them immediately begun to play some soft airs on instruments between a lute and a guitar, which they accompanied with their voices, while the others danced by turns. This dance was very different from what I had seen before,[27] nothing could be more artful, or more proper to raise *certain ideas.* The tunes so soft, the motions so languishing, accompanied with pauses, and dying eyes, half-falling back and then recovering themselves in so artful a manner, that I am very positive the coldest and most rigid prude upon earth, could not have looked upon them, without thinking of *something not to be spoke of.*[28] I suppose you may have read that the Turks have no music but what is shocking to the ears, but this account is from those who never heard any but what is played in the streets, and is just as reasonable as if a foreigner should take his ideas of the English music from the *bladder* and *string,* and *marrow-bones* and *cleavers.*[29] I can assure you, that the music is extremely pathetic. 'Tis true, I am inclined to prefer the Italian; but perhaps I am partial. I am acquainted with a Greek lady who sings better than Mrs. *Robinson*[30] and is very well skilled in both, who gives the preference to the Turkish. 'Tis certain they have very fine natural voices, these were very agreeable. When the dance was over, four fair slaves came into the room with silver censors in their hands, and perfumed the air with amber, aloes-wood, and other rich scents. After this, they served me coffee upon their knees, in the finest japan china, with *soucoups*[31] of silver gilt. The lovely *Fatima* entertained me all this time, in the most polite agreeable manner, calling me often *Uzelle Sultanam,* or the Beautiful Sultana, and desiring my friendship with the best grace in the world, lamenting that she could not entertain me in my own language.

When I took my leave, two maids brought in a fine silver basket of embroidered handkerchiefs, she begg'd I would wear the richest for her sake, and gave the others to my woman and interpretress. I retired thro' the same ceremonies as before, and could not help fancying I had been some time in Mahomet's paradise,[32] so much I was charmed with what I had seen. I know not how the relation of it appears to you, I wish it may give you part of my pleasure, for I would have my dear sister share in all the diversions

LETTER XLII.

To the Countess of—.
(excerpt)

I am well acquainted with a Christian woman of quality who made it her choice to live with a Turkish husband, and is a very agreeable sensible lady. Her story is so extraordinary I cannot forbear relating it, but I promise you it shall be in as few words, as I can possibly express it.

She is a Spaniard, and was at Naples with her family, when that kingdom was part of the Spanish dominion. Coming from thence in a *Feloucca* accompanied by her brother, they were attacked by the Turkish Admiral, boarded, and taken; and now; how shall I modestly tell you the rest of her adventure? The same accident happened to her, that happen'd to the fair Lucretia so many years before her, but she was too good a Christian to kill herself as that Heathenish Roman did.[33] The Admiral was so much charmed with the beauty and *Long-suffering,* of the fair captive, that, as his first compliment he gave immediate liberty to her brother, and attendants, who made haste to Spain and in a few months, sent the sum of four thousand pound sterling as a ransom for his sister. The Turk took the money, which he presented to her, and told her, she was at liberty, but the lady very discreetly weighed the different treatment she was likely to find in her native country. Her Catholic relations as the kindest thing they could do for her, in her present circumstances, would certainly confine her to a nunnery for the rest of her days. Her Infidel lover, was very handsome, very tender, fond of her, and lavished at her feet, all the Turkish magnificence. She answered him very resolutely, that her liberty was not so precious to her as her honour, that he could no way restore that but by marrying her, she desired him to accept the ransom as her portion, and give her the satisfaction of knowing no man could boast of her favours, without being her husband. The admiral was transported at this kind offer, and sent back the money to her relations, saying he was too happy in her possession. He married her, and never took any other wife, and (as she says herself) she never had reason to repent the choice she made. He left her some years after, one of the richest widows in Constantinople, but there is no remaining honourably a single woman, and that consideration has obliged her to marry the present Capitan Bassa (i.e. Admiral) his successor.[34] I am afraid that you will think my friend fell in love with her ravisher, but I am willing to take her word for it, that she acted wholly on principles of honour, tho' I think she might be reasonably touched at his generosity, which is often found amongst the Turks of rank.

Notes

1. Harrowby MSS 253-54.
2. The printer's copy, from which the first edition of the *Travels* was set, was identified in 1948 by Penelope Morgan, Reference Librarian, Hampstead Public Library. Now held at the Holborn Public Library.
3. *Horace Walpole's Correspondence with Sir Horace Mann,* vol. 6, ed. W.S. Lewis et al. (New Haven: Yale University Press, 1960), 84.
4. Montagu: "the city itself is very large, and extremely populous, here are hot baths, very famous for their medicinal virtues." (Letter 25, my text). Grundy: "Sofia's medicinal hot springs come at about 46.8°C" (*Lady Mary Wortley Montagu: Comet of the Enlightenment* [Oxford: Oxford University Press, 1999], 137).
5. Aaron Hill (1685-1750), *A Full and Just Account of the Present State of the Ottoman Empire* (London, 1709): "The Second sort of *Bagnio*'s, are those Publick Places where for payment of a *Penny* or *Three-half-pence,* they are wash'd *Politely,* and supplied with *Linen,* or whatever else the *Bath* requires, by the diligent Attendance of *appropriated* Servants . . . These *Publick Baths* are very common in their largest Cities . . ." (50). In the *Catalogus Bibliothecae Kingstoniae* (*CBK*), the printed catalogue of the library owned by Montagu's father.
6. Hill: "To every *Mosque* belongs a *Bagnio,* which as, much frequented by the *Tur[k]ish People,* is esteemed a Building of no small Importance; these are *often,* tho' not *always,* built of Marble, *Square* and *Spacious,* all divided into many and convenient Chamgers, *Sweating Rooms* and *Cooling Baths,* which each makes use as he finds himself inclin'd; in every one of these Divisions *different Cocks of Brass* admit what quantity of Water *hot* or *Cold* they think convenient" (132).
7. John Milton, *Paradise Lost,* iv: 314-318: "For contemplation he and valour form'd, / For softness she and sweet attractive Grace. . .Nor those mysterious parts were then conceal'd, / Then was not guilty shame, dishonest shame / Of natures works." (London: Jacob Tonson, 1711), 106-7. In *CBK.*
8. Titian, or Tiziano Becellio (ca. 1485-1576), Venetian painter. Guido Reni (1575-1642), Bolognese painter.
9. Hill: "Their Motions carry a *Peculiar Grace* in an *Easy* and *unaffected* Freedom of behaviour, the Native Cha[r]ms of an *Amorous Softness* appear unfeignedly in *every look,* while *every Step* bears somewhat of an Air not altogether free from a *Majestic Gravity,* and yet entirely Govern'd by an *uncommon Easiness*" (110).
10. Charles Jervas (1675?-1739), who painted Montagu in as shepherdess, ca. 1710. Grundy characterizes Jervas's studio as "a meeting-place for the worlds of fashion and the arts, for sitters and customers" (91-92).

11. In Halsband "skirt," but the manuscript letterform is *h,* and stays were a kind of waistcoat; see the *Oxford English Dictionary* (OED) and Johnson's *Dictionary.*
12. The Countess of Mar, Frances (Pierrepont) Erskine, was Montagu's younger sister. No genuine correspondence between Montagu and her sister survives for the period 1716-1718.
13. Aphra Behn, *The Emperor of the Moon* (London, 1687). Doctor Baliardo asks Harlequin if, on the *Lunar Mundus,* the women drink, gamble, or scheme as they do in London. Harlequin replies yes, and the Doctor responds, "Just as 'tis here" (III.i).
14. Hill: "The numerous *Mahometans* are *like our selves* divided into *Good* and *Bad,* according to the Lessons of their different Educations, or the contrary Impulses of a *vicious Soul,* or one inspir'd with a sublime and generous love of Vertue" (76).
15. Hill: "Whoredom, and Adultery, too much encourag'd by the *Christian*'s Practice, and alleviated commonly by the *palliating* Cloak of a pretended *Gallantry,* are seldom found among the *Turks;* the Punishments so great, and obviously known, and Difficulties of obtaining Female Correspondence so insuperably hazardous, deterr them from attempting such *Illegal Love,* and the little value which they hold their Women at, the Slavery that poor unhappy Sex are there subjected to, and the unbounded Liberty their Laws afford 'em in promiscuous Use of *Wives* and *Concubines,* excite no search of such unvalued Pleasure" (80).
16. Paul Rycaut: "The delightful Fields of *Asia,* the pleasant Plains of *Tempe* and *Thrace,* all the Plenty of *Egypt* . . . the Tributary Principalities of *Moldavia* and *Walachia, Romania, Bulgaria,* and *Servia,* and the best part of *Hungary,* concur together to satisfie the Appetite of one single person . . . the Grand Signior, in his sole Disposal and Gift they remain" (*The Present State of the Ottoman Empire.* . . . [London: J. Starkey & Brome, 1668], 2:2).
17. Rycaut: "The next to the Vizier *Azem,* or the first Vizier, are the several *Beglerbegs* (which may not unaptly be compared to Arch-dukes) . . . the Grand Signior bestows them three Ensigns [insignia]. . .this is to distinguish them from *Bashaws* who have two Ensigns; and the *Sanizch-beg,* who hath also the name of *Pascha,* and hath but one" (2:23).
18. Rycaut: "There are amongst the *Turks* three degrees of Divorce . . . The second . . . the Husband is compelled to make good her *Kabin*" (2:75).
19. Rycaut: The other great Officer, is the *Testerdar* or Lord Treasurer, who receives the Revenue of the Grand Signior, and pays the Souldiery, and makes other publick Disbursements" (2:26).
20. Rycaut: "The Prime Vizier, called in *Turkish* Vizier *Azem,* is as much as chief Counsellor; he is sometimes termed the Grand Signior's Deputy or Representative, or *Vicarius Imperii,* because to him all the Power of the Sultan is immediately devolved . . . he . . . can, without the formality and process of Law, remove all Obstacles and Impediments which hinder the free sway of his Government" (2:20).
21. Rycaut: "The Black Eunuchs are ordained for the service of the Women in the Seraglio; as the White are to the attendance of the Grand Signior, it not seeming a sufficient Remedy by wholly dismembring them, to take the Women off from their inclinations to them, as retaining some relation still to the Masculine Sex; but to create an abhorrency in them; they are not only castrated, but Black . . ." (2:18).
22. Rycaut: "The state and Greatness the Prime Vizier lives in, is agreeable to the Honour of him whom he represents, having commonly in his Court about 2000 Officers and Servants . . ." (2:20).
23. Montagu: "This set of men are equally capable of preferments in the law, or the church, those two sciences being cast into one, a lawyer and a priest being the same word . . ." (Letter 27).
24. A dish usually consisting of meat cut in small pieces, stewed with vegetables and highly seasoned. *OED,* 2nd ed., s.b.
25. Rycaut: "The whole *Turkish* Militia then is of two sorts; one that receives maintenance from certain Lands or Farms bestowed on them by the Grand Signior; others that receive their constant pay in ready money . . . Those of the second sort, paid out of the Grand Signior's Treasure, are *Spahees, Janizaries* . . . the Lieutenant General of the *Janizaries* called *Kiahaia-Begh* . . ." (2:82, 91). Montagu: "the Grand Signor with all his absolute power, as much a slave as any of his Subjects, and trembles at a Janizary's frown" (Letter 30).
26. Apelles (ca. 375-305 B.C.), Greek artist credited with having created classical Hellenism.
27. Grundy: "The evening's entertainment was what is now called belly-dancing" (149).
28. Rycaut: "The doctrine of *Platonick* Love hath found Disciples in the Schools of the *Turks,* that they call it a Passion very laudable and vertuous, and a step to that perfect Love of God . . . in reality this Love of theirs, is nothing but libidious Flames each to other . . . This Passion likewise reigns in the Society of Women; they die with amorous Affections one to the other . . . these

Darts of *Cupid* shot through all the Empire, especially *Constantinople,* the Seraglio of the Grand Signior, and the Apartments of the Sultans" (2: 16-17).

29. Hill: "The *Turkish* Nation . . . are altogether Strangers to the melting Strains of Vocal Harmony, nor understand the charming use of those delightful Instrucments, whose elevating sounds have unresisted Power to move the Souls of dying Men, and make the poor desponding Wretch forget his Sorrows, and erect *with Joy* his drooping Head, to hear the *soft,* and *tunefull* Call, that lulls his Cares, and huses *for a while* has loud Misforuntes" (73). A detailed, uncomplimentary description of Turkish music and muscial instruments follows.
30. Anastasia Robinson, afterwards Countess of Peterborough (ca. 1692-1755), opera singer. Grundy notes that in 1721 "two leading singers were lodging in Twickenham, both altos: Anastasia Robinson and the top castrato Senesino. Robinson moved in Lady Mary's social circle . . . Lady Mary took a passionate delight in music" (225-26).
31. French, nf., meaning *saucers.*
32. John Donne, "To his Mistris Going to Bed" (pub. 1669), 20-21: "Thou Angel bring'st with thee / A heaven like Mahomets Paradise." *The Elegies and The Songs and Sonnets,* ed. Helen Gardner (Oxford: Clarendon Press, 1965), 15.
33. The virtuous wife of Collatinus Tarquinius. Livy: "*Sextus Tarquinius* was bewitched and possessed with wicked wanton lust, for to offer violence and villanie unto *Lucretia* . . . *Tarquinius* in great pride and jolitie, that he had by assault won the fort of a womans honor, departed thence . . . shee stabbed her selfe to the heart, and sinking downe forward, fell upon the floore readie to yeeld up the ghost . . ." *The Romane Historie Written by T. Tivius of Padua,* trans. Philemon Holland (London, 1600), 40-41. In *CBK.*
34. Rycaut: "Captain *Pascha,* or as the *Turks* call him, General of the White Seas . . . is Admiral of the Grand Signior's Fleet, and commands as far as the *Turkish* Power by Sea extends" (2:25).

A WOMAN TRIUMPHS: FROM *TRAVELS OF AN ENGLISH LADY IN EUROPE, ASIA, AND AFRICA* (1763) BY LADY MARY WORTLEY MONTAGU

Rebecca Chung

> I have made strict observations and enquirys on the Health and manner of Life of the Countrys in which I have resided, and have found little Difference in the length of Life.
>
> —Lady Mary Wortley Montagu, letter to her husband, 1752

In 1809/10, John Cam Hobhouse toured Turkey with the poet Byron. At some point Hobhouse inscribed his 1790 copy of Lady Mary Wortley Montagu's *Travels* with this representation of her:

> The merit of these letters has been much overrated—There is certainly a great deal of ease in the air and style, but as must be expected in easy writing of all kinds, no great correctness nor purity. What renders the collection peculiarly agreable is that the letters were all of them written really to the author's friends without any immediate view of publication, though, as her ladyship kept copies of them it is to be supposed she did think it possible that they might be collected by her friends, and collected in such a defective form as to render an authentic copy not only useful but necessary—The letters from Turkey have by all readers been considered the most entertaining and, indeed, [ ] writing of the whole appointment. But they please me less from the knowledge I have gained that her representations are not to be depended upon. Some of her assertions none but a *female* traveller can contradict, but what a *man* who has seen Turkey can controvert, I am myself capable of proving to be unfounded—From what I have seen of the country, and from what I have read of her book I am sure that her Ladyship would not stick at a little fibbing; and as I know part of her accounts to be altogether false I have a right to suppose she has exaggerated other particulars—One cannot fail to discover Lady M W M's ruling frailty in these letters—She wished be considered a striking beauty, and as I heard did not very much discourage the story spread abroad of an amour between her & Sultan Mustapha—Her son, a graceless dog who turned Turk, publickly professed himself to be the child of the Grand Signor—Yet she was not very beautiful, and soon began the repair of her charms by paint which she laid on so thick that it was scraped from her face & bosom with a knife—The charge brought against her, it must be confessed rather in an unmanly way, by Mr Pope of filthiness was very well founded. Horace Walpole from her own good authority used to tell of her that upon her leaving some apartments at Florence lent to her by the Grand Duke of Tuscany, they were so insufferably odorous from her habitual nastiness that they were obliged to be perfumed & aired from some time before they could be inhabited by another lodger—[1]

Hobhouse is wrong about many facts. The letter-units that make up the *Travels* were not sent to anyone; Montagu's *Travels* survive not as a transcription of letters, but as a fair copy manuscript for an epistolary travelogue. Montagu intended the *Travels* for print publication, actively contrived for them to be published posthumously, and successfully circumvented her family's efforts to suppress them. Very few of Montagu's letters from her 1716–18 travels survive, and those that do show that Montagu altered "authentic" content: rewriting it, or reassigning it from its real-life recipient to another one. Sometimes that new recipient was wholly fictional. Finally, Hobhouse's version of Montagu is too old. Montagu was a 27-year-old mother of a three-year-old boy when she went to Turkey, and she finished writing the *Travels* by age 35, in 1724. Montagu's estrangement from Alexander Pope must have been in the works while she wrote, but the breakdown that generated lasting hostility between them had not yet happened. Horace Walpole was 28 years younger than Montagu, born in 1717 while she was in Turkey.

Hobhouse worked from insufficient information, and so his critique of Montagu can run through its logic of *ad feminam* attack with little qualification. Her authority was on his mind; Hobhouse published *A Journey through Albania, and other provinces of Turkey in Europe and Asia, to Constantinople* in 1813.[2] Hobhouse's published remarks about Montagu are comparatively restrained:

> Lady M. W. Montague, whose book is so commonly read that you will scarcely pardon me for quoting rather than referring to it, talking of the Arnoots, says, in her agreeable manner—"These people, living between Christians and Mahometans, and not being skilled in controversy, declare that they are utterly unable to judge which religion is best, but to be certain of not entirely rejecting the truth, they very prudently follow both. They go to the moscks on Fridays, and to the church on Sundays, saying, for their excuse, that they are sure of protection from the true Prophet; but which that is, they are not able to determine in this world."
>
> This may have been true in the days of our accomplished countrywoman, but I could not learn that there is now to be found an instance of so philosophical an indifference, or rather of so wise a precaution. However, it is certain that the Christians, who can fairly be called Albanians, are scarcely, if at all, to be distinguished from the Mahometans. (147)

Which better represents Hobhouse's opinions about her? Was he more candid in his unpublished notes? More thoughtful in his printed assessments?

Through both citations emerges Hobhouse's reluctant concession to Montagu's authority as an observer in Turkey, especially her authority as observer of women—despite relishing the gossipy insults he heard about her, and despite his inability to replicate her observations for himself one century later. Hobhouse writes his own contact experiences through his textual representations of, and literary engagement with, Montagu. His annotations document the sources for that representation: gossip attributed to literary men. They also suggest that Hobhouse saw himself in competition with Montagu; he attempted to defeat his internalized representation of her through gossip about her gender, her sexuality, her hygiene, and her observational reliability. Hobhouse's fixation on Montagu's *Travels* confesses something like an inferiority complex; Hobhouse was experiencing some anxiety of influence. In neither annotation or published reference does Hobhouse feel able to attack Montagu's literary skill. He ventriloquizes her exaggeratedly.

In any case, the *Travels* have maintained their canonical status among Anglo-American texts of contact in part because of Montagu's easy writing, agreeable manner, and accomplishments as an Englishwoman. Paradoxically, Montagu uses the same source techniques in her own writing as Hobhouse did for his—but a record of her literary sources has survived instead of her anecdotal sources. For this reason, rereading the *Travels* through Montagu's sources is in some sense as simple as deconstructing Hobhouse through his notes—and politically, as treacherous. But through these sources, Montagu produces in the *Travels* their convincing performance of masculine authority that disturbed Hobhouse, disturbed Montagu's contemporaries, and has disturbed Anglo-American culture generally, for nearly two hundred years.

Montagu's performance of authority (via her sources and style) is separate from her authority as a participant in contact. Thus, Hobhouse's impulses were fair: he viewed her literary power skeptically, and he separated Montagu's own will to power from her observations. His evidence and methods parody the actual situation, but they are (as Donna Haraway has said about romance novels) wrong on the right subject. Hobhouse accused Montagu of desiring a superficial physical beauty, and what she produced was observation imbued with the traces—the textual trappings, to a skeptical mind—of high literary beauty: allusions, translations, transcriptions of Greek and Latin, a polished middle prose style, and a neoclassical symmetry to the structure of the text. As writer of the *Travels,* Montagu is a celebratory participator in the practices of being well-dressed and well-read:

> for me that am in arrear at least two months news, all that seems very stale with you, would be fresh and sweet here. Pray let me into more particulars, I will try to awaken your gratitude, by giving you a full and true relation of the novelties of this place, none of which would surprize you more, than a sight of my person as I am now in my Turkish habit, though I believe you would be of my opinion that 'tis admirably becoming.[3]

In a critical mind, this passage should elicit complex responses. Is Montagu simply flaunting her writing skills as she gives her "full and true relation of the novelties" she saw? Or is she placed—"in her Turkish habit"—in empathy with the native women? Subalternity, tropicopolitanism, or cultural capitalism?

In evincing ambivalence, in expressing both anxiety and desire, Hobhouse is not alone. Feminist and postcolonial readers of the *Travels* are also concerned about Montagu's position of contact observer as feminist pose, and can be said to line up on either side of suspicion or sympathy. In 1994, Devoney Looser put the matter directly:

> We might now justifiably ask what it is that we have celebrated and how we have constituted our sisterhood. Even the scholarship published on Montagu and Orientalism leaves us with unanswered questions about the correlation of "progressive" gender politics to those of race, class, and action. Especially in regard to Montagu's Turkish Embassy letters, the verdict is not yet in.[4]

Looser's concern is useful, and suggests to me that sometimes critics practice fairness too absolutely. In so doing, they reproduce the well-meaning, useful, yet inequitable logics of liberal ethical oppositions. Hobhouse argues *ad feminam* unfairly, but this can be accounted for without dismissing his impulse to critique. Hobhouse had inadequate information and a competitive motive. He understood Montagu only through a set of textual representations. These came to him shaped by conditions of access to facts, the perceptual conditioning that shapes Montagu's own self-representation in the *Travels,* and the perceptual conditioning habitual to Montagu's observers. Hobhouse's distance from Montagu is like ours: it is historical, ideological, and representational.

In 1965, Robert Halsband's edition of Montagu's *Complete Letters* printed this transcription of Montagu's visit to the bathhouse at Sophia:

> The Lady that seem'd the most considerable amongst them entreated me to sit by her and would fain have undress'd me for the bath. I excus'd my selfe with some difficulty, they being all so earnest in perswading me, I was at last forc'd to open my skirt and shew them my stays, which satisfy'd 'em very well, for I saw they beleiv'd I was so lock'd up in that machine that it was not in my own power to open it, which contrivance they attributed to my Husband.[5]

But Montagu's holograph reads differently, by one letter:

> The Lady y^{t} seem'd the most considerable amongst y^{m}:, entreated me to sit ‸^{by} her, + would fain have undress'd me for ye bath. I excus'd my self wth some difficulty, they being all so earnest in perswading me, I was at last forc'd to open my **shirt** + shew y^{m} my stays, wch satisfy'd 'em very well, for I saw they beleiv'd I was so lock'd up in y^{t} machine y^{t} it was not in my own power to open it, wch contrivance they attributed to my Husband.[6]

Montagu's *h* letterform is not ambiguous, nor does it resemble her *k*. All other editions that used the holograph as copy text—the first edition, the 1803 James Dallaway edition, the 1837 Lord Wharncliffe edition—print *shirt.* Both the *Oxford English Dictionary* and *Johnson's Dictionary* confirm that *stays* were worn above the waist, as "a kind of stiffe waistcoat made of whalebone, worn by ladies."[7] This discrepancy complicates every study of Montagu that has relied on her observations of women in the baths, because scholars tend to read this description closely.

Representations condition the experience of contact. Like most current readers of the *Travels,* I read them first in the Halsband edition. In July 1999 I transcribed *skirt* while working with a microfilm of the holograph, and I did not catch the error until September, when I was in England using the holograph itself. When I changed textual forms (from microfilm to manuscript), and working locations (from the U. S. to England), my perceptions were somehow less conditioned by habit, and a one-consonant discrepancy between representation and materiality came into view.

It looks like another text conditioned Halsband's perceptions of Montagu's manuscript—the 1820 Singer edition of Joseph Spence's *Anecdotes.* Spence met Montagu in the early 1740s, when she traveled, alone, to Rome. She was in her early fifties.[8] In his 1956 biography of Montagu, Halsband cites not Montagu's *Travels,* but Singer's edition of Spence's anecdotal account of a conversation with Montagu:

> In her Embassy Letters she painted a ravishing picture of the beautiful bathers, who when they saw her corset believed she had been locked in it by her jealous husband. She related this delightful anecdote to Spence in a less formal version:
>
> The first time she was at one of these baths, the ladies invited her to undress, and to bathe with them; and on her not making any haste, one of the prettiest run to her to undress her. You can't imagine her surprise upon lifting my lady's gown, and seeing her stays go all round her. She run back quite frightened, and told her companion, "that the husbands in England were much worse than in the East, for that they tied up their wives in little boxes, of the shape of their bodies." She carried 'em to see it. They all agreed that 'twas one of the greatest barbarities of the world, and pitied the poor women for being such slaves in Europe.[9]

So perhaps Montagu did open her skirt, or even remove her gown in the Sophia baths, but chose not to say so in her publication version of the bathhouse visit, and wrote *shirt.* On the whole, Halsband transcribed words accurately (punctuation less so); even with all the awkward abbreviations in the holograph, I found only 12 positive spelling errors.[10] Given this level of accuracy, the best explanation for Halsband's error is perceptual conditioning. Perceptual conditioning perhaps also explains why Halsband did not catch the error in copyediting or proof stages, why so many commentators on the *Travels* have chosen to interpret what Montagu does with her skirt, and why editors continue to choose Letter 26 for their anthologies—all without checking their work against the manuscript, or at least earlier editions of the *Travels.* In fact, everyone who has reckoned with Montagu's visit to the baths—Halsband, Spence, Montagu, and even the artist who produced the one known eighteenth-century illustration of Montagu in the baths (see Figure 2)—has struggled with the gender-marked, sexually-charged signifiers *skirt, shirt,* and *gown.* The illustration depicts Montagu simultaneously naked and clothed, because the artist has left the fabric texture and drape of the bodice almost undrawn. The artist emphasizes the shape and musculature of Montagu's body instead. Our own anxious and voyeuristic reckoning with this moment in the *Travels* has long-standing editorial and ideological precedents.

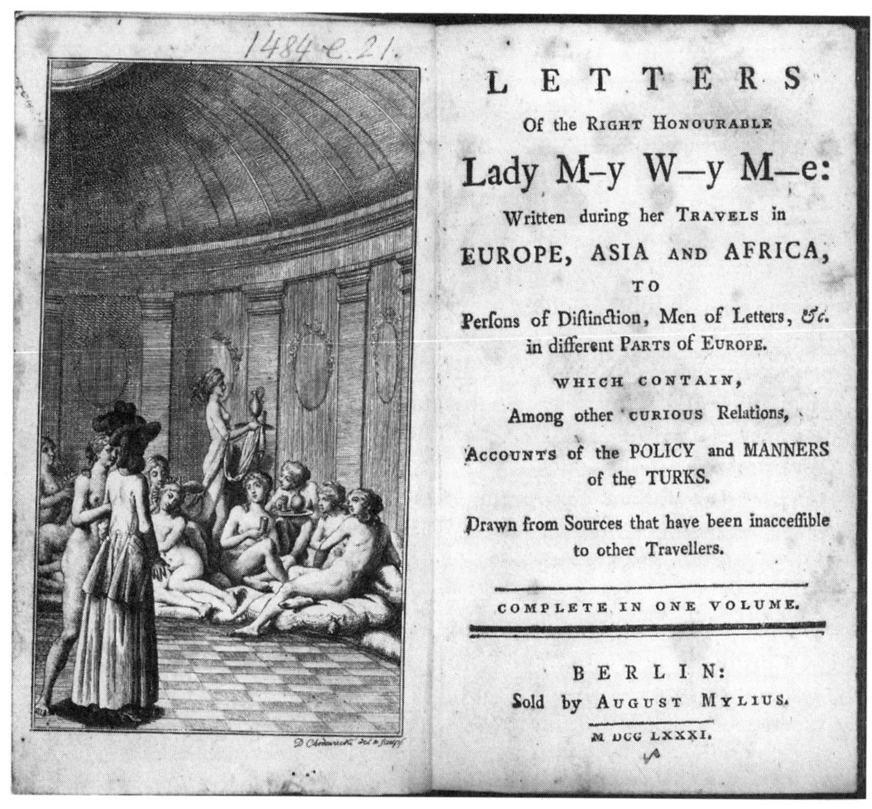

L E T T E R S

Of the RIGHT HONOURABLE

Lady M–y W–y M–e:

Written during her TRAVELS in

EUROPE, ASIA AND AFRICA,

TO

Perſons of Diſtinction, Men of Letters, &c.
in different PARTS of EUROPE.

WHICH CONTAIN,

Among other CURIOUS Relations,

ACCOUNTS of the POLICY and MANNERS
of the TURKS.

Drawn from Sources that have been inacceſſible
to other Travellers.

COMPLETE IN ONE VOLUME.

BERLIN:
Sold by AUGUST MYLIUS.
M DCC LXXXI.

Figure 2. Lady Mary Wortley Montagu visiting the baths in Sophia, Bulgaria, as described in Letter 26. The engraving emphasizes the musculature of Montagu's upper body instead of the folds and texture of her clothing, making her appear as if naked, or "shirtless." Frontispiece of a 1781 Berlin edition of the *Travels of an English Lady in Europe, Asia, and Africa.*

But it is equally important that descriptive details persist through all versions of the *Travels*. Montagu figures her stays as a device used by her husband to lock her up. Montagu is positioned, and positions herself, as a successful negotiator of potentially humiliating challenges both to her sexual agency and her understanding of political discretion. She imposes on herself multiple double-binds that will condemn her if she turns Turk by becoming naked, if she stays in her inhospitably English stays, or if she displays cunning by doing neither. To some degree, she does all of these things, as Srinivas Aravamudan observes; she complies with all that is demanded of her in this moment.[11] In the end, the persistent image here is of an Englishwoman producing material proof for her culture's symbolics of patriarchal enslavement, and the evidence against corsets is sobering. No twentieth-century woman could have an eighteenth-century English body. Eighteenth-century women wore corsets from adolescence; the bodices deformed or even broke ribs, and they weakened the back muscles to the point where back muscles could not comfortably support the weight of a woman's own torso and limbs.[12]

Montagu's corset is, cross-culturally, evidence of an ethnic truth about being female in eighteenth-century England. Hence, I suspect, comes some of the disturbed unease Hobhouse articulates about truths appearing *where no man can controvert,* and his discursive effort to control for himself the truths her body tells. Montagu's account registers unease as well. The first sentence of the letter puts her geographically inside the Ottoman empire, and psychologically within the safety of "another world":

> The first Sofas were covered with cushions and rich carpets, on which sat the ladies, and on the second their slaves behind 'em, but without any distinction of rank by their dress, all being in the state of nature, that is, in plain English, stark naked, without any beauty or defect concealed. Yet there was not the least wanton smile, or immodest gesture amongst 'em. They walked, and moved with the same majestic grace which Milton describes of our General Mother.[13]

Onto Orientalism Montagu superimposes Milton's paradise. From her frame of reference, East has merged with West, Turk with Christian, the Arabian Nights with the Fountain of Domestic Sweets, prose with poetry, and Other with self. But in the context of the entire description of the Sophia baths, Montagu's transference to the Turkish women around her changes to alterity from identity, to threat from salvation, to paradise lost from paradise, to tempting Other from loving Mother.

But in a March 1740/41 letter to his mother, Spence reported a fuller version of the conversation. This version was generally unavailable until 1966 (and so to Halsband), when the James Osborn edition of Spence's *Anecdotes* appeared. Osborn records both the anecdote and this excerpt as the anecdote's source (Singer did not):

> The Turkish Ladies, you know, are a sort of Prisoners: they have very little Liberty; & their chief place of their meeting & conversing together, is at the Womens Baths. Lady Mary went thither; & says she never saw finer shap'd women, than the Turkish Ladies; tho' they never wear Stays. Their make is more natural; & really more beautiful, than that of the Ladies with us. The first time she was at one of these Baths, the Ladies invited her to undress, and to bath with them: and on her not making any haste, one of the prettiest run to undress her. You can't Imagine her surprize upon lifting my Lady's Gown, & seeing her Stays go all round her. She ran back quite frighten'd; & told her companions "That the Husbands in England were much worse than in the East; for that they ty'd up their Wives in little Boxes, of the shape of their Bodies": She carried them to see it; they all agreed that twas one of the greatest Barbarities in the World; & pitied the poor Women, for being such slaves in Europe."[14]

If Osborn's attribution is correct, then Spence edited out of his for-publication version Montagu's most radical observation, that English women would be very free, and very beautiful, if they never wore stays at all.

In any case, Spence's letter supplements his anecdote. The excisions he made when revising letter into anecdote bring into view his attraction to commonplace patriarchal conflations of aesthetics, politics, and gender ideology: of female wit, female nakedness, and female cross-cultural humiliation. Ironically, Montagu did something like this too. Spence put the richer, less witty description in a familiar letter. Montagu at age 35, remembering an experience had at 28, turns the beauty of the Turkish women into an aesthetic statement, not a political one (as Elizabeth Bohls has observed): "There were many amongst them as exactly proportioned as ever any goddess was drawn by the pencil of Guido, or Titian, and most of their skins shiningly white, only adorned by their beautiful hair divided into many tresses hanging on their shoulders braided either with pearl or ribbon, perfectly representing the figures of the graces" (Letter 26).[15] At age 51, Montagu compared Turkish and English women more directly.

Now, what if Montagu *did* undress in the Sophia baths? Then both Spence and Montagu have attempted to turn political transgression into civic discretion. In doing so, they make shifts from private to public, to which they link shifts from coterie genres to commercial printing, from detail-rich documentary rambles to the art of prose, from something resembling Hakluyt's *Traffiques* to something resembling Addison's *Spectator*. Hobhouse, again, is on the right subject: *as must be expected in easy writing of all kinds, no great correctness nor purity.* Montagu's travel prose does not ramble. It vibrates with the complexity of its political and aesthetic negotiations, and her diaries are burned. Small spaces separate woman from feminist, *shirt* from *skirt*, and antithetical cultural critique from decorative, paradox-laden Augustan conversational wit. These spaces—between critique and conversation, between politics and aesthetics, between power and submission, between materiality and its perception through representation—are the contested epistemological spaces in the *Travels*, and the contestants are patriarchy and feminism.

Hobhouse, Halsband, Spence, and Montagu herself all have altered her observational and literary authority at its source: the writer's character, her manuscripts, and her body. But they did not see women in the Ottoman empire in 1717, nor have we; Montagu did. The persistence of factual and textual alterations suggests that the reception history of the *Travels* may be conditioned to permit them. The alterations misrepresent, to varying degrees, Montagu's own material presence in history as a participant in contact, which survives only through artifacts shaped into representation: Montagu's manuscripts, her talk, and her actions.

The persistence of nonauthorial and erroneous alterations to Montagu's own records of contact suggests that the *Travels*'s reception history may be conditioned to permit them. Patriarchal biases may have distracted travelers and scholars from better attention to the primary materials relating to Montagu's *Travels.* But I would also argue that all this work, including my own work, responds to a sexist anxiety implicit in much of Montagu scholarship: that the uses Montagu made of her gender and sexual agency while in Turkey determine her authority as contact participant and cross-cultural authority on the condition of women. The consequence has been to pay more attention to representations of Montagu herself, particularly when she locates herself in erotically-charged scenes, than to the scenes of contact themselves and the texts through which

Montagu represents them. The remedy is reobjectification of the restorative kind: to take Montagu's contact representations seriously enough to subject them to the same multiple interpretive practices, the same level of sustained and informed scrutiny as has been given to male authors whose skills are comparable to Montagu's and whose writings have been canonical longer.

Spence records another anecdote about Montagu's observations of women in Turkey:

> The ladies at Constantinople used to be extremely surprised to see me go always with my bosom uncovered. [She had frequent disputes with them on that subject.] It was in vain that I said that everybody did so among us, and added everything I could in defence of it. They could never be reconciled to so immodest a custom, as they thought it, and one of them after I had been defending it to my utmost, said, "o my sultana, you can never defend the manners of your country, even with all your wit! But I see you are in pain for them, and shall therefore press it no farther." (311)

Like the first anecdote, this one foregrounds a situation that links Montagu's sexual agency to her ethics, her ethnicity, and her freedom. Montagu has again put herself in a crisis situation involving cross-cultural disputes over female body display practices and female freedom. In the anecdote, all the women are in some sense complicit with their version of patriarchal enslavement; the Turkish women are afraid to bare their bosoms at home, and Montagu is afraid to clothe hers abroad.

Conversational

Montagu's *Travels* never discuss this dispute. No defense of uncovered bosoms appears anywhere else in Montagu's surviving papers; 185 years later, repeating the phrase still invites comedy. Archivally, this anecdote supplements Montagu's observations of women in Turkey. The most reliable information in it is the fact of the conversation, its topic, and its attraction to issues linking gender, sexuality, and patriarchal domination across cultures. When Montagu was in her late twenties, she wore low-necked gowns in an Islamic culture: "I thought I should very little satisfy her curiosity (which I did not doubt was a considerable motive to the invitation) by going in a dress she was used to see, and therefore dressed myself in the court habit of Vienna, which is much more magnificent than ours" (Letter 33). The Turkish Muslim women were tolerant, but also willing to express their discomfort (and perhaps suppress fascination). Montagu defended her practice through explanation, but she also did not mind that her interlocutors were unconvinced, that they saw her persist in wearing low necklines, or that they felt affronted by her choice. Yet Montagu did not articulate the complexities of this cross-cultural conflict in the *Travels.* She represented the conflict instead by material and chronological disjunction: articulating it through two genres (conversation and print) and deploying the terms of the conflict at two different times (while she lived, and after she died). She told Spence what she did while she lived, *and* she kept the conflict out of her manuscript.

Montagu was also older when she spoke to Spence. Montagu had moved away from England, and all England was moving away from the literary forms conventional during the Age of Satire. What would a 1740s version of Montagu's *Travels* have looked like? At 27, Montagu was a wholly westernized subject entering non-westernized territory. She was coping both with the shock of contact and a limited and culturally localized capacity for self-critical utterance. These limitations, and the ideological dichotomies they produce, are the effect of class privilege and gender oppression simultaneously. They need to be read accurately as the consequence of aristocratic birth and education, yet Montagu used her semiotic inheritances resourcefully for the purposes of anti-patriarchal feminist critique. To put one of Gayatri Spivak's invaluable observations another way, the subaltern *tries* to speak. Rey Chow has noted, to deny subjects the language of their subjectivity only on the grounds of ideological contamination is inhumane:

> Although the point that we must not be trapped within dichotomies is a familiar one, many of us, especially those who experience racial, class, or gendered dichotomies from the unprivileged side, are still within the power of dichotomization as an epistemological weapon. The above kind of interrogation slaps me in the face with the force of a nativist moralism, precisely through a hierarchical dichotomy between West and East that enables my interrogators to disapprove of my "complicity" with the West. Such disapproval arises, of course, from a general context in which the criticism of the West has become mandatory. However, where does this general critical imperative leave those ethnic peoples whose entry into culture is, precisely because of the history of Western imperialism, already Westernized? For someone with my educational background, which is British colonial and American, the moralistic charge of my being "too Westernized" is devastating; it signals an attempt on the part of those who are specialists in "my" culture to demolish the only premises on which I can speak . . . what is left out is precisely the material reality of a Westernized subjectivity that is indelibly present in the non-Western intellectual's entrance into the world.[16]

Montagu's discreditors, advocates, and finally Montagu herself have all been working the tenuous and potentially manipulable spaces between observable realities and their textual representations. In the *Travels,* Montagu leveraged her observational authority by speaking through the most authoritative texts available to her. She used the materiality of her experience in Turkey to displace linguistic and political manipulations of gender power at home. Thus, to fully understand the formation of her authorial voice via other texts, let us turn to the writing of Bishop Gilbert Burnet, a mentor for Montagu.

Burnet was a cleric with a mixed record for civic service; among other things, he blocked funding for an all-women's residence proposed to Queen Anne by Montagu's friend Mary Astell. But Burnet's mentorship of Montagu was fully literary, political, and doctrinal. Isobel Grundy describes Bishop Burnet as one of "two elderly bishops she corresponded with before her marriage, both men of social conscience: Burnet, an eminent Whig politician, opponent of pluralism and champion of the poverty-stricken lower clergy. . . . Both men were Latitudinarians, ecumenicals who wished to see a liberal, progressive Church of England opening her doors to Protestants of every sect; they might have contributed toward making the young Lady Mary an egalitarian."[17] Burnet died in 1715; Montagu set out for Turkey in 1716.

In his 1688 travelogue, *Some letters containing an account of what seem'd most remarkable in travelling thro' Switzerland, Italy, and some Parts of Germany,* Gilbert Burnet published this account of his journey from Paris to Lyons:

> As I came all the way from *Paris* to *Lions,* I was amazed to see so much misery as appeared, not onley in Villages, but even in big Towns, where all the marks of an extream poverty, showed themselves both in the Buildings, the Cloaths, and almost in the looks of the Inhabitants. And a general dispeopling in all the Towns, was a very visible effect of the hardships under which they lay.[18]

Montagu wrote this about her journey from Lyons to Paris:

> The air of Paris has already had a good effect on me, for I was never in better health, though I have been extreme ill, all the road from Lyons to this place, you may judge how agreeable the journey has been to me, which did not need that addition to make me dislike it. I think nothing so terrible as objects of misery, except one had the God-like attribute of being capable to redress them, and all the country villages of France shews, nothing else. (Letter 49).

Montagu internalized Burnet's travel prose: its epistolary form, its arrangement into itinerary, its choice of topics, its religious and political views, its paragraphing, and its habits of expression. Montagu's close imitation of Burnet's attitudes suggests either an uncritical or conflict-free absorption of his theology, civic, and domestic attitudes:

> Burnet: "[The Bishop] also told me the other legend of King *Lucius*'s sister *S. Emerita,* who was burnt there, and of whose veil there was yet a considerable remnant reserved among their reliques: I confess I never saw a relique so ill disguised, for it is a piece of worn linnen cloath lately washt, and the burning did not seem to be a Month old; and yet when they took it out of the Case to shew it me, there were some there that with great devotion rubb'd their beads upon it." (55)

> Montagu: "I have been to see the churches here, and had the permission of touching the relics which was never suffered in places where I was not known. I had by this privilege the opportunity of making an observation, which I doubt not might have been made in all the other churches, that the emeralds and rubies, that they show round their relics and images are most of them false, though they tell you that many of the *Crosses* and

> *Madonas* set round with them stones have been the gifts of Emperors, and other great Princes, and I don't doubt but they were at first jewels of value, but the good fathers have found it convenient to apply them to other uses, and the people are just as well satisfied with bits of glass." (Letter 6)

Reading Burnet and Montagu next to each other, a critic can feel momentarily disarmed, her sophisticated theorizing being pulled right out from under her. In the obvious and uncomplicated sense, Montagu was working from example.

But unlike Hobhouse, Montagu's prose is not anxious about influence. Montagu's textual attachment to Burnet looks less oedipal than relational-differentiational, and this extends into her relationships with the Mother and Other of her contact experience: Turkish Islamic women. Montagu's stylistic preferences owe less to her publicized relationship with Alexander Pope than to her quiet attachment to Gilbert Burnet, and are better understood not through a Bloomian agonism but through Nancy Chodorow's feminist countermodel of relationality:

> Differentiation happens *in relation to* the mother, or to the child's primary caretaker. It develops through experiences of the mother's departure and return, and through frustration, which emphasizes the child's separateness and the fact that it doesn't control all its own experiences and gratifications . . . separateness is defined relationally; differentiation occurs in relationship: "*I*" am "*not-you*." Moreover, "*you,*" or the other, is also distinguished.[19]

I am suggesting for Montagu's habits of perception a psychic economy, inspired in part from my reading of Mary Jo Kietzman's essay as a postcolonialist analogue to a Chodorowian psychological model. Kietzmann's central observations for the *Travels* are *cultural fluidity, multiple alterity,* and *transfer* of identity positions:

> Montagu's letters may be read as an ethnography of Turkish women's culture that does not represent culture as an order, an Other, or a fixed world of any sort but, instead, conveys the fluidity of a culture whose women seem remarkably able to accommodate a multiplicity of alterities into their social fabric and recommended critics shift from models derived from Edward Said's alterist Orientalism to Sara Suleri's notion of transfer. . . . My reading of Montagu's letters will show that they neither reproduce nor simply complicate the eurocentrism and authoritarianism of Said's model but replace it by representing moments of cultural confrontation in which self and Other do not remain fixed in polarized positions but are rewritten through discursive and social interaction. . . . Even commentators such as Lisa Lowe who seek to show how *The Turkish Embassy Letters* complicates a monolithic Orientalism can do little more than point out Montagu's occasional divergences from traditional British Orientalism because they lack a critical language able to describe ways of seeing and writing Turkish women in terms other than those derived from an alteritist model.[20]

Yet while Kietzman makes her case through feminist and postcolonial discourse analysis, I argue for relationality on materialist grounds, on biographical, editorial, and other artefactual evidence. Either way, the possibility of relational modeling is one agon-fixated writers like Hobhouse could be inclined to miss.

The textual trace of Montagu's differentiation from Burnet is the formality of her prose: the political trace, her feminism. Burnet's descriptions of women are outsider descriptions: "As for their Wives they are bred to so much ignorance, they converse so little, that they know nothing but the dull superstition on Holy-daies, in which they stay in the Churches as long as they can, and so prolong the little liberty they have of going abroad on those daies, as Children do their hours of play. . . . I was told that they were the insipidest creatures imaginable" (121). Burnet does not side with either patriarchy or feminism; he does not appear to be attracted to the controversy. Burnet's prose, unlike Montagu's, does not rely on antithesis, chiasmus, isocolon, or comparative prosodic forms; his travel prose is neither witty nor formally expressive. Burnet's sermons are both, which suggests Burnet found more energy to stylize when he wrote on religious topics than when he wrote on travel. In contrast, sermons were not one of Montagu's characteristic genres at all.

Instead, Montagu wrote as a female insider:

> As to their morality or good conduct, I can say like Harlequin, 'tis just as 'tis with you, and the Turkish ladies don't commit one sin the less for not being Christians. Now I am a little acquainted with their ways, I cannot forbear admiring either the exemplary discretion, or extreme stupidity, of all the writers that have given accounts of 'em. 'Tis very easy to see, they have more liberty than we have, no woman of what rank soever, being permitted to go in the streets, without two *Muslins,* one that covers her face all but her eyes, and another that hides the whole dress of her head, and hangs half way down her back, and their shapes are wholly concealed by a thing they call a *Ferigee,* which no woman of any sort appears without. This has strait sleeves that reaches to their fingers ends, and it laps all round 'em not unlike a riding-hood, in winter 'tis of cloth, and in summer plain stuff or silk. You may guess how effectually this disguises them, that there is no distinguishing the great lady from her slave, and tis impossible for the most jealous husband to know his wife when he meets her, and no man dare touch or follow a woman in the street. (Letter 29)

Montagu's attraction to English feminism created an historical innovation for the travelogue genre. Montagu thematizes throughout her *Travels:* not only by representing individual relationships with her correspondents, but also through her cross-cultural comparisons of civic life, the condition of women, religion, and literature.

When Montagu's feminist response to contact defied explicit articulation, it registered through aesthetic means, and multiple eighteenth-century literary forms. Like her contemporaries, Montagu used form expressively. Montagu's *Travels* mimic two kinds of printed books of letters simultaneously. The *Travels* follow both the style and page layout of the collected book of familiar letters, as does Aphra Behn's *Love Letters Between a Nobleman and His Sister* (1684); and the epistolary divisions of travel literature and chorography, like Gilbert Burnet's *Travels,* or Daniel *Defoe's Tour of the Whole Island of Great Britain* (1724-26). Like Fielding's use of "Palladian" chapter organization for *Tom Jones* (1749), Montagu's *Travels* use the number and organization of letters expressively. Fifty-two letters suggest a calendar year, and also reference Aristotle's recommendations for the duration of epic and tragedy: "Epic poetry . . . [has] no fixed limit of time, whereas Tragedy endeavours to keep as far as possible within a single circuit of the sun."[21] Through this figuration of time, the *Travels* reference other literary genres structured with reference to time and space: pastoral, epic, tragedy, diary, newsletter and newspaper, periodical essay.[22] Finally, Montagu's *Travels* have succeeded, on the scale of *Robinson Crusoe,* as a fabricated text—a narrative in epistolary form read and edited as transcribed letters. Montagu's imitation of letters fooled Hobhouse and have fooled many subsequent readers; to a cynical mind, Montagu's joke has had an impressive run, as long as that of the travel narratives of either Defoe or Jonathan Swift.

Notes

* Grateful acknowledgments to: Lord Harrowby and the Harrowby Manuscripts Trust, Columbia University Library, University of Chicago Library, the British Library, David Bevington, Alice Schreyer, Sandra Macpherson, Isobel Grundy, and Roger Chartier.

1. *Letters of the Right Honourable Lady M—y W—y M—e* (London: Thomas Martin, 1790). British Library, 1477. b. 29, p2r-v.
2. London: Cawthorn, 1812.
3. Letter 26, my text.
4. Devoney Looser, "Scolding Lady Mary Wortley Montagu? The Problematics of Sisterhood in Feminist Criticism," *Feminist Nightmares, Women at Odds: Feminism and the Problem of Sisterhood,* ed. Susan Ostrov Weisser and Jennifer Fleischner (London: New York University Press, 1994), 45.
5. *The Complete Letters of Lady Mary Wortley Montagu,* ed. Robert Halsband, 3 vols. (Oxford: Clarendon Press, 1965-66), I: 314.
6. Harrowby MSS 253: 187-188, italics mine.
7. Samuel Johnson, *A Dictionary of the English Language,* 4th edition, 2 vols. (London, 1773), n. s. *stays.* For help with this reference, thanks to Thomas Bonnell.
8. Isobel Grundy, *Lady Mary Wortley Montagu: Comet of the Enlightenment* (Oxford: Oxford University Press, 1999), 430-441.
9. Robert Halsband, *The Life of Lady Mary Wortley Montagu* (Oxford: Clarendon Press, 1956), 68.
10. In Halsband, *Complete Letters,* vol. 1. Transcription error given first:
    p. 293, possibly Audience/Audiences; the word runs to the end of the page
    p. 302, the/yt [that]
    p. 314, skirt/shirt
    p. 359, Woolen/Woollen
    p. 387, possibly expectation/expectations; the script is ambiguous
    p. 341, freedoms/freedom
    p. 378, or children/or even children
    p. 383, leaded/headed
    p. 385, possibly tales/tale
    p. 397, Shewing/Showing
    p. 403, shewing/showing
    p. 441, Waters/Water
11. Aravamudan: "In continuing to maintain herself in full dress, she appears both dignified and

ridiculous, imprisoned by her own culture. She has pulled off a brilliant improvisation, successfully negotiating the Scylla of offending her Turkish hosts and the Charybdis of scandalizing her English readers. By banking on Turkish cultural misapprehension, she narrowly escapes the sacrifice of her English virtue. However, doubly bound as her body is by her lingerie, she is also in a fictional double bind. . . . she has exposed herself, ever so slightly, to the English gaze by revealing a glimpse of her underwear. The tantalizing readability of her straitjacketed body to the Turkish ladies is all too readable within this fictionalized scenario, providing a further contrast with the enigmatic and wide-open scenario, providing a further contrast with their exultant freedom in the *hammam*. The focus shifts to her: she is the fetish for the female gaze at the bath, and for the mixed gaze back in England." "Lady Mary Wortley Montagu in the *Hammam:* Masquerade, Womanliness, and Levantinization," *English Literary History* 62:1 (1995): 83-84.

12. Londa Schiebinger reproduces Thomas von Soemmerring's 1785 illustration of female skeletal deformation from corset wearing, and she discusses the eighteenth-century European controversy over medical drawings of male and female skeletons. *The Mind Has No Sex? Women in the Origins of Modern Science* (Cambridge, MA: Harvard University Press, 1989), fig. 38, 191-200.
13. My unpublished text.
14. Joseph Spence, *Observations, Anecdotes, and Characters of Books and Men Collected from Conversation* (Oxford: Clarendon Press, 1966), 311-312.
15. Bohls: "Likening the beautiful bathers to prestigious European works of art—Milton's Eve, the nude paintings of Guido and Titian, and the frequently painted classical motif of the Three Graces—is the crux of Montagu's ingenious rhetorical strategy. Such comparisons, by invoking contemporary aesthetic thought, in particular the concept of disinterested aesthetic contemplation, reinforce Montagu's claim that these Turkish women are neither "wanton" nor "immodest." By the early eighteenth century a consensus was beginning to emerge, articulated by British aestheticists like Addison, Shaftesbury, and Francis Hutcheson, that the aesthetic gaze must be sharply distinguished from ways of looking which incorporate what Kant (whose aesthetics owe much to this British tradition) would later call "vested interest"—"practical" needs and desires such as hunger, sexual lust, acquisitiveness, and so on. If works of art are, by definition, not objects for prurient regard, then Montagu's aesthetic comparisons should, at least to some extent, de-eroticize her readers' imaginary gaze and block the crassly sexualized representations of Withers's and Dumont's lascivious crew. "Aesthetics and Orientalism in Lady Mary Wortley Montagu's Letters," *Studies in Eighteenth-Century Culture* 23 (1994): 180.
16. Gayatri Spivak, "Subaltern Studies: Deconstructing Historiography," in *In Other Worlds: Essays in Cultural Politics* (New York: Routledge, 1988), 197-221. Rey Chow, "Violence in the Other Country: China as Crisis, Spectacle, and Woman," *Third World Women and the Politics of Feminism,* ed. Chandra Talpade Mohanty, Ann Russo, and Lourdes Torres (Bloomington: Indiana University Press, 1991), 97.
17. Grundy, *Lady Mary Wortley Montagu: Comet of the Enlightenment,* 87-88.
18. n. p. , 1687, 3-4.
19. Nancy J. Chodorow, *Feminism and Psychoanalytic Theory* (New Haven: Yale University Press, 1989), 102.
20. Mary Jo Kietzman, "Montagu's *Turkish Embassy Letters* and Cultural Dislocation," *Studies in English Literature* 38:3 (1998): 538. Lisa Lowe, *Critical Terrains: French and British Orientalisms* (Ithaca: Cornell University Press, 1991). Elizabeth Bohls is also useful on Suleri's notion of transfer.
21. *Aristotle's Art of Poetry,* trans. Ingham Bywater (Oxford: Clarendon Press, 1940).
22. I must emphasize that no "No. 52" appears in holograph. There are two explanations. First, Montagu misnumbered in holograph by skipping from "21" to "23." Second, when text overflowed into the second of her quarto blank books, Montagu began renumbering: "Letter 1st," and so on. This will be a controversial footnote, but my argument here is that critics need to take seriously the persistence of an imaginatively suggestive structure, despite a conscious slip in counting, the number of letters in the *Travels* and, finally, their sequence.

PART II:

Travelers to India

SECTION FIVE

FERNÃO LOPES DE CASTANHEDA

Fernão Lopes de Castanheda (?-1559) was born in Santarem, Portugal, as the son of Lopo Fernandes de Castenheda, the first "ouvidor"—a special judge of a ministry or tribunal—of Goa. In 1528 he accompanied his father to India with the armada of Nuno da Cunha. Little is known of his tenure. He was apparently commissioned by the Portuguese monarch, João III, to write a history of the Portuguese discovery of India. Books one through seven of the *História* appeared between 1551 and 1554, and were soon translated into French and Spanish. Book eight appeared posthumously in 1561.

The composite translation below combines the Portuguese source (*História do descobrimento e conquista da índia pelos Portugueses,* edited by M. Lopes de Almeida, 2 vols. [Porto: Lello & Irmão, 1979]) with the florid but energetic 1582 English translation by Nicholas Lichefield, published as *The First Book of the Historie of the Discoverie and Conquest of the East Indias.* Despite its looseness, Lichefield faithfully conveys the tenor and spirit of Castanheda's *História.* Insertions between angular brackets mark Lichefield's additions to the source text. Spellings have been modernized.

—Shankar Raman

From the Prologue dedicated to King João III of Portugal (pp. 3-4)

Men are greatly obliged to historians, O great and powerful King our Lord; and especially [so indebted] are Princes for whom it appears especially important that one compose history—a thing so beneficial for human life, which teaches us what we should do and from what we should flee. [For an history] concerns Princes even more than other men. . . . Whatever a Prince does affects all those who are under his governance, since upon his being good or bad depends the well-being of all who belong to his republic. It is therefore necessary for a Prince to be more virtuous, knowledgeable and prudent than all other [people] and for him to learn [such] things there is no better teacher than history. And for this reason all famous Princes—Barbarians as well as Greeks and Romans—have been so given to reading histories . . . (3).[1] [K]nowing these benefits, I took upon myself the labour of composing [this history] of the discovery and conquest of India accomplished by the Portuguese, [both] under the mandate of the famous and well-fortuned King Manuel, your father, as through [the command] of God; and so that the notable and grand actions which were carried out with the help of [you], our Lord [the King João III], in this discovery and conquest would be spread throughout the world. Of all the which there has remained no memory other than in four persons, so that if they had died, all the same would have ended with them, the which would have been imputed to their great shame and rebuke. [A]nd [thus] being written [the memory of these deeds] would perdure forever as [is the case] with the achievements of the Greeks and the Romans, over which these [actions] of the Portuguese towards the Barbarians have a great and well-known advantage. For their conquests were all by land—like the conquests of Semiramis, of Ciro, of Xerxes, of great Alexander, of Julius Caesar and of other Barbarians, Greeks and Romans, . . . their doings are no more marvelled at than a dead lion in respect of one alive. [B]ut this of the Indies was done by sea, and that by your Captains, being upon the same a whole year and eight months, and at the least six months not along or near any coast, but by the bottomless and great Ocean Sea, and departing from the utmost limits of the Occident and sailing to [the limits of] the Orient, without discovering or seeing any other thing, but only the heavens and water, going round about all the Sphere—a matter never before attempted by any mortal man, nor even imagined by any to put the same in practise. Suffering greatly through hunger, thirst and other infirmities, besides [being] every day with those furious storms and rains, in danger a thousand times of their lives, they arrived in India, having passed those fears and dangerous troubles by sea, to find themselves in great and cruel battles, not with men that did fight only with their bows, arrows and spears as they did in Alexander's time, but with such as were stout and of a haughty stomach, and with men that were experienced in war, . . . [who had], besides their accustomed weapons, . . . ordinance . . . and . . . fire-workers more plenty that the Portingales. . . . [N]otwithstanding, the Portingales did give them continually the overthrow, although their strength was but small . . . (4).

Chapter 16: The Portuguese enter a temple in Goa, India, which they mistakenly assume to be a church. (pp. 46-47)

The Catual did carry him unto a certain Pagode of their idols, into which when they were entered, he told him the same was a church of great devotion, which the Captain General believed to be true and to be some church of the Christians,

[1] This paragraph is omitted by Lichefield, who substitutes for it an extended reflection on the necessity of historial writing to perpetuate and revivify the memory of past achievements. In general, Lichefield rearranges and amplifies aspects of Castanheda's prologue in order to foreground the role of the written word in memorializing colonial history.

and therefore he gave the more credit thereunto, the rather for that he saw that over the principal door thereof there hanged seven little bells, and afore the same was the mast of a ship upon the top thereof there stood a wethercock made likewise of wire. The church was as great as a good monastery and was made all of free stone, and covered or vaulted over with brick, which gave an outward show, as though within side it should be of very fair workmanship. Our Captain was very glad to see the same, for that he thought himself to be among Christians, and entering within this church with the Catuall, they were received by certain men, naked from the girdle upward . . . , and upon their left shoulders they had a certain number of threads, which came under their right shoulders, much like the priests were wont to wear their stoles here amongst us. . . . These sprinkled some water out of a fountain . . . upon Vasco da Gama and the Catual and us . . . .

So going into this church, they saw many images painted upon the walls, whereof some there were that had great teeth, which appeared to be so monstrous that they were of an inch in length without their mouth. Others there were that had four arms, and therewith were so ill-favoured, that they seemed to be very devils, the which sight made our men stand in doubt whether the same were a church of Christians or no. Being come afore the chapel which stood in the midst of their church, they perceived that the same had a certain little roof, made much in the manner of a tower, the which was also built of free stone, and in a part of this roof there was a door made of wire, by the which a man might enter into it. The going up to the same tower was by a stair of stone; within this tower, which indeed was somewhat dark, was enclosed in the wall a certain image the which our men beheld a far off, for that they would not suffer them to go near the same . . . howbeit they made a sign to the image, naming the same our Lady, giving thereby to understand that it was her image. The Captain General supposing the same to be true, fell upon his knees with the rest of the company, making their prayers; but one whose name was John de Sala, being in doubt whether the same church were of Christians or not, for that he saw so monstrous images painted on the walls, as he fell on his knees said, *If this be the devil, I worship the true God,* the Captain General that heard him say so looking upon him laughing. The Catual and his company as they came before this chapel, did fall down flat upon the ground before them, and this they did three times, and afterword they arose and made their prayers standing.

LUIS VAZ DE CAMÕES

Luis Vaz de Camões (1524-1580) was born, probably in Lisbon, of a noble Galician family. His father, a sea captain, died at Goa after a shipwreck en route to India. Banished from Lisbon in 1546 for an unknown transgression, he became a soldier at Ceuta in Morocco. There, he lost an eye in battle, learnt the tactics of war, and developed a lifelong hatred of his Moorish enemies. Returning to Lisbon in 1549, he was forced to join the king's service in India three years later to avoid punishment for injuring a court official in a street brawl. He sailed to India in March 1553, as part of Fernão Alvares Cabral's armada, and did not return for 17 years. He narrowly missed being shipwrecked en route to Goa, and served on numerous military expeditions to Malabar, Ormuz, the Moluccas, and Macao. Legend (and his own boast) has him swimming to safety, clutching the manuscript of *Os Lusíadas,* after a shipwreck in 1559 off the Mekong River in Thailand. In 1567, Camões determined to return home, though succeeding only through the help of his friends, Heitor de Silveira and the historian Diogo do Couto. Os Lusíadas was published in 1572, and King Sebastian signified his pleasure at the epic by awarding him a small pension, Camões's primary source of income in his last years. He died of the plague in 1580, the very year in which Phillip II of Spain absorbed Portugal into the Spanish empire.

The portion of Luis Vaz de Camões's *Os Lusíadas* reprinted here are taken from the translation by Leonard Bacon (*The Lusiads* [New York: Hispanic Society of America, 1950]).

—Shankar Raman

Opening Invocation: Canto I, Verses 1-3 (p. 3)

Arms, and those matchless chiefs who from the shore
Of Western Lusitania began
To track the oceans none had sailed before,
Yet past Tapróbané's fair limit ran,
And daring every danger, every war,
With courage that excelled the powers of Man,
Amid remotest nations caused to rise
Young empire which they carried to the skies;

So, too, good memory of those kings who went
Afar, religion and our rule to spread;
And who, through either hateful continent,
Afric or Asia, like destruction sped;
And theirs, whose valiant acts magnificient
Saved them from the dominion of the dead,
My song shall sow through the world's every part,
So help me this my genius and my art.

Of the wise Greek, no more the tale unfold,
Or the Trojan, and great voyages they made.
Of Philip's son and Trajan, leave untold
Triumphant fame in wars which they essayed.
I sing the Lusian spirit bright and bold,
That Mars and Neptune equally obeyed.
Forget all the Muse sang in ancient days,
For valor nobler yet is now to praise.

Bacchus deceives the Portuguese voyagers in Mombasa: Canto II, Verses 10-12 (pp. 43-44).

False human guise and gear he wore anew.
Seeming most like a Christian, he had made
A splendid altar before the which he prayed.

There the high Holy Spirit painted fine,
As it were a portrait, had the artist placed,
With a white dove worked into the design
Above that phoenix sole, the Virgin chaste.
The twelve Apostles' company divine,
Their faces all sore troubled, had been traced
As when from heaven tongues all fiery broke,
And various languages forthwith they spoke.

And thither were the two companions led
Where Bacchus waited for them with his cheat.
And on their knees their prayers to God they said,
Who in the universe hath power complete
Perfumes in odorous Panchaia bred
Bacchus was burning, excellently sweet.
Thus at long last the false god came to do
His homage and must venerate the True.

King Manuel's Dream and Ganges' Prophecy: Canto IV, Verses 69-74 (pp. 152-53)

It seemed to him that he had climbed so high
He must have touched on the first sphere at last,
Whence many various worlds he could descry
And, of strange folk and savage, nations vast,
And near the place of Dawn's nativity,
As far away as farthest sight could cast,
'Mid the long range of huge and ancient mountains,
He saw spring forth two clear and noble fountains.

Wild fowl and savage beasts and monstrous kine
Live on those mountains where the forests grow.
Thousands of shrubs with trees combine
To block all passage where no man can go.
Those desperate heights, as enemies malign
Of any traffic, all too clearly show,
For not since Adam sinned until our day
Has human foot across them broke the way.

From the waters issued (or so fancy told)
Two beings, who, far-striding, near him drew (71. 1-2)
. . .
And one of them approached with weary air,
As who had had the longer way to go (72. 3-3)
. . .

He, as the person of chief dignity,
Thus to the King from far his thought made plain:
"Know a great cantle of this world shall be
Awarded to your crown and to your reign.
And we, whose fame so far abroad doth flee,
Whose necks to bow none ever could constrain,
Warn you 'tis time for you to make demand
For the great tribute given by our hand.

"I am illustrious Ganges. In the earth
Of Paradise I have my cradle true.
And this, O King, is Indus, he whose birth
Befell among the heights you have in view.
Fierce warfare is the price that we are worth.
But if your ends you constantly pursue,
Unfearing, after victories untold,
You'll bridle all the nations you behold."

ABRAHAM ORTELIUS

Abraham Ortelius (1527-1598) was born in Antwerp. At the age of 20 he joined the Guild of St. Lukes as illuminator of maps, while also dealing maps he imported from other countries. He traveled widely and was well-known throughout Europe, even becoming geographer to King Phillip II of Spain in 1573. He was responsible for a large number of important cartographic texts and especially for one of the earliest and most comprehensive atlases, the *Theatrum Orbis Terrarum,* first issued in 1570. The *Theatrum* was subsequently published in at least 26 editions in the author's lifetime, each time with a different number of maps since Ortelius was in constant contact with cartographers, navigators, and geographers, who suggested corrections and additions. He died in Antwerp in 1598.

Reprinted here is an anecdote from Abraham Ortelius' sixteenth-century atlas, *Theatrum Orbis Terrarum* (translated by John Norton [London: 1606]).

—Shankar Raman

Ortelius describes a forged prophecy about Portuguese dominion (p. 5)

Like to those Sibyllin verses, which (as Jacobus Navarcus writeth) were found at the foot of the promontory of the Moon (commonly called Rochan de Sunna) upon the Ocean sea-shore, ingraven upon a four-square pillar, in the time of Don Emmanuel, King of Portugal, to this or the like purpose:

> The stone with my stick letters, rowl'd shall be,
> When West the treasure of the East shall see.
> The Portugals and Indians (a thing admir'd)
> Shall truck their wares, on either part desired.

Howbeit, that these verses are not ancient, but graven in our times, nor part of Sybilles Prophecies, but counterfait; I was advertised (being in hand with the second edition of this my *Theatrum*) from *Rome,* by *Caesar Orlando,* a Civilian, in his letters, out of some printed works of *Gaspar Vaerius,* in which since that I myself have read same. And afterward I found it confirmed by *Amil Resende* in his antiquities of Portugal: namely, that in the time of *Dom Emmanuel,* King of *Portugal,* one Hermes Caiado of the same country, caused them to be ingraven and buried in the earth: and when he supposed that the marble began to corrupt with the moisture of the ground; pretending some cause of recreation, he invites his friends to a country house of his, near into which this fained Prophecy lay hid. Wherefore being all set at meal, in comes his Bailiff with the news, that his Labourers had by chance digged up a stone engraven with letters. They all immediately run forth, they read it, they admire it, they highly esteem it, and are ready to adore it, &c. See how apt *Caiado* was to delude his friends.

BACK TO THE FUTURE: FORGING HISTORY IN LUÍS DE CAMÕES'S *OS LUSÍADAS*

Shankar Raman

I. Remembering the Past

As with Othello's fateful "travelers history" which Desdemona "with a greedy ear" "devour[s] up,"[1] the Portuguese word *história* registers an ambiguity both felicitous and dangerous. While insisting upon a basis in historical events—like the "disastrous chances," "moving accidents," and "hair-breadth scapes" to which Othello refers—colonial *histórias* of discovery encompass both the narrower factual reconstructions of past events and their performative rehearsal as stories.[2] These aspects are necessarily intertwined. As Jacques Lezra puts it, "the 'form' of [the event's] appearing—the morphology of the culture or of the moment—'precedes' the event, which comes to form always and already in the shape of a sign, an event that 'means something'" (41).[3] This essay explores the narrative modes of remembering the past that accompanied and succeeded the Portuguese voyages of discovery to India and the east. It operates on the blurred line dividing histories from stories, colonial events from their narrative reshapings.[4]

I focus on one of the earliest literary works to deal explicitly with the colonial voyages eastward: Luís de Camões's 1572 epic *Os Lusíadas* [The Lusiads], which retells the story of Vasco da Gama's historic voyage to India in 1497-98. Transforming the historical narratives surrounding da Gama's exploratory voyage into the stuff of legend, the epic directs attention both to the initial stages of Portuguese colonialism and the representational strategies that shape our understanding of those events. Camões draws upon a wide range of sources: ancient history; literary predecessors; contemporaneous histories of colonial Portugal; as well as his own experience as a traveler to India, a servant and representative of the Portuguese *Estado da índia*. His *Os Lusíadas* synthesizes these diverse materials into the most durable representation—for the Portuguese at any rate—of the early years of European colonial expansion.[5]

From its opening invocation, Camões's text reveals a doubled allegiance: on the one hand, to the generic demands of the classical epic and, on the other, to the burden of recounting a contemporaneous history. The cohabitation of these two loyalties is not, of course, specific to Camões, for it bespeaks the mutual dependence of epic and historiography as generic forms. From the moment of its emergence, historical discourse shares with the classical epic a desire to preserve the traces of human action. Faced with the possibility of death and oblivion, it opens up a space in which those actions become visible again. By so doing, history, following the example of the epic, dispenses *kleos* or an everlasting glory. "Through [its] song of remembrance," as François Hartog puts it, "heroes are transformed into the men of old and thus represent a collective 'past'" (84).[6] Representing a collective past needs to be taken here in the strong sense of constituting a past as *the* collective past. Remembering heroic exploits brings into being a single and enduring collective identity, one constituted through, but transcending, individual action.

An affinity with the epic task of remembering and surpassing the historical past imbues Portuguese historiographical descriptions of their eastern ventures with a sense of the magnitude of these colonial achievements. Thus, the traveler and historian Fernão Lopes de Castanheda, dedicating in 1551 an early history of the *"descobrimento & conquista da India"* to the King João III (see Extract I, 3-4), emphasizes the need to preserve and perpetuate "the notable and grand actions" performed by the Portuguese discoverers "which have never been surpassed in valour, or even equalled, in any age or country." Alexander the Great's victories cannot be compared to the "noble deeds" performed by the Portuguese in India, the Lichefield translation dismissively adds, any more than "a dead lion can be likened to one alive." While Alexander's conquests were all by land, the Portuguese conquest of India required a voyage by sea of "a year and eight months . . . from the utmost limits of the Occident . . . to the limits of the Orient, without . . . seeing any other thing but only the heavens and water, going round about all the Sphere, a matter never before attempted by any mortal man, and nor yet almost imagined by any." And were this not enough, one had only to note the difference in the caliber of the enemy: while Alexander fought men armed with bows and spears, the outnumbered voyagers had to contend against opponents "as were stout and of haughty stomach . . . and experienced in war, [having] ordinance and fire-workers more numerous" than the Portuguese.

Nonetheless—and here Castanheda's awareness of the importance of his own undertaking becomes manifest—the "memory" of these glorious recent actions remains confined to four persons, "so that if they had died" all remembrance of these transactions "would have ended with them . . . to their great shame and rebuke." Struggling against time and the death of memory, Castanheda's *Historia* represents a way in which sixteenth-century Portugal shaped itself by remembering and rehearsing its colonial accomplishments.

If the Portuguese historian's concern with memory recalls the dependence of both antique and early modern historical writing upon the classical epic, his countryman Luís Vaz de Camões's *Os Lusíadas* aims even higher than these historical and literary progenitors: it takes up the epic task of remembrance precisely as a topos whose calculated inversion then modulates history into myth (see Extract II, 3). Not content merely to advocate memory, Camões calls instead for oblivion, calling for the erasure of a mythic past by a living history that has taken its place. Of course, the Portuguese rupturing of the past can only be meaningful or significant in relation to that past. Camões's assertive stand strategically calls epic and historical precedents to mind, holding before us the very things he urges us to forget. He evokes the *Aeneid,* the *Odyssey,* and the Macedonians in order to mark the limits of the thinkable and the achievable. By rejecting them he challenges, consequently, not just these limits but the possibility of worldly limits as such for the Portuguese who have tracked "the oceans none had sailed before" [mares nunca dantes navegados], who have sailed even beyond Taprobane's far limit [ainda além da Taprobana] (1. 1. 2-3, 3).

Yet, Camões's self-conscious engagement with his classical inheritance also reveals a tension between its impositions and the claims of history. To cite H. V. Livermore, for Camões "[t]he subject of heroic poetry is necessarily historical. The epic is not simply narrative, but historical narrative" (12).[7] At stake is the authority or legitimacy of the epic's narrative. Unlike the classical epic, historical discourse eschews the privileged epic relationship to the Gods (through the Muses). François Hartog emphasizes this shift in terrain entailed by the historian's assessment of points of origin "solely in terms of his own knowledge" (86). In the place of the Muses, historical writing such as Castanheda's installs "a new regime of authority" (90): that of the *histôr* who speaks in the name of history.[8] Similarly, rather than relay the inspiration of the Muses, Camões insists that *his* voice replaces theirs. *His* song shall "spread" [espalharei] the glorious deeds of the Portuguese far and wide, *his* "genius and art" [engenho e arte] (1. 2. 6-8, 3) will make us forget all "the Muse sang in ancient days."[9] Indeed, adopting the historian's task within the frame of his epic, Camões appears to transform a national colonial history into a mythic enterprise. Historical memory is eternalized in order to unveil the existing Portuguese nation as a living embodiment of the "Lusian spirit bright and bold, / that Mars and Neptune equally obeyed" (1. 3. 5-7, 3).

This forced conjugation of historical discourse and the classical epic remains, however, a fraught one. Camões's daring rejection of mythic authority, and the assertion of an equivalent status for his own historical voice, come at a price: he thereby subjects history's "truth" to the flux and uncertainty of human history. In contrast to those "foreign muses" [estranhas musas], who extol "empty deeds / Fantastical and feigned and full of lies" [vãs façanhas, / Fántasticas, fingidas, mentirosas], Camões insists that his epic celebrates an historical truth which "exceeds" [excedem] such dreams [sonhadas] and fictions [fabulosas]: "the great and true" acts of the Portuguese adventurers [verdadeiras vossas são tamanhas] (1. 11. 1-8, 5). He assures King Sebastian, to whom the text is dedicated, that he "shalt never see, for empty deed . . . [t]hy people praised" (1. 11. 1-4, 5). Yet Camões's very insistence betrays an uncertainty: lost with authority of the Muses is their guarantee that his own narrative is "true." By embracing history in order to make the Portuguese its privileged agent, *Os Lusíadas* itself falls into history's grasp; its own legitimacy is rendered suspect, dogged by the fear of being taken as "fantastical and feigned." Like Othello's "wondrous pitiful" and "passing strange" encounters with "[t]he Anthropophagi, and men whose heads / Do grow beneath their shoulders," the history of Portuguese exploration *Os Lusíadas* recounts seems less "a round unvarnished tale" (1. 3. 90) than a mighty narrative "conjuration" (1. 3. 92).

II. Exorcising the Past

If Vasco da Gama's voyage is truly to out-rival the journeys of his mythic predecessors, he must brave dangers exceeding those faced by his forerunners. Nowhere does Camões's concern with the legitimacy of his—and, by extension, Portugal's—colonial mission eastward become more evident than in the epic's recurring fixation on the dangers of deception. Fernão Lopes de Castanheda's widely disseminated *História* already sets the stage for Camões in an account of how the natives of the east African port of Mombasa received the Portuguese embassy. According to the historian, the Moorish inhabitants deliberately counterfeited an image of the Holy Ghost in order to entice the Portuguese to dock and come onto the land, where they were, one might say, to give up the ghost entirely. Forever suspicious, Vasco da Gama first sent ashore two sailors to verify whether the "white moors" of Mombasa were indeed the Christians they claimed to be. Entertained with much celebration by the putatively Christian merchants from India, these two guinea pigs returned to confirm joyfully what Vasco da Gama had hoped to find. The Portuguese captain therefore decided to lead his ships into the harbor where, unbeknownst to him, a small army of "moors" gleefully awaited in ambush. But fortune stepped in to save the fragile mission. An accidental reanchoring of his ships on the way in—interpreted by Castanheda as a sign of divine intervention—induced the hidden warriors into springing the trap too soon, allowing the Portuguese to raise anchor quickly and thus escape unscathed.

While Camões incorporates this incident into *Os Lusíadas,* he nonetheless significantly transforms Castanheda's largely matter-of-fact historical account. The epic's mythic villain is Bacchus, to whom Camões ascribes an unrelenting desire to undo the Portuguese colonial project. An erstwhile companion to Lusus—Portugal's mythic founder—and now a petulant and renegade god bent upon destroying the adventurers, Bacchus attempts time and again to deflect the mission from its goal by turning the different eastern peoples encountered by the Portuguese voyagers into implacable enemies arrayed against the voyage's success. Like the Antichrist to whom the epic often compares him, Bacchus relies on misrepresentation to undermine the Portuguese colonial mission, repeatedly staging deceptions intended to lure Vasco da Gama and his crew to their destruction. At Mombasa, Bacchus attempts to prevent the Portuguese conquest of the land he rules by simulating a Christian: "Seeming most like a Christian, he had made / A splendid altar before the which he prayed."[10] Camões elaborates on the planned deception by describing in detail the paper upon which—following Castanheda's lead—the Holy Ghost has been painted (see Extract II, 43-44). Seeing this replica convinces the Portuguese sailors that they have reached a Christian land.

The reference to Bacchus as "the false god" [o falso Deus] provides a significant clue to why Camões magnifies Castanheda's anecdote, since Camões stages here the crucial issue of idolatry, to engage thereby the problem of "true" representation in the colonial context. Camões not only supplements Castanheda's minimal reference to the painting of the Holy Ghost with detailed visual content—the dove, the phoenix-like Virgin, the Apostles with their troubled faces—but also underscores the likeness of the image through modifying phrases that point to its status as representation ("painted fine," "as it were a portrait"). The closeness of the representation to the real thing, to the truth it expresses, both deceives the Portuguese and—in an important reversal—leads to Bacchus's self-deception. The elaborate description of the painting and the "false" god's homage combine to evoke an ever-present danger surrounding this inaugural voyage:

that of being trapped into an idolatrous veneration of what the east reveals to them, and losing as a result their lives and their souls.

Camões's fascinated turn to idolatry registers, I wish to suggest, the epic's recurrent evocation of the distinction between the forged products of human history and the timeless source that they represent or signify, between artefacts created by man and the divine truth they intend to express. For idolatry emerges when the forged product counterfeits as the divine truth of which it is only the shadow or contingent form. The danger it epitomizes inheres ultimately in the structure of representation itself: in the possibility that the image designed to provide access to the divine source will instead take its place, becoming the very object that blocks access to what it represents. Seduced by the picture's beauty, the Portuguese explorers are deceived into taking the form of Christian truth for that truth itself. And only an additional—and undeserved—divine intervention can hold them from their deaths: in the form of Venus [a Deusa] and her "squadron of nymphs" who "save the Portuguese from that bad end," (2. 23. 7-8, 46) Camões includes Castanheda's emphatic assertion that the "Almighty, disposer of events" prevented the Portuguese ships entering the harbor.

Camões's use of the construct of idolatry harks back to a long-standing biblical concern regarding divine representation, captured most forcefully in the Old Testament story of the golden calf.[11] In sixteenth-century Europe, post-Reformation Protestant attacks on iconoclasm and graven images conferred upon the issue of idolatry an urgent presence, thrusting the problem of imagistic representation to the forefront of the theological debates of the Catholic Counter-Reformation. Himself a participant in a colonial venture deeply identified with Catholicism, Camões thus voices here an anxiety that expresses itself in an intensified focus on the idolatrous image. Rather than being conquerors of the east, the discoverers may be deceived by the east and conquered through that deception. Only by overcoming the seductive deceptions of idolatry can the colonial mission emerge triumphant.[12]

Indeed, the need for constant vigilance turns out to be a lesson that the Portuguese take to heart, as a subsequent, potentially idolatrous encounter underscores. When Vasco da Gama finally disembarks at Calicut in India to carry his embassy to its ruler, the Zamorin, the Zamorin's deputy first leads him into a temple, which is described to the voyagers as "a church of great holiness." Castanheda recounts in some detail this second encounter with putatively eastern "Christians" (see Extract I, 46-47). The 1582 Lichefield translation of Castanheda's *História* summarizes the Portuguese reaction to the images they see in the "pagode's" interior through a marginal note: "the general deceived, commits idolatry with the devil." But the Portuguese voyager John de Sala's reaction to the "monstrous images" adorning the temple emphasizes rather the converse. It reveals the Portuguese determination *not* to be gulled, *not* to be drawn a second time into an idolatrous veneration of the image. There is nothing extraordinary, of course, in the voyagers' interpreting what they see in this unknown land in terms of what they had expected to find. Not only were the Portuguese largely ignorant of Hindu religious practices, but they had also expected the voyage east to lead them to Prester John, the putative eastern Christian monarch who would (they thought) aid them in the struggle against the Moorish foe.[13] Nevertheless, even the most ingrained assumptions are likely to be shaken when belied by visible signs. Thus, aptly enough, John de Sala testifies to his rejection of the signifier in favor of the signified, even as he falls to his knees before the image: "If this be the devil, I worship the true God."

Vasco da Gama's laughter at de Sala's exaggerated caution suggests that Castanheda's anecdote aims to deflect the serious threat shadowing this first contact with alien religions and peoples. That the threat is indeed real becomes evident when we recognize that John de Sala's defensive inversion of a potentially idolatrous situation lurks behind Camões's own distinctive addition to that earlier confrontation at Mombasa: the ironic reversal of the concluding couplet in which the false god, kneeling, finally takes up his subject-position in the divine scheme, driven by his own "cheat" into cheating himself. Camões transfers de Sala's recognition of the gap between the image and what it represents onto Bacchus's equally contrived genuflection before the image of the Holy Ghost.

But the reassignment crucially inverts, too, the sense implied by the historical source. Beyond indicating a (Portuguese) Christian insight into the truth behind the visible, Camões appropriates the story ironically to expose the blindness of the pagan to the hidden truth, emphasizing thereby the power of Christian representation to convert intended evil into good. Precisely because it so faithfully represents he "who governs the world" [que o Mundo governava], the "cheat" prepared by Bacchus takes upon itself the power of that divine source—irrespective of what Bacchus intends it to mean. In "seeming most like a Christian," the false god becomes for an instant what he imitates: the true believer.

This episode further clarifies how idolatry functions in the context of Portuguese expansion eastward as a means of differentiating Us from Them, the European colonizers from those who are to be made their subjects. For Bacchus reveals against his intention the ideological frame the epic repeatedly insists upon: that the adequation of representation to its object is the preserve of Christian truth alone. The prior legitimacy of Christian doctrine asserts the inherent superiority of Christian representations of the world over the counterfeit and inauthentic images produced by the alien peoples encountered in the course of "discovering" the east. And the power of the Christian image is such that it usurps the possibility of native resistance. In the irony of Bacchus's homage to the Holy Ghost, idolatrous deception rebounds upon itself. The epic ironically deflects the false god's counterfeit image from its intended path in order to reiterate the prior legitimacy of the divine source.

And yet there remains a whiff of something forced about Camões's rhetorical one-upmanship. By mocking Bacchus's unwitting self-deception, the text no doubt cleverly undercuts the deception practiced by the demonic other. But the reversal works by mirroring the very structure of idolatry that it condemns: only by equating the counterfeit image with the truth that it represents/blocks can the false god be forced to "venerate the true." If, on the one hand, the epic's iconoclastic impulse directs attention to a gap between the demonic copy and its divine source, on the other, its rhetorical usurpation of Bacchus's cheat reinvests the counterfeiting artefact with a quasi-divine power. In so doing, the attack upon idolatrous representation itself partakes of the form of idolatry: the image must simultaneously be both false and true.[14] Consequently, the textual irony goes a step further than Camões seems to have intended: differentiating authentic Christian from feigned pagan representation finally uncovers a deeper idolatrous impulse that extends to embrace Christian image-making as well. For all the epic's protestations to the contrary, the presence of a counterfeit destabilizes the status of the original.

III. Futures Past: Prophecy and History

Camões's concerns in *Os Lusíadas* with distinguishing the true representation from the counterfeit illustrate deeper tensions within Renaissance historiography over the problems of origins and meaning. Discussing humanist methodology, David Quint posits a distinctive coupling of history and allegory as characteristic of early modern textual practice. Philological and textual scholarship directed at recovering the "original" texts from classical antiquity implicitly evoked an understanding of culture as a human creation whose meaning was determined by historical circumstances and the individual dispositions of its authors. However, Renaissance humanism shied away from the fullest consequences of such an insight: the impossibility of an absolutely fixed and authorized meaning in the face of history and historical change. When the "counterfeit" productions of human endeavor threaten to engulf the "original" text, an allegorical reading became necessary to recover an original dispensation of divine meaning that once authorized and continues to authorize the cultural sign.[15] Camões's abrupt reversal of Bacchus's counterfeit expresses the need for just such an "original dispensation" to stabilize the Christian image. He asserts an original power and truth conferred upon the image in order to block its appropriation by the colonial imitator.

This fraught effort to rescue the created image from the historical circumstances of its making resonates with the tale of "fained prophecy" that Abraham Ortelius recounts in the second edition of his atlas, *Theatrum Orbis Terrarum* (see Extract III, 5). As we shall see, this anecdote shares with Camões's *Os Lusíadas* a narrative strategy aimed at reinforcing the uncertain legitimacy of colonial history. At first blush, Ortelius's story of how an idle nobleman, Caiado, "accidentally" unearths a stone prophesying Portugal's colonial dominion seems no more than an amusing tale of an upper-class prank. Indeed, Ortelius appears to dismiss quite lightly Caiado's involved "discovery" of a prophecy underwriting an already achieved Portuguese dominion in the eastern seas ("see how apt *Caiado* was to delude his friends"). But let us not give in too readily to this gesture. After all, Ortelius makes it clear that he had been sufficiently "deluded" himself to check the veracity of the tale twice. And even after establishing the prophecy as "counterfeit," he still includes the anecdote in the forthcoming edition of his *Theatrum*. Why, one wonders, this interest in a falsified prophecy, an interest that prompts its inclusion, along with an assertion of its illegitimacy and a rather complicated narrative of its origins, in an atlas already in preparation?

We should first note that Ortelius echoes the oppositional structure of original to counterfeit we have located in Camões's *Os Lusíadas.* As is the case with Bacchus's cheat, Ortelius's anecdote explains away the counterfeit image, in this case as an elaborate hoax. But here, too, descriptive detail problematizes the final dismissal. The "four-square pillar" is presented as historical fact, supported by citation ("as Jacobus Navarchus writeth"), by description of its precise geographical location, as well as by a temporal signpost that coincides with the prophecy's prediction of its own discovery—"rowl'd when West sees the treasures of the East." The reversal signalled by "howbeit" begins to unravel the historical source, but only *after* the source has first been presented as factually accurate. Moreover, even as Ortelius scrutinizes the historical origin of the stone, he makes no attempt to refute the content of the prophecy itself—after all, even if the stone is inauthentic, a forgery, it is nonetheless true that it was "rowl'd" at the appropriate moment.[16] Like the image of the Holy Ghost, the counterfeit prophecy is thus allowed to stand *as if* it were authorized by the Sybil, despite the demonstration that the stone's claim to antiquity is a specious one.

Ortelius's invocation of prophecy also introduces, though, a different way of addressing the tension between divine truth and its (potentially idolatrous) human imitation. If the figure of the original aims in the Renaissance allegorically to connect the products of human history to a transcendent or divine source of meaning, then this linkage is achieved in this colonial context by rewriting history as prophecy. Prophecy performs the task of linking the time-bound to the timeless, of anchoring the instability of human meaning in the stability of a truth outside history. But links it, of course, at the proper time, when history is itself in the position to legitimate the prophecy which proleptically legitimates it. Prophecy and history thus reciprocally constitute each other's identity: the future as accomplished history testifies to the validity of that which had foreseen it, just as prophecy extends its mantle to guarantee the truth and inevitability of what is ostensibly to come.

Thus, while Ortelius's own historical excavation proves the material artefact a counterfeit, forged by man—"that these verses are not ancient, but graven in our times"—an allegorical reinterpretation that moves beyond the contingent origin of the physical sign is able to compensate for that historical contingency. The prophetic truth expressed by the verses reattributes the prophecy to another source, one outside the play of human history. By locating the confirmation of the stone's prophetic content in a completed history—that of Vasco da Gama's inaugural voyage—the atlas recovers the engraving's essential truth in the very face of the image's falsity.

Bearing in mind this circular strategy of legitimation, let us turn to a related moment of anterior justification in *Os Lusíadas* (not by any means the only one). The situation is as follows: after the travails of Mombasa, Vasco da Gama has at last reached the safe harbor of the African town, Melinde. Its ruler asks him to describe the west, Portugal and its history, and his own voyage thus far. Obligingly, Vasco da Gama begins "first [his] giant continent to explore, / And after tell the tale of bloody war" (3, 5-6, 82). The annals of Portuguese history that occupy the best part of two books finally reach the reign of King Manuel, whose dream it was to initiate the Portuguese voyage of discovery under the narrator, Vasco da Gama. Manuel dreams that he is approached by two ancient men who provide the impulse for the imperial enterprise, as well as prophesy its success (see Extract II, 152-153).

The ecphrasis of the dream is nicely poised between a culmination of Portuguese history—an aggressive expansionism that has reclaimed all Portugal from the Moors—and a new phase in the nation's history in which it moves beyond its borders to become a dynamic *imperium* in its own right. The moment of articulation, joining past to present, signals the origin of a new age posited as a break from other ages, all the while reinscribing this origin within the history of the Portuguese spirit that the epic celebrates. The legitimation of such a narrative derives precisely from its invocation of the topos of the prophetic source from whence the two rivers spring. The epic correlates thereby an image of a prophetic truth outside time and history with the historical demands of Portuguese imperialism. The coordination takes the form of a double movement that we have already teased out of Ortelius: the fulfilled prophecy authenticates the source, while the source itself provides the reference to an absolute and invariant truth outside human history. The historical events can thus be marked as *original*—that is to say, as a break from what has come before and therefore intrinsically inimitable—while their historical contingency can simultaneously be overcome by a transcendent and absolute source.

And the seductive power of the strategy is such that neither Ortelius nor Camões can let it go. In their own ways, they too must rehearse the counterfeits they describe. Repeatedly, through their own forgeries, they reach toward that relation that Caiado's tablet evokes: between an unalterable prophetic past and an unstable historical present.

IV. Forging History

But King Manuel's dream does more than simply echo the relationship between history and prophecy exemplified by Ortelius's anecdote: after all, the prophecy is uttered by Ganges, by the very figure personifying the Orient. The instigation for the voyage thus appears to be the Orient's voluntary offering of itself to the Portuguese king. Willingly acceding to its own subjugation, the Orient legitimates the Portuguese imperial project by offering a prophecy of its future. The dominance that the Portuguese are to win by force is thereby underwritten by the very entities upon whom this force is to be exercised. And conversely, the fact of dominance legitimates Camões's representation of the East as a place inviting its own colonization. *Os Lusíadas* thus narrativizes Portuguese imperialism constituting itself around a moment of (self-)negation, in which not only does the colonizing power assume the ability to represent and speak for the "Orient," but in addition, the "Orient" is reconstituted as an object that needs to be spoken for.

The legacy bequeathed by Camões's *Os Lusíadas* extends well into the nineteenth and twentieth centuries. As Michael Harbsmeier has argued, only early modern European civilization incorporated its own ability properly to describe and understand the other into the very definition of its own identity.[17] In this sense, Camões's epic already reveals a renarrativization of the (multiple) histories of the East. It increasingly assigns the possibility of an adequate representation of colonial lands to the European civilizers, and begins to replace "native" histories with its own representations of these histories. What takes place thereby is, to use Gayatri Spivak's term, a "worlding" of the East in accordance with the economic and territorial project of Portuguese colonialism.[18]

Spivak's notion of "worlding" indicates not only the process by which the colonized world is produced on the level of the signifier by the colonizing power[19] but, equally crucially, the way in which the inhabitants of that world are "worlded" or induced into producing of *their own accord* a determinate relationship between their selves and the colonial power. Such a relinquishing of subjectivity to the colonizer implies a prior understanding of the colonized worlds as "uninscribed earth," blank and homogenous spaces unmarked by history. This fictive construct is not an innocent one, related as it is to the historical processes through which the colonial powers consolidated their hold on the "discovered" lands. It already implies a prior ideological shaping of discovered lands as the basis for colonial expropriation in later periods.

Camões's *Os Lusíadas* brings into view the embeddedness of these putatively uninscribed spaces in an historical process of colonial consolidation. As I have suggested, the process has two dimensions: the ideological separation provided through the rhetoric of idolatry and the further usurpation of voice within the structure of a prophetic history. Bacchus's genuflection before the image of the Holy Ghost exemplifies the first of these: in his self-deception an imagined mode of colonial resistance is turned ineluctably against itself. King Manuel's dream undermines native resistance even more drastically in that it makes the voice of the Other the site upon which both the legitimacy of the colonial project and the accompanying colonial narrative are constituted. Through such representational means, the discovered lands and people are, as it were, emptied out, severed from their own distinctive pasts to be located rather differently in a colonial narrative.

The texts accompanying the Portuguese "discovery" of India reveal the ways in which the historical events of early modern colonialism took shape through colonizing narratives, stories that not only reinterpreted but also directed the actual practices of discovery. These subject-constituting strategies of colonial narratives delineate the space in which the diverse *material* practices lumped together under the term colonialism unfold. Hence, justified by the East's prophecy of its own subjugation are also the practical means through which Portuguese eastern dominion is to be achieved. The ends of imperialism required, after all, the continual and strategic use of violence. The sixteenth-century Portuguese chronicler Manuel de Faria y Sousa's description of how Vasco da Gama's return to Portugal was greeted, hints in passing at the violent realities underlying the process of "discovery":

> There were Publick Thanksgivings through the Kingdom for the good success of this [the first] Voyage; . . . And all mens expectation being raised with the glory of the Action and hope of ensuing Profit, it was now consulted

> how to prosecute what was begun, and resolved, that according to the disposition they had found in the People of those Countries there was more need of Force than Intreaty, in order whereunto thirteen Vesses [sic] of several sizes were fitted, and *Peter Alvarez Cabral* was named Admiral. (53)[20]

"There was more need of Force than Intreaty": a phrase tucked in a subordinate clause, as if its truth were evident to all. The fiasco of Vasco da Gama's first voyage—in which he lost 105 of the 160 men to sickness or disease and failed, largely through his own misinterpretations, to establish friendly trade with Calicut's Zamorin—hardly seems to justify that phrase. But the use of force was in fact an implicit assumption from the very beginning. Wresting control over trade from the Arab and Gujerati ships plying the Indian Ocean littoral required the continual exercise of violence: first to actually break into the existing spatial networks, and then to maintain them in a way that yielded the requisite results.

Thus, once the capture of most of the important Asian towns was largely complete, the Portuguese instituted a system of *cartazes* to regulate the movement of ships in the Indian Ocean. For a trading vessel to pass unmolested, the captain had to carry a pass or *cartaz* issued by the competent Portuguese authority in India (generally the governor or the captain of a fort). Though given freely, the pass carried numerous stipulations: it limited the amount of munitions a ship could carry, forced ships to trade only at a Portuguese fort, or compelled them to call at a Portuguese fort to pay duties on cargo before proceeding to their destinations. It also required the captain of a vessel to leave a cash security at the fort where the pass was issued, restricted the passengers he could carry—Turks and Abyssinian Muslims were not allowed—and the goods he could trade. Any ship without a *cartaz* could automatically be confiscated, and its crew either put to death or sent to the galleys (41).[21] The indignation at such high-handed actions is well-captured by a sixteenth-century Muslim chronicler:

> Now it should be known, that after the Franks [i.e., the Portuguese] had established themselves in Cochin and Cannanore, and had settled in these towns, the inhabitants with all their dependents, became subject to these foreigners, engaged in all the arts of navigation, and in maritime employments, making voyages of trade under the protection of passes from the Franks; every vessel, however small, being provided with a distinct pass, and this with a view to the general security of all. And upon each of these passes a certain fee was fixed, on the payment of which the pass was delivered to the master of the vessel, when about to proceed on his voyage. Now the Franks, in imposing this toll, caused it to appear that it would prove in its consequence a source of advantage to these people, thus to induce them to submit to it; whilst to enforce its payment, if they fell in with a vessel, in which their letter of marque, or pass, was not to be found, they would invariably make a seizure both of the ship, its crew and its cargo! (40)[22]

If earlier, the winds and tides had determined the passage of ships in and around the Indian Ocean, it was now the "discoverers" who regulated their movement through an administrative system designed to consolidate Portuguese control over existing trading and commercial networks.

On his second voyage to the Indian Ocean a few years later, Vasco da Gama eschewed even the pretence of friendly negotiation:

> The Admiral arrived the *12th* of *July* at *Quiloa,* having lost two Ships in bad Weather. He entered furiously, firing all his Canon, and battering the Town in revenge of the ill usage others had received from that King. But he to prevent his total ruin, came in a Boat to appease the Admiral, offering to be a subject, and pay a Tribute to King *Emanuel.* Thus the Storm was converted into Joy (64).[23]

Remarkable about Faria y Sousa's description of Quiloa's capture is the way a rhetorical statement of Portuguese mastery over nature merges into the brutal conquest of the town. The ships arrive in "bad weather," but with the conversion of the Quiloan king into a Portuguese vassal, the storm itself is "converted" into joy. Skillfully slipping from the natural power of the storm into the Portuguese "storming" of Quiloa, the historian not only asserts the equivalence between the Portuguese forces and the forces of nature, but also establishes the Portuguese as a force *over* nature, capable of "converting"—note, too, the monetary resonance of this word—the power of nature to their own ends.

In this movement, colonial history, having displaced myth, itself becomes mythical. The logic of such a reversal has, as we have seen, been present in Camões's *Os Lusíadas* all along. If we recall King Manuel's dream—which provided the impetus for Vasco da Gama's pioneering voyage—it was the East itself that had, in the shape of the Ganges and the Indus, underscored the necessity of colonial violence:

> Fierce warfare is the price we are worth.
> But if your ends you constantly pursue,
> Unfearing, after victories untold,
> You'll bridle all the nations you behold. (4. 74. 4-8, 153)

By "faining" a prophecy that "the strange folk and savage, nations vast" (4. 69. 4, 152) cannot themselves utter, the brutality of colonial history assumes the mantle of heroic inevitability.

Notes

1. William Shakespeare, *Othello,* ed. Gerald Eades Bentley (New York: Penguin Books, 1970), 1. 3. 139ff. The Arden Shakespeare's "travailous" diverges from both Folio ("travelers") and Quarto ("travels"), but usefully emphasizes the etymological link between "travel" and "travail," journeying and labour.
2. Aptly enough, Fernando Cristóvão's introduction to a recent compilation of thematic and bibliographic studies on the literature of journeying begins with "hesitations and ambiguities" that beset any determination of what comprises the subgenre of travel literature. The term implies the existence of a collection of texts sharing a turn to the voyage in search of themes, motives, and forms that together identify "an autonomous conjunction, distinct from other textual conjunctions. "It remains nonetheless difficult to establish criteria to define the scope of the cate-

gory. A stringent restriction to a relatively short list of national texts of broadly historical character—such as contemporary chronicles, descriptions of lands, voyage diaries, rotaries, and nautical guides—would seemingly contain the ambiguity of *história,* but only at the expense of excluding the *stories* of travel, the narrative forms in which historical reconstructions of events often truly became alive, imprinting themselves upon the cultural spaces of their origin. See Fernando Cristóvão, ed., *Condicionantes Culturais da Literatura de Viagens: Estudos e Bibliografias* (Lisbon: Edições Cosmos, 1999), 14-21.

3. Jacques Lezra, *Unspeakable Subjects: The Genealogy of the Event in Early Modern Europe* (Stanford: Stanford University Press, 1997), 41.
4. Indeed, for the case of Portugal, an apparently commonsensical working distinction between "primary" texts of an eyewitness character and "secondary" texts that reshape such accounts into an *história* would seem an artificial one. Besides the ones discussed in this essay, many of the most influential texts, such as Gaspar de Correia's *Lendas da Índia* and Diogo Couto's *O Soldado Practico,* are the works of writers who themselves traveled to India and spent many years living and experiencing Portugal's engagement with the east. Even if these writers do not describe their own voyages, their writings are shaped by a firsthand experience of and participation in a nascent colonial project.
5. On references to ancient history, see Frank Pierce, "Ancient History in *Os Lusiadas,*" *Hispania* 57 (May 1974): 220-30. Roger Stephen Jones compares Camões's use of literary and historical similes in "The Epic Similes of *Os Lusiadas,*" *Hispania* 57 (May 1974): 239-45. See also Ronald W. Sousa, "From Homer to the Testament: Toward a theory of Camões's Similes, " in *Camoniana California,* ed. Maria de Lourdes Belchior and Enrique Martinez-Lopez (Santa Barbara: Jorge de Sena Center for Portuguese Studies, 1985), 259-64. A measure of his success lies in the fact that *Os Lusíadas* has been embraced by the Portuguese as their national poem, "the prime literary representation," to cite Richard Helgerson, "of who they have been, who they are, and who they should be," in: *Forms of Nationhood: The Elizabethan Writing of England* (Chicago: University of Chicago Press, 1992), 162.
6. See François Hartog, "Herodotus and the Historiographical Operation," *diacritics* 22. 2 (1992): 84. The historiographical topos of remembrance can be traced back to the emergence of history as a new form of knowledge and practice in Herodotus's foundational *History.* As Hartog's discussion of the structure, vocabulary, and cadence of Herodotus's opening shows, Herodotus inaugurates a different way of describing events in the human world, but his point of departure remains the classical epic.
7. H. V. Livermore, *Epic and History in The Lusiads* (Lisbon: Comissão Executiva do IV Centenário da Publicação de "Os Lusíadas," 1973), 12.
8. Hartog, "Herodotus," 86 and 90.
9. David Quint, *Epic and Empire: Politics and Generic Form from Virgil to Milton* (Princeton: Princeton University Press, 1993), 117. Camões's epic is no less fantastic and fictional; but, as Quint convincingly argues, Camões's rationalist reduction of classical myth allows him "to assert a historicity and human truth for his fabulous classicising invention."
10. As Camões often equates Bacchus with the Antichrist or the devil, the god's prior possession of the east reaffirms that the Portuguese mission is not simply to find Christians but (forcibly) to make Christians, wresting the misled from the unbeliever's control. This imperative allows the colonial enterprise to be characterized as a rescue mission, saving the bedevilled colonial other from itself.
11. On the biblical prohibition of images, see Moshe Barrasch's *Icon: Studies in the History of an Idea* (New York: New York University Press, 1992).
12. At the same time, Camões's hysterical logic leads to an otherwise perplexing equation: he aligns the false pagan god, Bacchus, with the traditional enemies of Portugal and of Christianity, the implacable Islamic foes who threaten this inaugural voyage. The Islamic tradition would appear, of the monotheistic religions, to be the one most unambiguously opposed to any form of figural representation of the deity. This collapsing of an idolatrous pagan figure onto an antiidolatrous religious foe suggests that Camões's concern with mistaken veneration responds not only to the threat posed by the colonial other but also to an internal fear of failure that dogs the colonial venture.
13. According to Castanheda, when asked by natives of Calicut what they sought "at so great a distance form home," the Portuguese replied that they had come "in search of Christians and spices."
14. Moreover, the reinvestiture has no practical effects, since the deception actually functions just as

Bacchus had wished: the forced homage does not negate the idolatrous reaction of the Portuguese to the sight of the altar. Nor does it affect Vasco da Gama's decision to bring his ships into the harbor.

15. David Quint, *Origin and Originality in Renaissance Literature: Versions of the Source* (New Haven: Yale University Press, 1983), chapter 1 *passim*.
16. And neither does he point up the rewriting of Portuguese involvement in Indian Ocean trade that the prophecy undertakes: "The Portugals and Indians (a thing admir'd) / Shall track their wares, on either part desir'd." Whatever the numerous differences between the various narratives generated by Vasco da Gama's voyage, they all certainly cast doubt upon the claim that trade was *mutually* desired.
17. Michael Harbsmeier, "Early Travels to Europe: Some Remarks on the Magic of Writing," in *Europe and its Others: Proceedings of the Essex Conference on the Sociology of Literature,* ed. Francis Barker et al., 2 vols (Colchester: University of Essex, 1985), I: 72-88. The clear-cut oppositions Harbsmeier derives from eighteenth-century travel literature cannot be transferred without modification to the sixteenth-century Portuguese colonial experience. The sixteenth-century Iberian representation of other cultures developed within an already existing opposition between Christian and Moor, leading to a floating, indeterminate representation of non-Christian and non-Moorish peoples as interchangeably savages, heathens, gentiles, etc. Nevertheless, even at this stage I would suggest we can recognize the emerging contours of a binary opposition of west to east crucial to subsequent colonial representations.
18. Gayatri Chakrovorty Spivak, "The Rani of Sirmur," in Barker et al., eds., *Europe and Its Others,* I: 128-51. "Worlding" is thus simultaneously a "self-worlding" through which the colonial subject "freely" accepts the place made available by the colonizing power. Spivak develops the notion of worlding to capture the representational strategies deployed in the nineteenth-century British colonization of India. She suggests that the formation of the native subject in that colonial phase transpires against the background of an "uninscribed earth," a fiction that characterises the to-be-worlded world as unwritten, a space bearing no traces of the human activity that has until this point constituted it.
19. The material expansion of colonial relations requires, in other words, a production of the colonial subject as well, since, as Robert Young puts it, imperialism is not only a "territorial and economic but inevitably also a subject-constituting project. " In: *White Mythologies: Writing History and the West* (London: Routledge, 1990), 159.
20. Manuel de Faria y Sousa, *The Portuguese Asia,* trans. Capt. John Stevens (London, 1706), 53.
21. M. N. Pearson, *Merchants and Rulers in Gujrat* (Berkeley: University of California Press, 1976), 41.
22. From Sheikh Zeen-udDeen's *Tohfut-ul-Mujahideen,* quoted in Pearson, *Merchants and Rulers,* 40.
23. Faria y Sousa, *The Portuguese Asia,* 64.

SECTION SIX

JAN HUYGHEN VAN LINSCHOTEN

Jan Huyghen van Linschoten's *Itinerario, Voyage ofte Schipvaert van Jan Huygen van Linschoten naer Oost ofte Portingaels . . .* was first published in Dutch in 1596 in Amsterdam by Cornelis Claeszoon. Born to an innkeeper and his wife in the Dutch town of Enkhuizen in 1562 or 1563, van Linschoten traveled to Portugal in 1579, and from there, in the employ of Portuguese Archbishop João Vincente da Fonseca, to Goa, India, where he arrived in September of 1583 after some six months at sea. In Goa, van Linschoten committed to paper most of the materials that make up the *Itinerario.* It is assumed, however, that he made substantial revisions between the date of his return to Holland in 1592 and publication in 1596. It is certain that van Linschoten was aided in this process by the Dutch scholar and physician Bernardus Paludanus (Barent ten Broecke), who had befriended van Linschoten and who had himself journeyed to the Levant. Van Linschoten's voyages were limited to the East and the north, and the materials in the *Itinerario* dealing with Africa and the Western Hemisphere were based on sources other than his personal experience. The first edition of the *Itinerario* includes 36 plates based on drawings made by van Linschoten himself.

Sections of van Linschoten's text were anthologized by Richard Hakluyt and Samuel Purchas. The whole of the *Itinerario* was translated into English and published by John Wolfe in London in 1598. The Dutch text has gone through several editions and reprints by De Linschoten-Vereniging in the Netherlands. The *Itinerario* was also translated into Latin (1599) and French (1610).

The excerpts reproduced here are taken from a reprint of the first English translation of 1598. Jan Huyghen van Linschoten, *Itinerario. The Voyage of Jan Huyghen van Linschoten to the East Indies. From the Old English Translation of 1598,* vol. 1, translated by William Phillip (?), and edited by Arthur Coke Burnell (London Hakluyt Society, 1885). Van Linschoten rarely divides his prose passages into paragraphs, even when such breaks are clearly called for. I have silently inserted paragraph breaks where this seems appropriate.

—Ivo Kamps

From Chapter 29. Of the customes of the Portingales, and such as are issued from them, called Mesticos, or half countrimen, as wel of Goa,[1] as of all the Oriental countries

The Portingales in India, are many of them marryed with the naturall borne women of the countrie, and the children procéeding of them are called Mesticos,[2] that is, half countrimen. These Mesticos are commonlie of yellowish colour, notwithstanding there are manie women among them, that are faire and well formed. The children of the Portingales, both boyes and gyrls, borne in India are called Casticos,[3] and are in all things like the Portingales, onely somewhat differing in colour, for they draw towards a yealow colour: the children of those Casticos are yealow, and altogether [like the][4] Mesticos, and the children of Mesticos are of colour and fashion like the naturall borne Countrimen or Decaniins of the countrie, so that the posteritie of the Portingales, both men and women being in the third degrée, doe séeme to be naturall Indians, both in colour and fashion.

Their livings and daylie traffiques are to Bengala,[5] Pegu,[6] Malacca,[7] Cambaia,[8] China, and everie way, both North and South: also in Goa there is holden a daylie assemblie, as wel of the Citizens and Inhabitants, as of all nations throughout India, and of the countries bordering on the same, which is like the méeting upon the burse[9] in Andwarpe,[10] yet differeth much from that, for that hether in Goa there come as well Gentlemen, as marchants, and there are all kindes of Indian commodities to sell, so that in a manner is it like a Faire. This méeting is onely before Noone, everie day in the yeare, except Sondayes and holie dayes: it beginneth in ye morning at 7., and continueth till 9., but not in the heate of the day, nor after Noone, in the principal stréete of the Citie, named the straight stréete, and is called the Leylon,[11] which is as much to say, as an outroop:[12] there are certain cryers appointed by the Citie for ye purpose, which have of al things to be cryed and sold: these goe all the time of the Leylon or outroop, all behanged about with all sorts of gold chained, all kindes of costly Iewels, pearles, rings, and precious stones: likewise they have running about them, many sorts of slaves, both men and women, young and old, which are daylie sould there, as beasts are sold with us, where everie one may chuse which liketh him best, everie one at a certaine price. There are also Arabian horses, all kinde of spices and dryed drugges, swéet gummes, and such like things, fine and costly coverlets, and many curious things, out of Cambaia, Sinde,[13] Bengala, China, etc. and it is wonderfull to sée in what sort many of them get their livings, which every day come thether to buy, and at an other time sel them again. . . . (183–85)

There are some married Portingales, that get their livings by their slaves, both men and women, whereof some have 12, some 20, some 30, for it costeth them but little to kéepe them. These slaves for money doe labour for such as have néede of their help, some fetch fresh water, and sell [it for money] about the stréetes: the women slaves make all sorts of confectures and conserves of Indian fruites, much fyne néedle worke, both cut and wrought workes, and then [their master] send the fairest and the youngest of them well drest up with their wares about the stréetes to sell the same, that by the neatnes and bewtie of the said women slaves, men might be moved to buy, which happeneth more for the affection they have to the slaves and to fulfill their pleasure with them, then for any desire to the conserves or néedle workes: for these slaves doe never refuse them, but make their daylie living thereby, and with the gaines that they by that meanes bring home, their maisters may well kéepe and maintaine them. There are others that use exchanging of moneyes, and to buy money, as tyme serveth to sell it againe. . . . (185-86)

The Portingales and Mesticos in India never worke, [if they doe, it is] but [very little, and that] not often, but the most part of them live in such sort, as I have shewed you, although there are some handie crafts men, as Hat-makers, Shoemakers, Saylemakers, and Coopers: but most of them have their slaves to worke in their shops, and the maisters when they walke up and downe the stréetes, goe as proudlie as the best: for there one is no better than an other, as they think, the rich and the poore man all one, without any difference in their conversation, curtesies and companies. (187-88)

From Chapter 30. Of the Portingales and Mesticos, their houses, courtesies, marriages, and other customes and their manners in India

The Portingales are commonly served with great gravitie, without any difference betwéene their Gentleman and the common Citizen, or soldier, and their going, curtesies, and conversations, common in all thinges: when they go in the stréetes they steppe very slowly forwards, with great pride and vaineglorious maiestie, with a slave that carrieth a great hat or vaile over their heads, to keepe the sunne and raine from them. Also when it raineth they commonly have a boy that beareth a cloke of Scarlet or of some other [cloth] after them, to cast over them: and if it bee before noone, hee carrieth a cushin [for his maister] to knéele on when he heareth Masse, and their Rapier is most commonly carried after them by a boy, that it may not trouble them as they walke, not hinder their gravities. When they méete in the stréetes a good space before they come together, they beginne with a great Besolas manos,[14] to stoope their bodies, and to thrust forth their foot to salute each other, with their hattes [in their hands], almost touching the ground: likewise when they come into the Church they have their stooles ready, which their slaves have prepared for them: all that are by him that commeth in do stande [up], and with the same manner of bowing doe him great reverence, and if it chaunceth that any doeth him reverence (as the manner is) and that he to whom it is done doth not greatly estéeme thereof, so that he doeth him not the like [curtesie], they do altogether for that cause go after him, and cut his hatte in péeces, saying that he had disgraced the partie, wherein it is not for them to aske wherefore they shold so do, for it would bee the greatest shame in the world unto them if they should not revenge [so great an iniury]: and when they séeke to bee revenged of any man that hath shewen them discurtesie, or for any other cause whatsoever it bee, they assemble ten or twelve of their friends, acquaintance or companions, and take him wheresoever they find him, and beat him so long, that they leave him for dead, or very neare dead, or els cause him to be stabbed by their slaves, which they for a great honor and point of honestie so to revenge themselves, whereof they dare boast openly, but if they desire not to kil him, they baste him well about the ribs and all his body over with a thicke réede, as big as a mans legge, which is called Bambus, whereby for eyght days after and more he hath inough [to do to kéepe his bed], and sometime in that manner they leave him for deade. This is their common custome, and is never looked unto or once corrected. Also they use long bagges full of sand, wherewith they will breake each others limmes, and [for ever after] make them lame. . . . (193-95)

Againe the long travaile and great voyage maketh many [Portuguese]*[15] to stay in India, and to employ [their time to other trades], as they can best provide [for]* [themselves]. By these meanes the wars in India are not so hot, [nor so throughly looked into], neyther any other countries sought into or founde out, as at first they used to doe. Now they doe onely strive to get praise and commendation, and to leave a good report behinde: and now likewise they are all given to scraping, as well as the Viceroy, Governours, and others, as also the spiritualitie,[16] little passing [or estéeming] the common profit or the service of the King, but only their particular profits, making their account, that the time of their abode is but thrée years: wherefore they say they will not doe otherwise then those what were before them did, but [say that] others which come after them shall take care for all: for that the King (say they) gave their offices, to pay them for their services [in past times], and not for the profit of the commonwealth: therefore there is no more [countries] in India won or new found out, but rather heere and there some places lost, for they have enough to doe, to hold that they have alreadie, [and to defende it from invasion], as also that they doe scoure the Sea coastes, and yet many Marchants have great losses every years, by meanes of the sea rovers,[17] and, together with the evill government of the Portingales, and it is to be feared, it will bee worser every day, as it is evidently séene. (203-204)

From *Itinerario, The Voyage of Jan Huyghen van Linschoten to the East Indies. From the Old English Translation of 1598,* vol. 2, translated by William Phillip (?), and edited by P. A. Tiele (London Hakluyt Society, 1885).

From chapter 92. Of certaine memorable things in India during my residence there

In the same month of August there happened a foule and wonderfull murther within Goa,[18] and because it was done upon a Netherlander, I thought good to set it downe at large, that hereby men may better perceyve the boldnesse and [filthie] lecherous mindes of the Indian women, which are commonly all of one nature and disposition. The thing was thus, a young man borne in Antwarpe called Frauncis King,[19] by his trade a stone cutter,[20] was desirous (as many young men are) to see strange countries, & travelled unto Venice, where he had an uncle dwelling, who being desirous to preferre his cosin, sent him in the company of other Marchantes to Aleppo in Suria,[21] where the Venetians have great trafficke, as I saide before, there to learne the trade of merchandise, and specially to deale in stones, to which ende he delivered him a [great] summe of money.

This youth being in Aleppo, fell into company in such sort, that insteede of increasing his stocke, as his uncle meant he should doe, he made it lesse by the one halfe, so that when the other Marchants had dispatched their businesse, and were readie to depart for Venice, Frauncis King perceyving that hee had dealth in such sorte, that halfe his stocke was consumed, and spent in good fellowshippe, knew not what to doe, as fearing his uncles displeasure, not daring to returne again [to Venice unlesse hee carried as much with him as hee brought from thence]: in the end hee took counsell of some Venetians, with whome hee was acquainted, that willed him to goe with the Caffila[22] or Carvana, that as then was ready to go unto Bassora,[23] and from thence to Ormus in India, assuring him seeing hee had knowledge in stones, that hee might winne much profite, and thereby easily recover the losse that by folly hee had receyved: which would turne to his great benefite, and likewise no hurt unto his uncle. Which counsell hee followed, determining not to returne backe againe before hee had recovered his losse, and ioyned himselfe with certain Venetians, who travelled thether, and so went with the Caffila till they came unto Bassora, the best Towne in all those Countries, lying uppon the utter parte of Sinus Persicus,[24] that goeth towardes Ormus,[25] and from thence by water till they arrived in Ormus, where everie man set uppehis shoppe, [and began to sell his wares]: but Francis [sic]* King being young and without government, seeing himselfe so far distant from his uncle, made his account, that the money he had in his hands was then his owne, and began againe without anie foresight, to leade his accustomed life, taking no other care, but onlie to and make good cheare so long, till in the end the whole stock was almost clean [spent and] consumed, and beginning to remember himself [and to call to mind his follies past,] hee knewe not what course to take: for that to goe home again, he thought it not the best way, as wanting the meanes, and again he durst not shew himself in the sight of his Uncle. At the last he determined to travell unto Goa, where he understood he might well get his living, by setting up his trade, til it pleased God, to work otherwise for him, and so he came to Goa, and being there presently set up shoppe to use his occupation. But because he found there good company, that is to say, Netherlanders and Dutchmen,[26] that served there ordinarily for Trumpetters and Gunners to the Viceroy, who did daylie resort unto him: he could not so well ply his worke, but that he fell into his wonted course: which he perceiving, in the end determined to make his continuall residence in Goa, and set downe his rest, there to abide as long as he lived, seeing all other hope was cleane lost for ever, returning againe unto his owne countrie.

[At the same time,] among other strangers, there was one Iohn of Xena,[27] a French man, borne in Deepe,[28] that in former times was come into India, for drum unto one of the Viceroys, and having beene long in the countrie, was maryed to a woman of Ballagate,[29] a Christian, but by birth a More.[30] This French man kept a shoppe in Goa, where he made Drummes and other Ioyners worke, and withall was the Kings Oare maker for the Galleyes, whereby he lived in reasonable good sort. He had by his foresaid wyfe two sonnes and a daughter: and as strangers, [of what nation soever they be], use to take acquaintance one of the other, being so out of their owne countries (speciallie in India, where there are very few) and do hold together as brethren, which to them is a great comfort: so this Frauncis King used much to this French mans house, by whom he was verie much made of, and very welcome, as thinking thereby to bring him to match with his daughter, because of his occupation, which is of great account in India: because of the great number of [Diamants & other] stones that are sold in those countries: and to conclude, as the manner of India is, that when they have gotten a man in once, they will never leave him: he ceassed not with many promises and other wonderfull matters to draw Frauncis so farre, that he gave his consent thereunto, which afterwardes cost him his lyfe, as in the historie following you shall heare the true discourse.

To make short, they were maried according to their manner, the Bryde being but 11. yeares old, [very] fair and comelie of bodie and limme, but in villanie, the worst that walked uppon the earth: yet did her husband account himselfe a [most] happie man that had found such a wyfe, as he often times said unto me: although he was so ielous of her, that he trusted not any man, were they never so neere friends unto him: but he in whome he put his greatest trust & least suspected, was the worker of his woe. When he was betroathed to his wife, the father promised him a certain peece of money, and untill it were payde, he and his wyfe should continue at meat and meale in his father in lawes house, and should have a shop adioyning to the same, and whatsoever he earned should be for himself. [When all] this was done, and the matter [had] remained thus a long while, by reason that the father in law could not performe the promised summe, because their houshold increased, [it came to passe that] the old man fell into sicknesse and died, [and then] Frauncis King must of force pay his part towards the housekeeping, which he liked not of: & thereupon fel out with his mother in law; and on a certaine time made his complaint to me, asking my counsel therein, I answered him and said, I would be loath to make debate betweene Parents & Children, but if it were my case, seeing I could not obteyne my dowrie, I would stay no longer there, but rather hyre a house by my self, and keep better house alone with my wife, then continue among so manie, wher I could not be master. In the end he resolved so do doe, & with much adoe tooke his wife & Child wt. his slaves, and parted houshold, hyred himselfe a house, set up his shoppe, and used his trade so handsomely, that having good store of worke, he became reasonable wealthie. But his mother in law, that could not conceale her Morish nature, after ye death of her husband: whether it were for spight she[31] to her sonne in law, or for a pleasure she tooke therein, counselled her daughter to fall in love with a young Portingal Soldier, whom the daughter did not much mislike: which soldier was verie great in the house, and ordinarilie came thether to meat and drinke, and Frauncis trusted him [as well as if he had bene] his brother, in so much that he would doe nothing without his counsell. This Soldier called

Anthonio [sic]* Fragoso continued this beastlie course with Frauncis his wyfe, with the helpe of her mother, all the while they dwelt with the mother: and it is sayd, that he used her company before shee was maried, although shee was but young, which is no wonder in India: for it is their common custome in those countries to doe it, when they are but eight yeres old, and have the flight[32] to hide it so well, but that when they are maried, their husbands take them for very good maides. This [order of life] they continued [in that sorte] for the space of foure yeares, and also after that they had taken a house, and dwelled alone by themselves (for Antonio Fragoso kept his old haunt) and although Francis [sic] used continually to shut his chamber dore,[33] yet was this Portingall oftentimes hidden therein, he not knowing thereof, where hee tooke his pleasure of his wife.

At the last, one Diricke Gerritson of Enchuson[34] in Holland, being Godfather unto Francis [sic]* Kinges wife, comming newly from China, desired the said Francis [sic]* and his wife to come and dine with him at his house without the towne, and among the rest had mee, [minding to bee merrie and made us good cheare]: but because the honest Damsell Francis Kinges wife, made her excuse, that she might not with her credit come where Batchelers were, for that they had no such use in India, he desired mee to hold him excused till another time. They being there at this feast, with the mother in law, and her sonne, & their houshold of slaves that waited uppon them, as the maner of India is. After dinner was ended, and they well in drinke, they went to walke in the fieldes, where not far from thence there stood a house of pleasure, that had neyther dore nor window, but almost fallen downe for want of reparations, having on the backe side thereof a faire garden full of Indian trees and fruites: the house and garden Francis Kinges father in law had bought in his life time for a small peece of mony, for as I say, it was not much worth: thether they went, and caused their pots and their pans with meat and drinke to be brought with them, being minded all that day to make merrie therein, as indeede they did. In the meane time it was my fortune with a friend of mine [to walke] in the fieldes, and to passe by the house, wherein they were, not thinking any company had bin there, & going by, Francis King being all drunken, came forth and saw me, wherewith he ranne and caught me by the cloke, & perforce would have me in, & made me leave my companion, & so brought me into the garden, where their wives and his mother in law, with their slaves sate, [playing upon certaine] Indian instruments, being verie merrie: but I was no sooner espied by them, but the young woman presently went away to hide herselfe for her credits sake, according to their manner, as their manner is when any stranger commeth into the house.

Not long after supper was made ready of such as they had brought with them, although the day was not so far spent, and the table cloth was laid uppon a matte lying on he ground: for that (as I said before) there was neyther table, bench, window, nor doore within the house. The meate being brought in, every man sate downe, only Francis Kinges wife excused her selfe, that I had shamed her, and desired that she might not come in, saying for the time shee would eate there with the slaves, and although her husband would gladly have had her come in among us, thereby to shew that he was not iealous of her, [yet shee would not, so that] seeing her excuse he let her rest, saying shee is ashamed. While we sate at supper where the slaves served us, going and comming to and fro, [and bringing such thinges as we wanted] out of the place where this honest women was, her husband thinking shee had taken pains to make it ready, it was nothing so, for that while we were merry together, not thinking any hurt, in came Antonio Fragoso with a naked Rapier under his cloake; it being yet daylight, and in presence of all slaves, both theirs and mine,[35] without anie feare of us led her away by the hand into one of the chambers of the house, having neyther doore, flore, nor window, and there putting off the cloth that she had about her middle, (which he laid uppon the ground to keepe her from fowling of her body) not being once ashamed before the slaves, neyther fearing any danger, he tooke his pleasure of her: but if any mischance had happened, that any of the slaves had marked it and bewraied it, the said Anthonie had tenne or twelve souldiers his companions and friendes not farre from hence, which with a whistle or any other token would have slaine us all, and taken the women with him, which is their dayly proffit in India, but we had better fortune: for that hee dispatched his affaires so well with her, that wee knew it not, and had leysure to depart as he came, without any trouble, and she well pleased there with: and when the slaves asked her how shee durst bee so bold to doe such a thing, considering what danger of life shee then was in, shee answered them that shee cared not for her life, so shee might have her pleasure, and saying that her husband was but a drunkard, and not worthie of her, and that she had used the company of that fine lustie youth for the space of foure yeres together, and for his sake she said shee would not refuse to die: yet had she not then béene married to Francis King full foure yeares, neyther was shée at that time above fifteene or sixteene yeares of age. Not long after shee had done, shee came into the garden, and as it should seeme, had cleane forgotten her former shame, when she began both to sing and dance, shewing herselfe very merry: where with all the companie was [very well] pleased, specially her husband [that commended her for it]. When evening was come, every man tooke his leave, and departed to his lodging, and [when wee were gone] wee chanced by our slaves to understand the truth of the fact [before rehearsed], and what danger we had escaped: where at we wondred much, and Francis King himselfe began to be somewhat suspicious of the matter, being secretlie advertised of his wives behavior, but hee knew not with whome she had to doe, not once mistrusted this Anthonie Fragoso, thinking him to be the best friend hee had in all the world: yea, and that more is, hee durst not breake his minde to any but onely unto him, of whome in great secret he asked counsell, saying, that he understood, and had well found that his wife behaved her selfe dishonestlie, asking him what he were best to doe, and told him further, that he meant to dissemble the matter for a time, to see if hee could take them together, thereby to kill them both, which the other counselled him to doe, promising him his help and furtherance, and to bee secret therein, and so they departed.

Anthonio Fragoso went presently unto his [King's]* wife, and shewed her what had past betweene her husband and him, where they concluded uppon that which after they brought to passe, thinking it the best course to prevent him. Now so it hapned, that in an evening in the month of August, 1588, Francis King had provided a rosted Pig for supper in his own house: where unto he invited this

Anthonio Fragoso, and his [King's]★ mother in law, who as it seemeth, was of counsell with them in this conspiracie, and the principall cause of the Tragedy, although boldly shee denied it afterwards. They being at supper and very merrie, at the same time it was my chance to suppe in a certaine place with a Dutch painter, whether Francis King sent us a quarter of the Pigge, praying us to eate it for his sake, [and to be merrie:] he that brought it being one of his owne house. They had caused him to drinke of a certaine wine that was mingled with the Hearbe Deutroa,[36] thereby to bereave poore Francis of his wittes, and so to effect their accursed device: for as it appeared, hee that brought the Pigge came halfe drunke, and out of his wittes, whereby we perceaved that all was not well.

To conclude, the Hearbe beganne to worke, so that of force hee [King]★ must needs sleep, and the companie beeing departed, shee[37] shutte his trappe doore, as ordinarily he used to doe, and laid the key under his pillow, and went to Bedde with his loving wife: where presently hee fell on sleepe like a dead man, partly by meanes of the Deutroa, and partly because hee had drunke well. About eleaven of the Clocke in the night, Anthonie Fragoso all armed, and another good friend of his, not knowing (as hee confessed) what Anthonie meants to doe, came to the doore of Francis Kinges house, and knocked softly, and willed the slaves that slept below to open the dore: but they answered him their master was a bedde, and that the trappe doore was shut on the in side. Francis his wife that slept not, when she heard it, ran to the window and willed him to bring a ladder and clime up, which he presently did, and she holp him in, where shee took him about the necke kissing him, and bad him welcome, leading him in by the hand, where her husband slept, little thinking on the villanie pretended by his wife, and such as he held to bee his best friend, and to be briefe shee said unto him: There lieth the drunkard and the Heriticke, that thought to bring us to our endes, thereby to seperate us from our love and pleasures, now revenge your selfe on him if you love me, and presently hee thrust him into the body with his Rapier, cleane through the breast, so that it came out behind at his backe, and [being not content ther with,] gave him another thrust, that went in at the one side, and out at the other side, and so 4. or 5. thrusts more after he was dead: whereby the innocent man ended his dayes: which done they took all the stones & Diamonds that hee had of divers men to worke, as also to sell: which amounted at the least to the value of 40. Thousand Pardawes,[38] and tooke Francis his own Rapier & put it into his hand, as if they would make men beleeve that hee would have killed them, & that in their owne defence they had slaine him, but it as well knowne to the contrarie: for that the slaves being below, heard all that past. They tooke with them also the childe, being of two yeares old, and went out of the house, but they had not gone farre, but they left the childe lying at a doore in the street, where in the morning it was found, & althogh the slaves made a great noise at their mistrisses flight, & went to fetch the officers, yet they could not find them, for that night the murtherers went & knocked at ye Iesuits cloister, desiring them to take them in, & gave them the most part of the stones, saying they had slain the man in their owene defence, but the Iesuites would not receive them, although they tooke the stones, of purpose to give them againe to the owners.

In the morning it was knowne through all the towne not without great admiration,[39] and although they sought diligently in all places [where they thought or suspected them to lodge], yet could they not find them: but long after they were seened in the towene of Chaul, which is about thirty miles Northwarde from Goa, where they walked freely in the streetes, without any trouble: for there all was covered, and few there are that look after such matters, [though they bee as cleare as the Sunne].

The dead bodie lay in that sort till the next morning, & we Dutch men were forced to see him buried, for the mother in lawe woulde not give one peny towards it, making as though she had not any thing to doe with him, but holpe the murtherer both with money and victuailes, therewith to travell unto Portingal, and so he sayled in the fleete with us: for I saw him in the Island of S. Helena, as bold and lusty[40] as if no such matter had beene committed by him, and so arived in Portingall, not any man speaking against him: having also promised both the mother, and wife of Francis King, that hee woulde come againe with the Kinges pardon, and marrie her: which I doubt not of, hee once went about it.

And thus Francis King ended his travel: which I thought good to set downe at large, that thereby you may perceyve the boldness and inclination of the Indian women: for there passeth not one yeare, but that in India there are [at the least] twentie or thirtie men poysoned, and murthered by their wives, onely to accomplish their filthie desires. Likewise there are yearely many women killed by their husbands being taken in adulterie, but they care not a haire for it, saying with great boldness, that there is no pleasanter death then to die in that manner: for thereby (they say) they do shew that they die for pure love. And to shew that this honest women was not of this badde inclination alone, you shall understande that a brother of hers, being but fourteene or fifteene yeares of age, was openely burnt in Goa for sodome or buggery, which was done when Francis King and his father in lawe were living: yet could not Francis thereby bee warned to take heede of his wife and that kindred: for as it seemeth [it was Gods will][41] he should end his dayes in that manner. (204–15)

Notes

1. The capital of Portuguese India. Goa is located on the west coast of India.
2. Mestico should be Mestiço. The offspring of a mixed union between, in this case, a Portuguese man and a native Indian woman.
3. Casticos should be Castiços.
4. Tiele and Burnell, the editors of Hakluyt Society edition (1885), routinely insert language in square brackets to bring the English translation of the *Itinerario* closer to the Dutch original. I have preserved some, but not all, of these insertions.
5. A province in north-east India.
6. A town (now named Bago) in south-east Myanmar (formerly Burma).
7. A town on the Malay peninsula in south east Asia.
8. An Indian seaport near the mouth of the river Indus.
9. A meeting place of merchants for the transaction of business; an exchange (Oxford English Dictionary [OED]).
10. Antwerp, in today's Belgium.
11. Leylon. The word is adapted from the Portuguese "leilào," which means "auction."
12. Outroop. Probably derived from the Dutch word "uitroep," meaning "to call out." The OED lists the English translation of van Linschoten's *Itinerario* as the earliest usage of the word.

13. Sind (?): a region in the region of present-day Pakistan. Possibly it is the river Sinde, better known as the Indus (see van Linschoten, chapter 7).
14. Besalos manos: a flourish of the hand (in Spanish, *Beso las manos* means, I kiss the [your] hands).
15. Language in square brackets accompanied by an asterix signals an addition by the present editor, generally for purposes of clarifying van Linschoten's prose.
16 Spiritualitie: members of the clergy.
17. Sea rovers: pirates (possibly derived from the Dutch "rover," meaning "robber").
18. The capital of Portuguese India.
19. Frans Coningh (Dutch).
20. In the Dutch original, he is a diamond polisher ("diamant slyper").
21. Syria.
22 Caffila: from the Arabic kãfilah, meaning a group of travelers or caravan.
23. Bassora: Basrah (in present-day Iraque).
24 Sinus Persicus: the Persian Gulf.
25 Ormus: the Straight of Hormuz, which connects the Persian Gulf to the Gulf of Oman.
26. Dutchmen: Germans ("Duytschen" is the Dutch original).
27. Probably "du Chesne" (Tiele).
28. Dieppe, a town in Belgium.
29 Ballagate: in Persian "bãlã" means "above"; "gate" refers to the Ghãts mountain range (Terpstra). The woman hails from high up in the mountains.
30. A Moor.
31. Read "she took" (Tiele).
32. flight: imagination (but the OED does not list this meaning for the word before 1668).
33. The Dutch original states that King always locked the door with a "mael-slot" (a pad lock) before leaving the house.
34. Enchuson: Enkhuizen, a Dutch fishing town on the then Zuiderzee.
35. A poor English translation appears to suggest that van Linschoten owned slaves. But the Original Dutch clearly states that Fragoso entered the house "in presentie van alle slaven en slavinnen, so van Dirck Gerritsz. als the myne, ende van haer eyghen man. . . . "The Dutch "myne" (my) refers to "presentie" (presence) and not to "slaven en slavinnen" (male and female slaves).
36. Datura (Tiele): a herb that supposedly causes those who take it to sleep for a period of 24 hours.
37. Read "he" (Tiele).
38. Pardoa: A coin circulating in Goa, worth at the end of the sixteenth century 4s. 6d. (OED).
39. "astonishment" (verwonderinghe) closer to the Dutch original.
40. "presumptuous" (Tiele) or "arrogant" are closer to the Dutch original. The Dutch word, derived from the French présomption is "presumptioen."
41. Closer to the Dutch original is "as his misfortune and planet had ordained" ("dat zijn ongheluck ende planeet het also gheordineert hadde om alsoo to voleijnden").

COLONIZING THE COLONIZER: A DUTCHMAN IN *ASIA PORTUGUESA*[1]

Ivo Kamps

> Beeing young, [and living idlelye] in my native Countrie, sometimes applying my selfe to the reading of Histories, and straunge adventures, wherein I tooke no small delight, I found my minde to much addicted to see and travaile into strange Countries, thereby to seeke some adventure, that in the end to satisfie my selfe.
>
> —*Itinerario, Voyage ofte Shipvaert* (1596, English translation 1598)

So read the opening lines of the travel narrative of one of the most notable of early modern Dutch travelers, Jan Huyghen van Linschoten. Born circa 1563 as the son of an innkeeper and his wife, and enthralled by the childhood "reading of Histories, and straunge adventures" (1),[2] van Linschoten left his hometown of Enkhuizen in 1579 and embarked on a voyage of almost 13 years that took him, among other places, to Portugal, India, Terceira Island, and many places in between. Van Linschoten may have set out merely to feed his addiction "to see . . . into strange Countries," but the impact of his travel writings on the course of Dutch exploration makes it clear that he satisfied the curiosity and economic appetites of many of his countrymen as well. Because, at a time when Dutch exploration to the East was still in its infancy, he drew his countrymen's attention to the Indonesian isle of Java, which eventually became a crucial part of an expansive Dutch colonial empire, he is credited as one of the "causes of the rapid rise of the Dutch power in the Indies."[3] Charles McKew Parr, who lauds him as the Dutch Marco Polo, shows that van Linschoten's *Reysgeschrift van de Navigatiën der Portugaloysers in Oriënten* (*Pilot's Guide*), a text that describes in detail various sea routes to the East (and that became part of the *Itinerario*), "had a direct effect upon the discoveries and conquests made by the Dutch, English and French navigators in the Far East."[4] What is more, van Linschoten told his readers that the Portuguese were no longer the unassailable power everyone thought them to be. He described the poor discipline, horrendous conditions, and inexperience aboard Portuguese vessels. He described the dilapidated and undermanned state of Portuguese fortresses in the East, and he drew attention to the corruption and decadence that existed on all levels of government in Goa, the Portuguese capital of India.[5] Such information surely contributed to the Dutch decision to compete with the Portuguese in the East, and hence added greatly to the reduction of Portuguese influence in that part of the world. That the Dutch commercial efforts in the East against the Portuguese were successful beyond anyone's wildest expectations is a matter of historical record.[6]

The Dutch States General immediately understood the value of van Linschoten's knowledge and personal experience, and in 1596 the assembly rewarded him with a license to publish the *Itinerario* .[7] The public dissemination of the *Itinerario* must have been a severe blow to the highly secretive Portuguese whose "trading advantage," Karel Steenbrink observes, had been "due to the fact that their knowledge of this area was much more extensive than other Europeans.'"[8] But if van Linschoten's role as an early contributor to what grew into a massive Dutch colonial project has been understood largely in terms of the maps with sea routes and the location of fortresses and ports, I want to examine how van Linschoten's close treatment of Portuguese colonial order, including its approach toward the other, women, non-Christian religions, trade, slavery, prostitution, widow burning, and so forth, gave the Dutch important inside knowledge helpful in their efforts to displace the Portuguese in India and elsewhere. While certainly van Linschoten's maps and charts were concretely related to the emergence of Dutch colonialism, a more nebulous but perhaps equally vital aspect of the *Itinerario* is that it employs a double gaze to scrutinize *both* the colonizer and the colonized, both the Portuguese and the Indian.[9] One half of this gaze, the gaze directed by the colonizer at the colonized—in this case by the Portuguese or van Linschoten at the Indian natives—has become familiar to us from recent postcolonial criticism. The other half of van Linschoten's gaze, however, is possibly unique for it colonizes the Portuguese colonizing power itself. This second "act" of colonization, which is an act of writing first, and which is deeply embedded in the complexities of the late sixteenth-century European socio-political landscape, scrutinizes and defines the Portuguese in ways that the writings of colonizing powers routinely objectify the natives for the purpose of their domination. The *Itinerario,* therefore, offers us a literary act of colonization of the Portuguese colonizer perpetrated by the Dutch colonizer-to-be. Whether or not this literary act of colonization was directly translated into the policies and practices of the Dutch may well be impossible to ascertain—there are no extant statements by members of the Dutch States General or the Dutch United East Indies Company (VOC) asserting that van Linschoten's book shaped their social and economic policies in the East (though we can be assured that many of them read it). That the *Itinerario* and Dutch attitudes and practices in the early years of the seventeenth century often mirror each other is undeniable.

These mirrorings offer us a myriad of shared observations, ranging from views on miscegenation and reproduction to religion and commerce, but if there is any one overarching similarity that shapes both van Linschoten's perspective and the Dutch outlook, it is that the colonizer should always maintain a safe *distance* from the colonized people. The Portuguese, van Linschoten insists, had become too entangled in Indian life and culture to conduct an efficient, profitable, and dignified India policy.[10] Endless sexual, social, and cultural encounters with the native population had degraded the Portuguese (and to some extend also the Indians) so much that the very *difference* between them could be called into question. In large part, the intermingling between the Portuguese and Indian natives was a result of the former's political goals. The Portuguese emphasized government involvement and control. "The Portuguese sent nobles and government officials who were to provide administrative expertise for the governing of a territorial rather than economic empire."[11] In many instances, this meant exerting Portuguese influence over Indian ways of life, making exchanges between the two cultures a necessity.

The Dutch, on the other hand, were primarily motivated by a desire for profit,[12] and deliberately avoided exerting ideological, religious, and cultural influence. Instead of government officials, they "sent Merchants to Asia."[13] The Dutch East India Company, which was governed by the Heren XVII—an "all-powerful Board of Directors"[14]—was granted "vast commercial, military and political powers"[15] and operated with comparatively little oversight or intervention from the Dutch government. The Heren XVII insisted that the acquisition of territory was "a last resort to protect . . . economic interests; it was not in fulfillment of a Company or government goal."[16] In 1604, for instance, as part of a treaty with the Indian ruler of Calicut, the Dutch were offered a fortress that "would remain in the hands and under the dominion of the Dutch for all time."[17] However, the VOC, not interested in territorial expansion, refused the fortress.[18]

Van Linschoten's Double Gaze

The *Itinerario*'s act of double colonization (the colonization of the Indians as well as the Portuguese) is made possible by the double gaze that marks the text. The double gaze, in turn, is spawned by van Linschoten's peculiar position as a Dutch Catholic who grew up in a predominantly Protestant town and who found employ with the Portuguese and, indirectly, with the archenemy of the Dutch, the Spanish. First the gaze. In general, colonial discourse centers on a binary relationship between the colonizer and the colonized (what Abdul JanMohamed calls a "Manichean allegory")[19] that is characterized by a presumably unidirectional gaze. As E. Ann Kaplan has noted, the colonist's "gaze is active: the subject bearing the gaze is not interested in the object per se, but consumed with his (*sic*) own anxieties, which are inevitably intermixed with desire. . . . The sight produces anxiety which closes off process [reciprocity] at once. The object is a threat to the subject's autonomy and security and thus must be placed, rationalized and, by circuitous route, denied."[20] That agency of the gaze is unidirectional, is predicated on the assumption that the gazer *identifies* wholeheartedly with the colonizing power. In van Linschoten's text, this identification is not complete enough for the gazer and the colonizing power to be one. As a result, the author of *Itinerario* gazes as much at the Portuguese as he does on the Indians, and in the process the *power* of the gaze—the power to colonize—is *partially* transferred from the colonizer to the would-be colonizer, the Dutchman.[21]

Through the deployment of such concepts as "mimicry," "ambivalence," and "hybridity," Homi Bhabha has recently challenged the Manichean view of colonial discourse with considerable success. Bhabha shows, for instance, how, in an effort to control and civilize the other, the colonizer encourages the natives to mimic the colonizer. Such mimicry, however, easily turns into subtle mockery or parody of the colonizer and thus subjects the colonizer to a partial reversal of the gaze.[22] Bhabha's approach is especially powerful in revealing the reciprocal (yet unequal) exchanges that take place in all encounters between colonizer and colonized. Van Linschoten's text, too, achieves insights into this reciprocity, but it does so by different means. If Bhabha's concept of "mimicry" signals a partial reversal of the colonial gaze, we could say that van Linschoten redirects or doubles the gaze, not by allowing the Indians to see the Portuguese but by allowing his stereotypical and often harsh assessments of the natives *to reflect back onto* the Portuguese. As we see the natives, so we see the colonizers.

By all appearances, Jan Huyghen van Linschoten shared the colonialist ideologies of the Portuguese. Never does he express any reservations about the domination and exploitation of the people of India. Nowhere does he indicate that he thinks that Indians stand on equal footing with Europeans. He deems virtually all Indian religious practices to be superstitions, and he cannot get himself to judge those Indians who convert to Christianity as the equals of the colonizers. He portrays Indian women who are married to Portuguese men as dangerously treacherous and lecherous creatures whom the colonizer may call his own but whose appetites are beyond Portuguese control. And yet, van Linschoten's Dutch roots, coupled with a veneration of the Spanish and a certain contempt for the Portuguese, prevented him from identifying *zealously* with the Portuguese enterprise. Though the *Itinerario* does not paint its author as a person of unusual psychological complexity, it is clear that van Linschoten's background and travel choices make it difficult to fit him into any simple, coherent ideological mold. He was raised a Catholic in the overwhelmingly Protestant town of Enkhuizen, and identified himself as late as 1596 as a citizen of Haarlem,[23] a staunchly Catholic town in which he had been born but where he had not lived since early childhood. Although there is every indication that he was a patriotic Dutchman, it can be seen as a sign of his independence that, well into the Eighty Year War between the Dutch and the Spanish which had erupted in 1568 with William of Orange's revolt against Philip II, he left for Spain to learn Spanish and to school himself in international business practices. One way to look at this seemingly controversial expedition is to suggest that van Linschoten, a Catholic in a country of which large portions had been governed directly by or by fiat of Habsburg Spain (and to the satisfaction of many Dutchmen) since the second half of the fifteenth century, may not have perceived any conflict between his veneration for the Spanish and his loyalty to his Dutch neighbors. With incessant territorial shifts in the struggle with Spain, and the lack of both a central government and a single, unifying religion, it is clear that the Dutch Republic was fragile at best, and that any sense of nation was only just beginning to emerge. That said, by the late 1570s, many of van Linschoten's countrymen, especially the Protestants, who increasingly were defining their Dutchness in terms of their religion and their opposition to Spanish power, would beg to differ. He nonetheless arrived in Spain in 1579 and spent eight months in Seville, where his brother already resided. Eventually, in 1580, he made it into Portugal, shortly after Philip II took the crown of that nation by force. Parr imagines that van Linschoten may have watched King Philip review his troops "on the meadows of Calatraveja before giving order to plunge across the [Spanish-Portuguese] border."[24] Van Linschoten, he suggests, "must have been stirred to see the old Duke of Alva, so hated in Enkhuizen, as he led a force of twenty-six thousand" into Portugal. That van Linschoten's text treats Alva with a measure of respect is nothing less than remarkable (as a junior high school student in the Netherlands in the 1960s, I was still taught that Alva was evil incarnate) because in the Low Countries Alva's name had become "a byword for atrocity and cruelty."[25] His crimes against Dutch Protestants were enormous, among them the "killing [of] every man, woman, and child" in the town of Naarden, which served as a punishment for the town's slow surrender to Spanish troops.

Parr is right to note that van Linschoten apparently (but incorrectly) believed Philip's take-over of Portugal to be a legal matter.[26] Van Linschoten tells us that the last Portuguese King, Cardinal Dom Henrique, designated Philip of Spain as his heir in his last will and testament (3). But there is no indication in the *Itinerario* to support Parr's assertion that van Linschoten thought of Philip of Spain, as some Dutch Catholics did, as *Graaf* of Holland, "his hereditary liege lord, one whose principal ancestral castle was at Haarlem."[27] Certainly, if van Linschoten's words of admiration for the Spanish royal family should be understood in an unreservedly pro-Spanish context, they would have been excised by the Dutch printer and certainly would not have been tolerated by the States General, to whom van Linschoten turned for approval and to whom he dedicated the *Itinerario.* Parr is aware of this problem, but cannot explain why the passages were left in the final version of the text. It seems reasonable to conclude that while van Linschoten's Spanish sympathies are plain, they fall far short of a failure to acknowledge that after the departure of the governor-general, Robert Dudley, earl of Leicester, the United Provinces had, for all practical purposes, turned its back on the concept of foreign control over its affairs and sought to become an autonomous republic. By 1596, the year of the *Itinerario*'s publication, the Dutch Republic's economy was making great strides and, under the skillful leadership of Prince Maurits of Nassau, the Dutch military had taken control over territories in both the north and the east, driving the Spanish back on all fronts. However, the republic was yet far from stable, as there were still a number of important cities and territories that looked on Spain with far too friendly eyes.[28] Hence it is unthinkable that the States General, an assembly that was controlled by staunchly Protestant, anti-Spanish forces and dominated by the political engineer of Dutch independence, *Raadspensionaris* (chief magistrate) Oldenbarnevelt, would have granted van Linschoten a license to publish a text detrimental to its cause. Indeed, the publication of the *Itinerario* hastened the decline of the Portuguese and therefore diminished Spanish influence and profits as well. Van Linschoten was keenly aware of Spanish authority over *Asia Portuguesa,* as he himself tells us that his Portuguese employer there, João Vincente da Fonseca, was persuaded to become the "Archbishopricke of all the Indies" by none other than Philip II himself (8-9).[29]

All this is to show that van Linschoten clearly occupied a subject position that is not easily reducible to an official ideology or nationalist agenda. Though Dutch by birth, van Linschoten was interpellated—to use Althusser's term—by a set of discourses and ideologies that exceed any narrow sense of late sixteenth-century Dutchness. The practical consequence of this is that van Linschoten himself always functions as "other" or "outsider" in his own narrative, as belonging neither to the Dutch nor the Spanish, nor the Portuguese. And since he obviously did not identify with the inhabitants of India, he occupies a subject position outside the familiar, binary colonizer/colonized dynamic.[30]

John Wolfe, the Englishman who published the *Itinerario* in England in 1598, offers in his preface a metaphor that aptly captures the detached perspective that van Linschoten is trying to achieve in his text. Wolfe draws on a passage from Lucian's *Dialogues* to illustrate the aeronautical, omniscient perspective that van Linschoten's text supposedly offers. Indeed, Wolfe's example prompts the familiar image of the early modern ethnographers, "whose veracity as eye-witnesses . . . privileges their act of *seeing* as a mode of unmediated access to experience, while implicitly ignoring the natives' own ability to *see*."[31] Underworld figures Charon (a boatman, like van Linschoten) and Mercury (like van Linschoten a messenger), Wolfe tells the reader, receive permission from Pluto to satisfy their desire to view the world "and the *Actions* of men therein."[32] The pair leave Hades and proceed to stack several mountains on top of each other—Parnassus on top of Aetna, both on top of Ossa, and all three on top of Pelius—in order to achieve the desired aerial panorama on the world. Here, from the top of the world, "hauing setled themselves, they did at leysure and pleasure take a view not onely of the *Seas,* and *Mountaines,* and Cities of the world, but also of the Inhabitants thereof, together with their Speeches, Actions and Manners." In just this metaphorical fashion, Wolfe submits, van Linschoten observes otherness from a safe distance. "Speech, actions, and manners" are grasped without a need to be among those who speak and act. The observer obviously takes pleasure in seeing the other, but prefers to remain aloof. Despite the "altitude" of this metaphor, the gazer remains connected to the earth and, though far away, is not out of reach. But Wolfe is plainly not satisfied with this metaphor of aloofness, for he goes on to describe how in a different *Dialogue* Menippus borrows wings—one from an eagle and one from a vulture—and, "hauing fastened them to his body," soars above mount Olympus, and then upward past the moon, the sun, and "the Habitation of Iupiter and the rest of the Gods," only there to rest and discover "all the world and euery particularity thereof, to the end he might more freely and like a *Scoggan* taunt and scoffe at the Actions of men in their seuerall kinds."[33]

Van Linschoten is of course neither a winged observer nor the Styx's ferryman, but to associate the divine heights reached by his classical counterparts with his vantage point grants the Dutchman a godlike relationship to the other. To be sure, Wolfe does not elevate van Linschoten over the Christian God, but this does not seem to diminish his powers of panoptic sight. In fact, the classical stories resorted to by Wolfe seem a safe, nonblasphemous way of attributing superior, godlike powers to the colonizer's gaze. Van Linschoten's text aspires to live up to these metaphors by the sheer breath of descriptions, but also through its author's obvious restraint when it comes to editorializing or moralizing. Clearly, for van Linschoten, a large part of showing himself superior to the other is to describe the other in cool-headed, (seemingly) nonjudgmental, and, at times, even (seemingly) sympathetic fashion. Hand in hand with this sense of superiority in Wolfe's metaphors goes a sense of distance, of a vast space, between the colonizer and colonized. This distance, we will see, is absolutely crucial to van Linschoten's implied vision of efficient and effective colonization. Indeed, he holds that a loss of distance between colonizer and colonized erodes the colonizer's superiority, and that the rapprochement between the Portuguese and the Indians is a fundamental reason for the Portuguese's decline.

Miscegenation

The *Itinerario* considers marriage to be a site of clear and present danger to the colonizer, for it obliterates the distance between the colonizer and the colonized, and greatly reduces the difference between them. The problem, as one might suspect, lies with the native and *mestiço* women—with both their desirability and their treachery. The "desire and derision"[34] that typify and structure the colonist's encounter with the other in general is brought into sharp focus in van Linschoten's descriptions of these women. In great detail, he depicts how, when the women go to church or on a visit to a friend, they "put on costly apparell, with bracelets of gold, and rings upon their armes. . . . [Their] clothes are Damaske, Velvet, and cloth of gold, for silke is the worst [thing] they doe ware" (206). Within the house, van Linschoten voyeuristically continues, they

> goe bare headed with a wastcoate called Baju, that from their shoulders covereth [their] navels, and is so fine that you may see al their body through it, and downewardes they have nothing but a painted cloth wrapped thrée or foure times about [their] bodies. These clothes are very faire, . . . all the rest of the body is naked without any hose. (206)

The Dutchman peers through the cloths of women and penetrates the walls of their houses to share with his readers an erotically charged gaze at the native female body. But derision for these same, desirable women manifests itself virtually every time they are mentioned in the *Itinerario.* A single but typical passage will suffice here.

> Manie and most [of the Portuguese soldiers in India] have their chiefe maintenance from the Portingales and Mesticos wives, as also the Indian Christians [wives], which doe alwaies bestow liberall rewardes and giftes [uppon them] to satisfie [and fulfill] their unchaste and filthy desires, which they know very well how to accomplish, and secretly bring to passe. (201)

Note that not even conversion to Christianity can deliver the native women from a life of sin. The irony, of course, is that van Linschoten also goes to great length to convince us that the wives are utterly inaccessible to men other than their Portuguese husbands. "The Portingales, Mesticos, and Indian Christian women in India, are little [séene abroad], but for the most part sit still *within* the house, and goe but *seldome* forth, unless it be to church, or to visit their friends, which is likewise but verie little, and when they goe abroad, they are well provided *not* to be seene, for they are carried in a Pallakim covered with a mat or [other] cloth, so that *they cannot be seene*" (205; emphasis added). The jealousy of Portuguese husbands is so extreme that

> they will never bring any man into their houses, how speciall a friend [soever] hee bee, that shall sée their wives and their daughters, unlesse it bee some gossip or any other married man with his wife in companie. When [the women; mine] will goe together to some place . . . they are alwaies well guarded by their slaves. . . . If any man commeth to the doore to aske for the master [of the house], presently the wives and their daughters run to hide them, and so leave the man to answer him that standeth at the dore: likewise they suffer no man to dwell within their houses, where the women and daughters bee, howe néere kinsman [soever] he be unto them, being once 15. Yeares of age, not their owne sons. . . . (208-9)

These anxious measures, however, van Linschoten remarks, are far from adequate in controlling these women, for they "are very luxurious and unchaste" (209) and resort to stealth and stratagem to fulfill their desires. Nearly all wives have "one or two of those that are called soldiers, with whom they take their pleasures: which to effect, they use al the slights and practises [they can devise, by] sending out their slaves and baudes by night, and at extraordinary times, over walles, hedges, and ditches, how narrowlie [soever] they are kept [and looked unto]." To complete this picture of near-total sexual anarchy, van Linschoten observes that not even the fear of death can deter the appetites of the women, for "so they are sacrificed for love, which they thinke to be a great honour" (212). They hold that "there is no pleasanter death" than to die for (adulterous) love (vol. 2, 215).

In their dealings with their Indian wives, Portuguese husbands seek to achieve a Foucauldian panopticism, but if we are to credit van Linschoten's assertions of the wives' astonishingly resourceful and relentless sexual transgressiveness, they obviously fail miserably. The husbands' panoptic gaze, its intense surveillance, does not discipline them in the least. If anything, it spurs the women to greater infidelity. Van Linschoten himself, on the other hand, succeeds absolutely as a panopticist: he sees the unseeable. He sees what goes on inside the houses—from which he is by his own admissions barred. He knows the stratagems of lustful Indian women, he knows how they use their slaves and bawds to arrange sexual liaisons. He knows it's done by night, and he knows how they climb over hedges and ditches and walls; he knows how they drug their husbands with *dhattûra* (210), even though he is of course not *actually* there to see any of it. To him, the women are as transparent as the thin, gauzy garments they wear. By seeing all without being seen himself, van Linschoten is like Charon and Mercury. Van Linschoten's panopticism does not discipline the Indian and *mestiço* women, but his panoptic *writing* disciplines the Portuguese and their colonial ways by reporting to the world everything in sight. Soaring above the gods with a wing from an eagle and a wing from a vulture, the colonizing gaze "discouere[s] all the world and euery particularity thereof [and] taunt[s] and scoffe[s] at the *Actions* of men in their seuerall kinds."[35] Unlike the jealous Portuguese who unproductively try to discipline their Indian and *mestiço* wives, van Linschoten's panoptic gaze accomplishes, in the words of Foucault, "hierarchical surveillance, continuous registration, perpetual assessment and classification."[36] Portuguese husbands seek unsuccessfully to be masters of their homes through the objectification and containment of their wives, but van Linschoten, who generally portrays himself an ally and admirer of his Portuguese employers, uses his panoptic powers to objectify the Portuguese "insidiously . . . to form a body of knowledge about these individuals, rather than to deploy the ostentatious signs of sovereignty."[37]

The native Indian and *mestiço* women, furthermore, even though they are objectified and disparaged in van Linschoten's account, achieve a powerful form of sight: they are *seeing* other men. They are seeing men that are invisible to their anxious and controlling husbands. In other words, the women reverse the colonial gaze, turning it on their lovers—and by reflection also on their husbands, whose inability to master their wives' sexuality betrays an impotence that highlights (both metaphorically and literally) their weakening grip on India. The only eyes that see all—eyes that peer into houses and bedrooms and into the dark of night—belong to van Linschoten, who amply illustrates that the proper colonial gaze—the one that *controls* the other—is one that maintains distance and difference from the other. Close engagement of the other makes the colonizer visible to the other (and to whomever is watching the spectacle), and undermines the binary relationship with the other, resulting in a loss of control.[38]

Although van Linschoten's comments serve as a critique of Portuguese men, there is no doubt that Dutch men stand to suffer equally if they marry native women. In the cautionary excerpt of the *Itinerario* included in this volume, Dutchman Frans Coningh (Francis King) becomes the hapless murder victim of his native wife and her Portuguese lover, Antonio Fragoso.

Reproduction

Loss of distance and difference is, not surprisingly, most visible is the reproductive consequences—both biological and social—of miscegenation. Van Linschoten speaks at some length on the subject:

> The Portingales in India, are many of them marryed with the naturall borne women of the countrie, and the children procéeding of them called Mesticos, that is, half countrimen. These Mesticos are commonlie of yelowish colour, notwithstanding there are manie women among them, that are faire and well formed. The children of the Portingales, both boyes and gyrls [which are] borne in India, are called Casticos, and are in all things like [unto] the Portingales, onely somewhat differing on colour, for they draw towards a yealow colour: the children of those Casticos are yealow, and altogether [like the] Mesticos and the children of Mesticos are of colour and fashion like the naturall borne Countrimen of Decaniins of the countrie. So that the posteritie of the Portingales, both men and women being in the third degrée doe séeme to be naturall Indians, both on colour and fashion. (184)

What is remarkable about this passage is not merely that the Portuguese "race" is erased in a union with the racial other but also just how this erasure occurs. One strain in van Linschoten's account makes the familiar claim that when a Portuguese man engages in miscegenation, and that his children do the same, then all traits—racial and cultural—that identify the person as Portuguese will have disappeared from the grandchild or the great grandchild. The racial progression runs from Portuguese to *mestiço* to Indian. Van Linschoten's argument is, however, not *only* a biological argument. Notice that the *castiços,* the children born of two Portuguese parents in India, suffer a similar fate. *Castiço* children "are in all things like the Portingales," except that they *already differ in color* from their Portuguese parents and thus begin to resemble the Indians, *without* the need for a biological link. *Castiço* children resemble native Indians simply by virtue of being born in India. Now the role of biology is not entirely clear when van Linschoten observes that "the children of those Casticos are yealow, and altogether [like the] Mesticos." Are these children the offspring of *castiço-castiço* marriages or from unions between *castiços* and *mestiços* or *castiços* and Indians? The language of the passage clearly suggests that the union is between *castiços,* and if this is indeed what van Linschoten means then he is proposing a logic of racial degeneration without miscegenation. It runs as follows: from Portuguese to *castiço* to *mestiço* to Indian. No matter how ludicrous the reasoning may seem, it replicates the logic of the first stage of the argument regarding the incipient resemblance between *castiços* and Indians. What is more, this racist logic also fits the racial logic of the Portuguese, who themselves discriminated against their hybrid offspring. *Castiços* and *mestiços* were commonly raised by Indian slaves, adopting their ways, and, as M. N. Pearson points out, they "were regarded with considerable suspicion by the [Portuguese] elite."[39] "The Portuguese reserved for themselves the upper-level political positions, excluding as far as possible even Indian-born Portuguese, let alone *mestiços.*"[40] What seems foremost at issue here for the Portuguese is a form of cultural contamination. Van Linschoten's account, however, collapses the arguments about miscegenation and cultural contamination into a single story of racial degradation, and, ultimately, of racial erasure.

One assumes that van Linschoten disapproves of miscegenation, but there is nothing explicit in his account that confirms this. His tone is matter-of-fact and does not reveal any particular anxiety or regret about the erasure of the Portuguese from the reproductive process or about the racial triumph of the colonized over the colonizer. Without as much as a pause or a transition, van Linschoten's narrative abandons the question of racial biology and plunges into a discussion of commercial habits in and around Goa, the capital of Portuguese India. At first, matters of race and commerce seem decidedly unrelated, but as the narrative develops and takes up the issue of slavery, commerce becomes linked to social leveling and decline and, ultimately, to a regrettable loss of biological difference.

After detailing several noncontroversial aspects of daily markets and the goods that are offered for sale, van Linschoten relates to the reader that the "Gentlemen" and "marchants" that buy and sell the "Indian commodities" (184), "have running about them, may sorts of [captives and] slaves, both men and women, young and old, which are daylie sould there, as beasts are sold with us" (185). Van Linschoten's tone is casual and does not signal any disapproval of the practice until we get to the end of the quoted passage where he compares the Goan slave trade to a Dutch cattle auction.[41] The extent of van Linschoten's outrage is difficult to measure (it may be quite minimal), but he does assert a moral difference between "us" and "them," between the Dutch and those who treat human beings as the Dutch do their animals. Furthermore, he seems to suggest that it is the *mestiços* and their offspring who are engaged in the slave trade.[42] If this is indeed what van Linschoten is saying then he associates biological decline (*mestiços*) with a morally/socially degrading practice (*mestiços* engage in slavery, but the Dutch never would). When, however, we read beyond the passage that introduces the practice of slave-owning, it becomes clear that not only the *mestiços* but "some married Portingales" themselves—without the "excuse" of degenerative hybridity—"get their living by their slaves, both men and women, whereof some have 12, some 20, and some 30, for it costeth them little to kéepe them" (185-86). What is more, van Linschoten states that these kind of markets are "holden in all places in India, where the Portingales inhabit" (185). Van Linschoten again does not explicitly moralize the practice, but his association of the Portuguese with a practice that is first associated with their racial inferiors, the *mestiços,* and that is morally dubious (the Dutch would never do it), has to be taken as a condemnation, however circumspect, of his Portuguese employers.

Furthermore, van Linschoten goes on to point out that these Portuguese and *mestiço* slave owners hire their slaves to anyone who can afford them for any kind of work, including prostitution (186). Life in India thus turns the Portuguese into slave owners and pimps: miscegenation but also cultural and commercial forms of exchange with the other lead to the deterioration of the colonizing race. Significantly, when van Linschoten views native others in their own milieu, he often grants them a definite nobility and dignity but when they become integrally connected to the Portuguese they become a corrupting and ruinous force. This tendency is also apparent when van Linschoten translates his gaze into a series of visual representations, as he did in the drawings reproduced in the *Itinerario.* This tendency is also evident in the drawings that van Linschoten himself contributed to the *Itinerario.* The native Indians in Figure 3 are presented as powerful, self-reliant, and are placed in classical poses in full sunshine. The Portuguese nobleman in Figure 4, surrounded by slaves and servants, appears effeminate, in need of shade, and dependent on his entourage. The slaves and servants in Figure 4, even though some of them carry weapons, also in no way resemble the Cochin warriors in Figure 3. Contact with the Portuguese has diminished them.

The upshot of this daily cross-contamination is that the "Portingales and Mesticos in India never worke" (187). They "walke up and downe the stréetes, goe as proudlie as the best: for there one is no better than an other, *as they think,* the rich and the poore man all one, without any difference in their conversations, curtesies and companies" (188; emphasis added). The "as they think" interpolation makes it clear that van Linschoten begs to differ. The Dutchman never explicitly states that proper social hierarchy ought to be observed, but this

Figure 3. The king of the Cochins, seated on an elephant and accompanied by members of his nobility, which are called Nairos. From *Itinerario: Voyage of Schipvaert van Jan Huygen van Linschoten naer Oost ofte Portingaels . . .* (1596).

leveling among the Portuguese and their descendents in India bothers him enough to return to the matter in his next chapter. Here the leveling process is taken indoors, as we are told that in their homes "[t]he Portingales are commonly served [by their slaves] with great gravitie, without any difference betwéene the Gentleman and the common Citizen, [townsman] or soldier, and in their going, curtesies, and conversations, common in all things" (193). However, the apparent lack of difference among men hardly leads to simple equality among them; indeed it disguises a most snobbish arrogance that, in turn, hides intense insecurity about their position in Portuguese India: "when they go in the stréetes they steppe very [softly and] slowly forwards, with a great pride and vaineglorious maiestie." In the streets, when the Portingales encounter one another, they do each other great "reverence" (194)—"they thrust forth their foot to salute each other, with their hattes [in their hands], almost touching the ground"—but if the receiver of such an act of courtesy "doeth him not the like [curtesie]," they will "go after him, and cut his hatte in péeces." In the eyes of the Portuguese, such acts of revenge are not only acceptable, they are required (194). To forsake revenge is to assume "the greatest shame [and reproach] in the world." Indeed, the man who escapes merely with the loss of his hat is lucky, for van Linschoten observes that revenge is typically also exacted by gathering a dozen or so friends and beating the offender "so long [together], that they leave

Figure 4. Members of the Portuguese nobility and government move about the streets of Goa in this fashion. From *Itinerario* (1596).

him for dead, or very neare dead, or els cause him to be stabbed by their slaves, which they hold for a great honor and point of honestie so to revenge themselves, whereof they dare boast [and bragge] openly [in the stréetes]" (194-95). That, as far as van Linschoten is concerned, these and other practices are endemic to the Portuguese way of life in India is clear because, despite their common occurrence, they are "never looked unto or once corrected" (195). Without a doubt, the mere fact of *mentioning* that these practices are systematically ignored by the Portuguese government in India indicates that van Linschoten would expect—and would expect his reader to expect—such instances of organized brutality (that hardly resemble an acceptable way for a gentleman to react to a violation of his honor) to be investigated and punished.

Although the country from which van Linschoten hailed was less obsessed with rigid divisions of social classes than were most European nations, the Dutchman looks in the *Itinerario* with pronounced ambivalence upon the lax attitude of the Portuguese in this area. He appears critical of Portuguese social leveling because it shows the corrosive effects of life in India upon the very structure and hierarchy of Portuguese culture.[43] Van Linschoten does not say that this social leveling is bad per se but (one suspects) that such leveling conflicts sharply with his personal need for social order and also with the fundamental needs of the Portuguese to live out their fantasies of superiority (which are evident in their absurdly intense and violent response to the slightest affront to their honor).

Indeed, van Linschoten finds that many who come to India from Portugal quickly shed their respect for values that they are made to respect back in Portugal. Touching on the question of why the Portuguese empire in India is no longer expanding, van Linschoten points out that all who come to India—the underpaid soldiers in service of the king, but also "the Viceroy, Governours, and others" (203)—soon turn to providing for themselves ("scraping [and catching]") instead of seeking "the profit of the commonwealth." "[F]or that the King (say they) gave them their offices, [thereby] to pay them for their services." The parenthetical "say they" makes clear that the view is widespread, for sure, but hardly the one van Linschoten believes the colonizers ought to embrace. And, again, van Linschoten does not attribute this decay in order and discipline to a few bad apples in the colonial barrel. In large measure, he lays the corruption and decay at the feet of "the evill government of the Portingales" (204).[44] Just as sexual hybridity erases the Portuguese as a cultural presence, so a shared pursuit of economic gain brings about a social leveling that cheapens the authority of the Crown and brings down the empire from within. It seems that simply "being there" spells the end of cultural and biological domination.

Insatiable Lust

It is not only through marriage and procreation that Indian and *mestiço* women present a legitimate threat to the Portuguese. *Any* sexual encounter with a native woman is potentially dangerous, even deadly. Van Linschoten identifies native women as a primary cause for why so many Portuguese men are dying of the "bloody Flixe" (dysentery). The reason they catch dysentery is because they are undernourished and "use much company of women, because ye land is naturall to provoke them therunto . . . for although men were of iron or stéele, the unchaste [life] of [a] woman, with her unsatiable lustes were able to grinde him to powder, and swéep him away like dust" (236-37). The exotic "land" itself provokes the desire in Portuguese men, and the insatiable women drive the men to such extreme performance that sickness and death result. Whether it is out of personal revulsion or out of a desire to admonish the men to keep their distance from native women is not clear, but van Linschoten's association of sex and dysentery is disturbingly appropriate to his vision of colonization. The diarrhea that is symptomatic of dysentery causes dehydration of the body; it squeezes the life force out of it and reduces the body to dust. In other words, sexual union with a native body constitutes contamination and leads to the colonizer's body trying to expel that contamination and, with it, its own anima.[45] Just as Portuguese husbands are incapable of disciplining and controlling their Indian and *mestiço* wives' sexual appetites, so Portuguese men are impotent to resist the seductive wiles of India and its women because they cannot discipline their own bodies. Moreover, the malfunction of the colonizer's organs results in a type of leakage that signals the breakdown of the borders between the colonizer's body and the Indian environment. Or, as Bakhtin puts it in a different context, "The object transgresses its own confines, ceases to be itself. The limits between body and world are erased, leading to the fusion of the one with the other and with surrounding objects."[46] The colonizer who fails to discipline his body literally leaks out and into the Indian earth, and returns to dust. Significantly, van Linschoten observes that the Portuguese have only one, completely ineffective, way of treating dysentery, which is bloodletting. The Indians and heathens, on the other hand, "do cure [themselves] with hearbes, Sanders, and other such like oyntments" (236).

The country and its women, van Linschoten delineates, defeat the Portuguese colonizer. Again, although the initial thrust of this account seems to be an indictment of the women, van Linschoten's double gaze reflects what is wicked about the women onto Portuguese who are shown to be feeble and vulnerable. Without the ability to resist or cure themselves, the proud colonizers are helpless. The Bakhtinian echoes can be summarized by suggesting that in their sexual encounters with native women, the colonizers suffer "debasement, uncrowning, and destruction," and are returned to the earth.[47] However, the logical consequence of this uncrowning—the "crowning" of the women—is only the penultimate outcome of this type of encounter. Van Linschoten hardly speaks on the women's behalf. The victory crown, it seems, is reserved for van Linschoten's readers—the Dutch would-be colonizers who are given ample reason to understand that the successful colonizer should prefer aeronautical distance to close involvement.

One could argue that at least some Dutch officials took van Linschoten's tales about miscegenation and procreation to heart, though it appears that his advice concerning keeping foreign concubines was largely ignored.[48] In sharp contrast with the Portuguese government, the East Indies Company, established in 1602, actively discouraged intermarriage with native Indian women. Jan Pieterz. Coen, governor-general of Dutch settlements, urgently requested of the VOC a supply of Dutch women. Coen "pointed out that adult women were needed immediately for the sailors and soldiers who wanted to settle as free citizens in the Dutch East Indies."[49] Other officials agreed that this was the only way to establish "reliable settled Dutch colonies in the East."[50] Because not enough women were willing to journey to the colonies, the preferred policy had to be abandoned. But even then, the Heren XVII "went so far as to urge Dutchmen to marry only high caste women of good social standing."[51] Following a somewhat different tactic, Rijkloff van Goens, the governor of Sri Lanka, did promote mixed marriages by offering a bonus of three months pay to those who married a native Indian or Indo-Portuguese woman, "but stipulated that the daughters of these marriages should marry Netherlanders, 'so that our race may degenerate as little as possible.'"[52] Whether the policies advocated by the Heren XVII and Goens had their origins in the *Itinerario* is of course debatable, but a seventeenth-century Dutchman reading van Linschoten's descriptions of the dire consequences of miscegenation in Portuguese India may well have arrived at just those policies.

Sati

A remarkable doubleness enters the narrative when van Linschoten comes to describe *sati,* the practice of widow-burning. In a manner typical of these accounts, van Linschoten relates that while the friends of the deceased Brahman are singing songs of praise for the dead husband, the widow distributes all her jewelry among her friends and then, "with a chéerefull countenance, she leapeth into the fire" (249; see Figure 5).[53] This is not the context to assess the veracity of van Linschoten's description, other than to say that modern scholars are rightly skeptical of a widow's cheerful countenance.[54] What I am interested in is what function this description serves in the narrative. For one thing, it allows van Linschoten to ascribe a far greater wifely devotion to the Brahman wives than he allows those women married to Portuguese men. It shows that some Indian men—even in death—are capable of disciplining their wives, but that Portuguese men are not. Yet the chapter on *sati* starts with the plain assertion that "The Bramenes are the honestest and the most estéemed nation among [all] the Indian heathens" (247). The Brahmans are first and most noble among an inferior race, but apparently not worthy of *direct* comparison to the Portuguese. To be first among heathens is still to be inferior to the colonizers. But, obviously, this implicit hierarchy, which is fundamental and structural to the logic of Portuguese colonial discourse, is partly overturned by van Linschoten's treatment of wifely obedience.

But then van Linschoten's narrative takes another turn in its attempt to explain the reasons for the practice of *sati:*

> The [first] cause . . . why the women are burnt with their husbandes, was (as the Indians themselves so say), that in time past, the women (as they are

Figure 5. As by law, the body of the Brahman is burned, and, out of love for him, his wife allows herself to be burned with him. From *Itinerario* (1596).

> very leacherous and inconstant both by nature, and complexion) did poyson many of their husbands, when they thought good . . . thereby the better means to fulfill their lusts. {When the king perceived this, he} ordayned, that when the dead bodies of men were buried, they should also burne their wives with them, thereby putting them in feare, and so make them abstaine from poysoning of their husbands. (251)

Suddenly, the Brahman wives closely resemble the native wives of the Portuguese (250–51). What this contradictory account of Brahman wives allows van Linschoten to do is exercise his double gaze: it allows him to assert the superiority of the Brahman wives (who cast themselves in the fire with a smile on their faces and are therefore properly disciplined) over the wives of the Portuguese (and therefore also holding these women up as paragons of wifely devotion for a European readership),[55] while also condemning those same wives by bringing them down to the level of the stereotypical, lustful, deceitful, murderous native wife.

If we recall that, according to van Linschoten, "there is no pleasanter death" for the Indian wife of a Portuguese than to die for adulterous love, we can see how a lapse in (ordinary) logic reveals the contorted logic of colonial discourse: the wives prefer dying for love over all other things, and yet van Linschoten's reason for *sati* is that it should discourage women who wish to engage, or to continue to engage, in love relations with other men from killing their husbands. How could *sati* be a deterrent? These same women who cheat on their husbands without any regard for their own life, cheerfully jump into the flames out of devotion for their dead husbands. Logic would dictate that the laughter ("lachende"; see Dutch original, vol. 2, p. 18) with which they hop into the fire is not an expression of their devotion to the husband at all but a sign of their happiness at dying for the sake of adulterous love.

Religious Encounters

Religious encounters between colonizer and colonized constitute another possible form of contamination. As far as van Linschoten is concerned, it's not so much that the Portuguese exercise excessive or not enough influence over religious life in India but that any involvement in this area is likely to be counterproductive. In the town of Goa (which is located on an island on the west coast), van Linschoten tells us, the Portuguese grant the "Indians, Heathens, Moores, Iewes, Armenians, Gusarates, Benianes, Bramenes, and of all Indian nations and people" (181) remarkable religious freedom. The authorities allow everyone to "hold" his own religion "without constrayning any man to doe against his conscience." Hailing from a predominantly Protestant town, the Catholic van Linschoten appears to admire this religious tolerance. The Portuguese, however, have placed some limits on religious freedom. The natives should refrain from

> their ceremonies of burning the dead, and the living, of marrying and other superstitious and develish inventions, [which] they are forbidden by the Archbishop to use . . . openly, or in the Iland, but they may fréely use

> them upon the firme land, and secretly in their houses, [thereby] to shunne [and avoid] all occasions of dislike that might be given to Christians, which are but newlie baptised. . . . [H]e that is once christened, and is after found to use any heathenish superstitions, is subiect to the Inquisition, what so ever he be, or for any point [of religion] what so ever. (181-82)

The ramifications of this freedom, which is obviously not based on respect for non-Christian worship, are intriguing. The freedom to worship consists of permission to practice behind closed doors or on the mainland, away from Goa where newly baptized Christians might take offense. Longtime Christians are apparently able to handle exposure to "develish inventions" without succumbing to "heathenish superstitions." Clearly, *contact* with other religions must be avoided to prevent religious contamination. Separation must be strict. What is more, once a person—regardless of race—has converted to Christianity s/he is answerable to the Inquisition, and there is no return to the former "freedom" of worship. A temporary change in a person's faith permanently changes that person in the eyes of the Catholic church. Once a Catholic, always a subject of the Inquisition. In other words, the (one-sided) cultural exchange of religion between the colonizer and the colonized produces a subject that is permanently fettered to, and a responsibility of, the colonizer. The conversion of Indians to Catholicism, while a good thing in general, according to van Linschoten, does bring unpleasant obligations with it for the colonizer. Significant space in the *Itinerario* is dedicated to the at times sympathetic description of the "ceremonies and superstitions" and religious artifacts of various Indian religions (see, for instance, chapter 44). Van Linschoten surely intends his readers to enjoy these descriptions and to learn from them. But knowledge about the religions of the other should be obtained as if one were a cautious sightseer, because intervention comes at a price. Whatever the Portuguese convert or touch may be lost forever, whether it is a person or an object. With admiration and an eye for delicacy of detail, van Linschoten describes a religious settlement around the "Pagode of the Elephant." The place with its cisterns for water is "very curiously made, and round about the wals are cut and formed, the shapes of Elephants, Lions, tigers, and a thousand other . . . cruel beasts [which] are so well . . . cut, that it is strange to behold" (291). But then he adds, sadly and critically, that "These Pagodes and buildings are now whollie left overgrowne, and spoyled, since the Portingales has it under their subiections." Even neglect constitutes a destructive exchange under colonial domination.

A final example, dealing with the treatment of non-Christian religions, will give us a concrete glimpse of how van Linschoten views himself vis-à-vis the dynamic between the Portuguese colonizer and the colonized. He tells us how he and a Portuguese friend, while walking about town, decided that it would be nice to visit a "Mahometicall Church" (287) and to observe "their manner of service." At guard at the door, a Moor, however, denies them access to the mosque unless they remove their shoes. This, van Linschoten and the Portuguese man refuse to do. The guard is nice enough to open "some of the windowes" to allow them to see in. Noticing that the hall of the temple is empty except for the worshippers, the Portuguese man asks "[for] their God and their Saintes which they used to pray unto." The "Moore answered him, that they used not to pray to stockes and stones, but to the living God, which is in Heaven, and said that the proude Portingale Christians, and the Heathens were all of one Religion, for that [they] prayed to images . . . and give them the glorie onely appertaineth to the living God." Now it is not clear from the narrative why the sudden change occurs in the guard. He seemed at first inclined to be patient with the visitors, allowing them to look into the hall even after they refused to observe the proper customs. The Portuguese's question also seems innocent (ignorant) enough. Or is the question taken as an implicit slander on the Moslem religion? Is the guard's attack on Portuguese Christians a delayed response to the refusal to remove footwear, or does the question about statues and saints suddenly bring into focus the sharp difference between the respective religions and therefore between colonizer and colonized? Or is it possible that van Linschoten has suppressed from his account either the tone—an insulting tone?—in which the Portuguese man addresses the guard (the refusal to remove the shoes can be read as an insult), or perhaps he has altogether censored offensive statements made by his friend. We cannot know the answers to these questions (except to say that the guard's insulting comments are at odds with the encounter as described). What we do know is how the Portuguese man responds to the guard.

> [T]he Portingale was so angrie, that he began to chide [and make a greate noyse, and to give] him manie hard words, so that there had growne a great quarrel, had it not bene for me, that got him to hold his peace, and [so] brought him away, and [let] the matter rest in that sort. (288)

Van Linschoten then concludes the discussion by noting that although the Portuguese trade heavily with the Moors and allow them to live in their midst, the Moors "are their most deadly enemies," who do the Portuguese great mischief, not least by preventing Indians from being converted to Christianity. There is obviously a lot going on here. Let me just say that van Linschoten's swipe at Moors and their "mischief" may well serve to justify his friend's anger, but the fact remains that the Portuguese man, who is a Catholic just like van Linschoten, loses his temper with a racial and social inferior and nearly causes a riot that could have cost them dearly. And it is the Dutchman, who should have been just as offended by the guard's comments as the friend, who keeps calm and saves the day. Van Linschoten portrays himself as the man who separates the feuding parties and who separates the cultures without getting involved any further than to palliate his friend's error in judgment by slandering Moslem Moors.[56]

To be sure, the author thanks God at various turns and points out that the religions of the Indians are mere superstitions, but he never suggests that the conversion of the natives is a priority or even of any great consequence. Indeed, when it comes to the conversion of Indian women to the Christian religion, van Linschoten makes it very clear that this does not curtail their promiscuous, deceitful, and murderous ways. The Brahman wives, he suggests, display greater devotion to their husbands than do the converts. The Dutch East India Company certainly shared van Linschoten's outlook. The company, though run for the most part by Calvinists, studiously refrained from promoting religion in any way whatsoever, preferring a focus on profit over a desire to proselytize. The VOC "did not count a chaplain as an essential member of a fleet's complement."[57] In fact, the Heren XVII, the governing body of the VOC, did not contain any provisions regarding religion in its charter. It was not until 1609 that the role and presence of preachers was discussed, and then not to proselytize but primarily to tend to "the spiritual needs of their [Dutch] staff, if only as a means of maintaining discipline and morale."[58] The Catholicism of the *Itinerario* seems not to have been an issue for the States General (a decidedly Protestant body). Indeed, the text's Catholicism may have been too diffuse to notice. Wolfe writes that the *Itinerario* may serve "for the dispersing and planting of true Religion and Ciuill Conuersation therein."[59] Undoubtedly, the reference to "true Religion" is to Anglicanism, not to Catholicism. Wolfe's comments should probably be read as a sop to the pious. Furthermore, the *Itinerario* hardly recalls the Spanish obsession with spreading the influence of Rome, and seems entirely unconcerned with planting any religion. In this respect, too, van Linschoten separates himself from the Spanish colonizer, whom he admired so much.

There is every indication that van Linschoten's subject position changed during the 13 years of his voyage. When van Linschoten leaves Holland in 1579, he leaves a country that has only just begun to extricate itself, with very limited success, from the colonizing forces of the powerful Habsburg empire. I would never suggest that the Spanish Netherlands was a colony of Spain in the same fashion that India was a Portuguese colony, but it is worth noting that the Duke of Alva, who ruled the Spanish Netherlands as governor-general from 1567 to 1572, was widely known for his "xenophobic bigotry" and "[h]is deeply suspicious attitude towards the Netherlands nobility and population [which] was tinged with scarcely veiled contempt."[60] Van Linschoten himself, therefore, is a hybrid product of an unequal exchange between unequal cultures. Then, in 1580, he enters Portugal on the heels of a conquering Spanish army. Yet, in order to feed his desire to travel, he seeks employment with the conquered, the Portuguese. And, again, while Portugal was not a colony of Spain in the way that India was a colony of Portugal, van Linschoten's own (erroneous) account of the rightness of the Spanish takeover and his countless comments regarding the weakness and corruption of the Portuguese in India make it very clear that he thought of the Portuguese as a justly colonized people. Indeed, as I have tried to argue, the *Itinerario* tries to do exactly that: colonize the Portuguese. But it does not perform this act of colonization from the Spanish point of view. The Spanish may have been the original colonizers (in van Linschoten's eyes), and they may have seemed superior (in van Linschoten's eyes) at the time of his departure from Enkhuizen and during his stay in Portugal and Spain, but the further we get into the *Itinerario,* the more we realize that, as George Masselman observes, van Linschoten feels that "the Portuguese ha[ve] lost much of their former prowess,"[61] and that their moment has passed. His perspective is neither Portuguese nor Spanish, nor does he present his text to the Portuguese for the betterment of their possessions in India, or to the Spanish for the purpose of disciplining the Portuguese. Instead, he offers it to the Dutch States General, who, as I argued earlier in this essay, use the *Itinerario* as a foundational document to guide Dutch (colonial) ventures in the East. Also, while it is difficult, for reasons discussed, to label van Linschoten's perspective in the *Itinerario* as fundamentally Dutch, there is no doubt that the ruthlessly efficient, nonproselytizing, profit-driven approach to the encounter with the other perfectly typifies the Dutch colonial enterprise. By 1688, at the height of the Dutch colonial empire, both the southwestern and southeastern coasts of India and the isle of Ceylon were firmly under Dutch control.[62]

One of the interesting ironies of van Linschoten's narrative is that while the exchanges between the colonizer and the colonized in *Asia Portuguesia* are always far more complex than any strict binary model can account for, van Linschoten also is a Manichean colonist who believes that separation and distance are absolutely required if the colonizer is to be successful. Such distance, the distance of Charon, Mercury, and Menippus, is an illusion. Economic hybridity, it seems, inevitably leads to social and biological hybridity. But that does not stop van Linschoten from pursuing a colonizer's dream. For, if nothing else, the illusion of *absolute* difference between colonizer and colonized must be maintained as long as possible in order for *any* difference to appear real and legitimate.

Notes

1. I am grateful to Deborah Barker, Karen Raber, Joe Ward, and Jay Watson for their helpful comments on this essay.
2. Throughout this essay quotations from van Linschoten's text are taken from the 1598 English translation (*The Voyage of John Huyghen van Linschoten to the East Indies,* 1598,vol. 1, ed. Arthur Coke Burnell, vol. 2, ed. P. A. Tiele [London: Hakluyt Society, 1885] of the Dutch original, published in 1596 (*Itinerario, Voyage ofte Schipvaert van Jan Hughen van Linschoten Naer Oost Ofte Portugaels Indien, 1597-1592.*, vol. 2, ed. H. Kern and H. Terpstra [The Hague: Martinus Nijhoff, 1955]). The English translation often inserts words or phrases that are not in the Dutch original. The editors of the 1885 reissue of van Linschoten's text saw fit to mark all such insertions with square brackets. I have retained these brackets to alert the reader to discrepancies between the English translation and the Dutch original. For literal translations of the passages in question, see the footnotes in Burnell and Tiele or the Dutch original. Whenever I had to insert a word or phrase inside a quotation I have enclosed it with {}. As all but one of my quotations from the *Itinerario* come from volume 1 of the 1885 edition, I have identified the single quotation from volume 2 with a volume reference but have sufficed to identify the others in my text with a parenthetical page number alone.
3. P. A. Tiele, "Introduction," in *The Voyage of John Huyghen van Linschoten to the East Indies,* Tiele, vol. 2 (London: Hakluyt Society, 1885), p. xxxvii.
4. Charles McKew Parr, *Jan van Linschoten: The Dutch Marco Polo* (New York: Thomas Y. Crowell, 1964), p. xvi. Parr also offers two contemporary quotations testifying to the importance of the *Itinerario* (pp. xxvi-xvii). The first Dutch merchant fleet that sailed to India relied on van Linschoten's *Reysgeschrift* to reach its destination (Tiele, "Introduction," p. xxxvi; Parr, *Linschoten,* p. xxvi; C. A. Davids, *Zeewezen en Wetenschap: De Wetenschap en de Ontwikkeling van de Navigatietechniek in Nederland tussen 1585-1815* [Amsterdam: Bataafsche Leeuw, 1985], pp. 70, 100-102). W. J. Balen expresses similar sentiments in *Naar de Indische Wonderwereld met Jan Huyghen van Linschoten,* 2d. ed. (Amsterdam: Amsterdamsche Boek and Courantmaatschappij, 1946), pp. 7-8, as does Owen C. Kail, *The Dutch in India* (Delhi: MacMillan India, 1981), p. 11.
5. See George Masselman, *The Cradle of Colonialism* (New Haven: Yale University Press, 1963), p. 73. Also see Parr, *Linschoten,* p. xxviii.
6. Describing Dutch successes in the East, M. N. Pierson observes that "As early as 1620 the Dutch

were importing more into Europe than the Portuguese had done in the last quarter of the sixteenth century. The Portuguese at this time returned 2,000 or 3,000 tons of shipping a year to Europe, by the middle of the century the Dutch returned 10,000. These figures become even more significant when we remember that the Portuguese had had a monopoly, while the Dutch did not. Even in the first decade of the century, from 1600 to 1610, the Dutch returned 75 ships to Europe, the Portuguese only 27. . . . A belated and feeble attempt to imitate the English and Dutch methods, that is by setting up a trading company, . . . failed miserably" (*The Portuguese in India* [Cambridge: Cambridge University Press, 1987], pp. 140-41). See further, Om Prakish, "Euro-Asian and Intra-Asian Trade: The Phase of Dutch Domination, 1600-1680," *European Commercial Enterprise in Pre-Colonial India* (Cambridge: Cambridge University Press, 1998), pp. 175-210.

7. Van Linschoten in turn dedicated his text to this political body. In 1594, van Linschoten was also rewarded with a commission to embark on the first of two naval expeditions that sought passage to the East along a northeastern route. Upon returning from the first voyage, van Linschotenwas honored with a request to come to The Hague and to deliver in person his report to Prince Maurits of Nassau and Johan van Oldenbarnavelt, the leaders of the United Provinces. See A. van der Moer, *Een Zestiende-Eeuwse Hollander in het Verre Oosten en het Hoge Noorden: Leven, Werken, Reizen en Avontured van Jan Huyghen van Linschoten (1563-1611).* (The Hague: Martinus Nijhoff, 1979), p. 12. But a later request (in 1610) for a "modest annual pension" was denied by the States General (Parr, *Linschoten,* p. xlv).
8. Karel Steenbrink, *Dutch Colonialism and Indonesian Islam: Contacts and Conflicts 1596-1950,* trans. Steenbrink and Henry Jansen (Atlanta: Rodopi, 1993), p. 25.
9. Technically, the historical period with which the present essay deals is often referred to as *pre*-colonial, preferring to reserve the term "colonialism" for the post-Renaissance period. I believe, however, that the practices of the Dutch and, especially, those of the Portuguese in the late sixteenth century carry within them the crucial seeds of conquest, domination, and exploitation that characterize the period of British control over India. Terms such as "early colonialism," "emerging colonialism," or "precapitalist colonialism" might be more accurate than "colonialism." They are, however, cumbersome terms, and I have opted simply to use "colonialism," with the understanding that when I use the term I'm invoking a mind-set and cluster of practices that are nothing more than the uncertain beginnings that eventually paved the way for the vast state apparatus of the British in India. Or, to put the matter slightly differently, but directly in the context of travel writing, "the transcendent traveler's gaze is the colonial gaze." As Indira Ghose observes, "the opposition between self and other set up in travel writing is, like all binary oppositions, grounded on an implicit (and sometimes explicit) hierarchy" (*Women Travellers in Colonial India: The Power of the Female Gaze* [Delhi: Oxford University Press, 1998]), p. 9.
10. On this question, see Pierson, *The Portuguese in India,* pp. 139-40. Interestingly enough, van Linschoten is not the last person to believe that the Indian environment and its people contributed to Portuguese "corruption" and "military ineffectiveness" (Pearson, *The Portuguese in India,* p. 140).
11. Ann Bos Radwan, *The Dutch in Western India, 1601-1632: A Study of Mutual Accommodation* (Calcutta: Firma KLM Private, 1978), p. 4.
12. For a succinct discussion of the VOC as a financial institution, see Jonathan I. Israel, *Dutch Primacy in World Trade, 1585-1740* (Oxford: Clarendon Press, 1989), pp. 67-73.
13. Radwan, *The Dutch in Western India,* p. 4.
14. For a succinct description of the governing structure of the Company, see Om Parakash, *European Commercial Enterprise in Pre-Colonial India* (Cambridge: Cambridge University Press, 1998), pp. 72-76. Also see Israel, *Dutch Primacy in World Trade,* pp. 69-73.
15. Israel, *Dutch Primacy in World Trade,* pp. 70-71.
16. Radwan, *The Dutch in Western India,* p. 4.
17. Kail, *The Dutch in India,* p. 31.
18. Kail, *The Dutch in India,* p. 31; see also Radwan, *The Dutch in Western India,* pp. 3-4.
19. Abdul JanMohamed, "The Economy of Manichean Allegory: The Function of Racial Difference in Colonialist Literature," *Critical Inquiry* 12 (1985): 59-87, 60.
20. E. Ann Kaplan, *Looking for the Other: Feminism, Film, and the Imperial Gaze* (New York: Routledge, 1997), p. xviii.
21. For the debate over the gaze, see MacLean, this volume; David Spurr, *The Rhetoric of Empire: Colonial Discourse in Journalism, Travel Writing and Imperial Adminstration* (Durham, N.C.: Duke University Press, 1993), pp. 16-23, *passim;* Mary Louise Pratt, *Imperial Eyes: Travel Writing and Transculturation* (London: Routledge, 1992); Indira Ghose, *Women Travellers in Colonial India: The*

Power of the Female Gaze; and Jyotsna G. Singh, *Colonial Narratives/Cultural Dialogues: "Discoveries" of India in the Language of Colonialism* (New York: Routledge, 1996), pp. 1-47.

22. Homi Bhabha, *The Location of Culture* (New York: Routledge, 1994), pp. 85-92.
23. See van Linschoten's "signed" dedication to the Dutch States General of the Dutch edition of the *Itinerario.*
24. Parr, *Linschoten,* p. 29.
25. Jonathan I. Israel, *The Dutch Republic: Its Rise, Greatness and Fall, 1477-1806* (Oxford: Clarendon Press, 1995), p. 178.
26. Parr, *Linschoten,* p. 35.
27. Parr, *Linschoten,* p. 34.
28. See Israel, *The Dutch Republic,* pp. 241-57.
29. Jan Pieterz. Coen, governor-general of Dutch settlements, was worried that Spain and Portugal, "united under one crown," could bar Dutch "access to the spices of Southeast Asia" (Masselman, *Cradle of Colonialism,* pp. 308-9). Ironically, it was also the unification of Spain and Portugal that drove the Dutch to travel to the East, because it led to the Spanish closing the formerly free port of Lisbon to the Dutch, who had formerly bought their spices there (Masselman, *Cradle of Colonialism,* pp. 70-71).
30. I disagree with Parr's suggestive hypothesis that van Linschoten was a Dutch spy (Parr, *Linschoten,* p. xvii). Instead, I concur with Masselman's assessment that van Linschoten probably "had no ulterior motives in collecting the wealth of information that eventually appeared in his *Itinerary*" (Masselman, *Cradle of Colonialism,* p. 71). Van Linschoten's perspective flows, it seems, from his background and unique position in a particular historical moment when, for a Dutchman, the boundaries of nationalism were blurred and in flux.
31. Singh, *Colonial Narratives,* p. 21.
32. John Wolfe, "To the Reader," in *The Voyage of John Huyghen van Linschoten to the East Indies,* 1598, vol. 1, ed. Arthur Coke Burnell (London: Hakluyt Society, 1885), p. xlvii.
33. Wolfe, "To the Reader," p. xlviii.
34. Bhabha, *The Location of Culture,* p. 67.
35. Wolfe, "To the Reader," p. xlviii.
36. Michel Foucault, *Discipline and Punish: The Birth of the Prison,* trans. Alan Sheridan (1975; New York: Vintage, 1979), p. 220.
37. Foucault, *Discipline,* p. 220.
38. Needless to say, from a practical standpoint it is impossible to "run" a country as a spectator from on high.
39. Pearson, *The Portuguese in India,* p. 95.
40. Pearson, *The Portuguese in India,* p. 108.
41. Having spent some time there, van Linschoten must have been aware that Portugal had an unusually high number of imported slaves among its inhabitants (see Pierson, *The Portuguese in India,* pp. 14-15). What is more, the VOC itself eventually resorted to the use of slave labor, although it resisted Jan Pietersz. Coen's requests in this score for some time (Masselman, *Cradle of Colonialism,* p. 343).
42. It is hard to say if van Linschoten is indeed holding the *mestiços* responsible, or if his prose is merely ambiguous, but his account of the market flows directly, without even a paragraph break, out of the discussion of reproduction (quoted earlier), which ends with the *mestiços.* The passage in question reads as follows: "so that the posteritie of the Portingales, both men and women . . . doe séeme to be naturall Indians, both in colour and fashion. Their livings and daylie traffiques are the Bengala, Pegus . . . and everie way, both North and South: also in Goa there is holden a daylie assemblie . . . as wel of the Citizens and Inhabitants, as of all nations throughout India . . ." (184; a gloss in the margin of the Dutch edition [vol. 2, p. 126] further strengthens the impression of the *mestiços'* role in the slave trade).
43. See also Pearson, *The Portuguese in India,* pp. 99-103.
44. It is a testimony to the judiciousness of van Linschoten's insights that Pierson, with the benefit of four centuries of hindsight and access to voluminous archives, comes to remarkably similar conclusions (140).
45. This explanation falls well within widely held Renaissance notions of bodily humors and their relation to sickness and health (see, for instance, Timothy Bright, *Treatise of Melancholie* [1586]; Andreus [Laurentius] Du Laurens, *A Discovrse of the Preservation of the Sight: of Melancholike Diseases. . . .* [1599]; Thomas Walkington, *Opticke Glasse of Humors* [1607]; for an overview, see

Lawrence Babb, *The Elizabethan Malady*[East Lansing: Michigan State University Press, 1951], pp. 1-41) with which van Linschoten was no doubt familiar.

46. Mikhail Bakhtin, *Rabelais and His World,* trans. Hélène Iswolsky (Cambridge, MA: MIT Press, 1968), p. 310; also pp. 317-18.
47. Bakhtin, *Rabelais,* pp. 372, 370.
48. Kail, *The Dutch in India,* pp. 147-48.
49. Heleen C. Gall, "'European' Widows in the Dutch East Indies: Their Legal and Social Position," in *Between Poverty and the Pyre: Moments in the History of Widowhood,* ed. Jan Bremmer and Lourens van den Bosch (New York: Routledge, 1995), pp. 103-21, p. 106. Gall points out that it was not until the early 1630s that this policy changed and marriage to native women was encouraged, but only after everyone realized that the women the VOC was sending "were the lowest of the low" (106), and that this was not going to change because women of more suitable character refused to make the journey.
50. Kail, *The Dutch in India,* p. 146.
51. Kail, *The Dutch in India,* p. 147.
52. Kail, *The Dutch in India,* p. 147.
53. Burnell observes that this is "the earliest precise account of the horrible rite now called 'Suttee'" (*The Voyage of John Huyghen van Linschoten,* vol. 1, 249 n. 7).
54. Lani Mani, "Cultural Theory, Colonial Texts: Reading Eyewitness Accounts of Widow Burning," in *Cultural Studies,* ed. Lawrence Grossberg, Cary Nelson, and Paula Treichler (New York: Routledge, 1992), pp. 392-405,399-400.
55. See Ania Loomba, *Colonialism/Postcolonialism* (New York: Routledge, 1998), p. 157.
56. For a brief treatment of the same passage, see Steenbrink, *Dutch Colonialism,* pp. 25-27.
57. Parr, *Linschoten,* p. xli
58. Kail, *The Dutch in India,* p. 157.
59. Wolfe, "To the Reader," p. lii.
60. Israel, *The Dutch Republic,* p. 156.
61. Masselman, *Cradle of Colonialism,* p. 73.
62. Israel, *The Dutch Republic,* p. 937.

SECTION SEVEN

EDWARD TERRY

Edward Terry was born in 1590 and educated at the Rochester School and Christ Church College, Oxford. In the spring of 1616 he accepted an engagement for a voyage to the Indies as one of the chaplains in the fleet commanded by Captain Benjamin Joseph. On the way out, the commander was slain in an encounter with the Portuguese, but they safely reached India's western shore on September 25, 1616. Sir Thomas Roe's chaplain had died a month earlier, and since Terry was well commended, he was engaged for the post. He later joined the ambassador near Ujjain and accompanied him to Mandu, where the Mogul Emperor Jehangir fixed his court until October of that year (1617), when he removed to Ahmedabad. In September 1618, the ambassador ended his stay in the court, resting for a few months in Surat before embarking for England on February 17, 1619. Thus, Terry had only seen a small area of western India—a fact one has to keep in mind while reading his generalizations about the land.

On his return, the clergyman went back to his Oxford college, where he probably wrote the result of his observations in India, as now reprinted. He presented this document in manuscript to the Prince of Wales, later King Charles I. It is not known how the manuscript came into the possession of the Reverend Samuel Purchas who published it in 1625 in his *Pilgrimes* (part ii, book ix, chapter 6).

After writing his account, Terry settled down as Rector of Great Greenford, near London, till his death in October 1660. During that time the East-India Company called on their former chaplain to preach before them on the occasion of the almost simultaneous return of seven ships from the East Indies. Delivered on September 6, 1649, it was printed under the title of *The Merchants and Mariners Preservation and Thanksgiving* . . . . Six years later, in 1655, Terry's account of his experiences first appeared as a separate volume entitled *A Voyage to East-India.* We have no evidence to suggest that Terry was consulted by Samuel Purchas, since in the preface of his own volume, he makes no reference to its prior appearance in Purchas's collection. It seems that the work attained some popularity since it was republished in a slightly condensed form in 1665 in a folio volume containing translations of *The Travels of Pietro della Valle*—from which I draw the excerpts below.

In 1777, a reprint of the longer 1655 edition was issued. Two Dutch editions of the narrative came out in the years 1707 and 1727, and a French translation appeared in 1663, and was reissued in 1669.

Edward Terry left an epitaph to his own life, inscribed under the portrait prefixed to his *Voyage to East-India* (1655 edition):

In Europe, Africk, Asia have I gone;
One journey more, and then my travel's donne.[1]

Selections from Terry's writings include excerpts taken from the 1665 version (included in Della Valle, Pietro. *The Travels of Pietro Della Valle into East-India and Arabia deserta, Whereunto is Added A Relation of Sir Thomas Roe's Voyage unto the East Indies.* London: Printed by J. Macock, 1665). Also included are excerpts from his 1649 *Sermon* to the merchants, which recapitulates some of his written observations.

—Jyotsna G. Singh

Extract I

Introductory Preface

A Description of the large Territories under the subjection of the Great Mogol

The twenty first [of September] We discovered the main Continent of *Asia* the Great, in which *East-India* takes up a large part. The twenty second, we had sight of Deu and Daman, places that lye in the skirts of *India,* principally inhabited and well-fortified by Portugals;[2] and the twenty-fifth of *September* we came happily to an Anchor in *Swally-Road* within the *Bay of Cambaia,* the Harbour for our Fleet while they make their stay in these remote Parts.[3]

Then after a long, and troublesom, and dangerous passage, we came at last to our desired Port . And immediately after my arrival there, I was sent for by Sir *Thomas Row*[4], Lord Embassadour, then residing at the *Mogol's* Court (which was very many miles up in the Countrey)[5] to supply the room of *Mr. John Hall* his Chaplain (Fellow of *Corpus Christi* Colledg in *Oxford*) whom he had not long before buried. And I lived with that most Noble Gentleman at that Court more than two years, after which I returned home to *England* with him. During which space of my abode there, I had very good advantage to take notice of very many *places,* and *persons,* and *things,* travelling with the Embassadour much in Progress with that King up and down his very large Territories.

And now, *Reader,* I would have thee to suppose me setting my foot upon the *East-Indian* shore, at *Swally* before-named. (344-45)

Extract II

Of the Soyl there, what it is, and what it produceth, &c.

This most spacious and fertile Monarchy (called by the Inhabitants *Indostan*)[6] so much abounds in all necessaries for the use and service of man, to feed, and cloath, and enrich him, as that it is able to subsist and flourish of it self, without the least help from any Neighbour-Prince or Nation.

Figure 6. A portrait of the Reverend Edward Terry from *Voyage to East- India* (1655 edition).

Here I shall speak first of that which Nature requires most, Food, which this Empire brings forth in abundance; as, singular good Wheat, Rice, Barley, with divers more kinds of good Grain to make bread (the staff of life) and all these sorts of Corn in their kinds, very good and exceeding cheap. For their Wheat, it is more full and more white than ours, of which the Inhabitants make such pure, well-relished Bread. . . .

The Countrey is beautified with many Woods and Groves of Trees, in which those winged *Choristers* make sweet Musick. . . .

Besides their Woods, they have great variety of fair goodly Trees that stand here and there single, but I never saw any there of those kinds of Trees which *England* affords. . . .

This Empire is watered with many goodly Rivers . . . the two principal are *Indus and Ganges;* where this thing is very observable (for they say there, that it is very true) that one pint of water of the *Ganges* weigheth less by one ounce than any other water in that whole great Monarchy . . . .

Besides their Rivers, they have store of Wells fed with Springs; and to these, they have many Ponds . . . some of them exceeding large, fill'd with water when that abundance of Rain falls . . . (358-64).

Extract III

What the chief Merchandizes, and most Staple, and other Commodities are, which are brought into this Empire.

The most Staple Commodities of this Empire are *Indico* and *Cotton Wool;* of that *Wool* they make divers sorts of *Callico,* which had that name (as I suppose) from *Callicut,* not far from *Goa,* where that kind of Cloth was first bought by the Portugals. . . .

In *Decan,* which bounds upon the *Mogol's* Territories South, (the Princes whereof are Tributaries unto him) there are many Diamond-Rocks, in which are found those most pretious of all other Stones; and they are to be sold in this Empire, and consequently to be had by those who have skill to buy them, and Money to pay for them. . . .

. . . the People here have some store of Silk, of which they make Velvets, Sattins, Taffataes, either plain, or mingled. . . .

The Earth there yields good Minerals of Lead, Iron, Copper, Brass, and (they say) they have Silver-Mines too; which (if true) they need not open, being so enriched from other Nations of *Europe,* and other parts, who yearly bring thither great quantities of Silver to purchase their Commodities . . . And this is the way to make any Nation of the world rich, to bring, and leave Silver in it, and to take away Commodities. (366-370)

Extract IV

Of the Inhabitants of East-India, who they are; Of their most excellent Ingenuity expressed by their curious Manufactures, their Markets at Home to buy and sell in, and their Trade abroad

The Inhabitants in general of all *Indostan* were all anciently Gentiles, called in general *Hindoes,* belonging to that very great number of those which are called *Heathens,* which take up almost two thirds of the number of the People who inhabit the face of the whole Earth. . . . There are some *Jews* (but they are not too many) here and there scattered and lost as it were, in those other great numbers of People . . . For the Inhabitants of *East-India* ever since they were subdued by *Tamberlain,* they have been mixed with *Mahumetans,*[7] which though they be by farr in respect of their number less than those *Pagans,* yet they bear all the sway, and command all in those Countries.

There are besides these, (now become as it were Natives there) a great number of *Persians* and *Tartars* (who are *Mahumetans* by Religion) that there inhabit, very many of which the Mogul[8] keeps for Souldiers to serve on Horse-back . . .

To those I have named of other Nations, (that are to be seen in *East-India*) there are besides some few almost of every people in *Asia,* and many Europeans of diverse parts (that use to stir from their own fires) to be found amongst them; and among that great variety of People and Nations there to be observed, I have taken special notice of divers Chinesaas, and Japanesaas there. . . .

There are some Jews here (as before I observed) whose stubbornness and Rebellion, long ago, caused Almighty God to threaten them . . . and scattered among all the Nations of the World.

Those ancient Satyrists, *Persius,* and *Juvenal,* after that most horrid act committed by them [Jews] in Crucifying our Blessed Saviour . . . call them . . . circumcised, Worms, vermin . . . *Marcus* the Emperour observing them well, concluded that they were a generation of men worse than savages or Canibals, to be even the worst of men, as if they were the very reffuse and dregs of mankind.

Yet we do believe, (because the Lord hath promised it) that he will find a time to call home this people [Jews] again to himself; when they shall receive honour above all the contempt they have been long under, after they shall see with sorrow, and with the eye of faith, Him, whom their Fore-fathers, out of ignorance, and despite, and unbelief pierced.

The Natives there [in East-India] . . . shew very much ingenuity in their curious Manufactures; . . . as in their Silk-stuffs which they most artificially weave, some of them very neatly mingled either with Silver or Gold, or both. . . .

The are excellent at Limning, and will coppy out any Picture they see to the life: for confirmation of which take this instance; It happened that my Lord Embassadour visiting the *Mogol* on a time, as he did often, presented him with a curious neat small oval Picture done to the life in *England.*[9] The Mogol was much pleased with it, but told the Embassadour withall, that haply he supposed that there was never a one in his Country that could do so well in that curious Art; and then offered to wager with him . . . that in a few days he would have two Copies made by that presented to him, so like, that the Embassadour should not know his own. . . . Two Copies taken from that Original were within few days after made, and brought and laid before the Embassadour, in the presence of the King; the Embassadour viewing them long, either out of Courtship to please the King, or else unable to make a difference twixt the Pictures being all exquisitely done, took one of them which was new made, for that which he formerly presented, and did after profess that he did not flatter, but mistake in that choice. The truth is, that the Natives of that Monarchy are the best Apes for imitation in the world, so full of ingenuity that they will make any new thing by pattern, how hard soever it seem to be done; and therefore it is no marvel, if the Natives there make Shooes, and Boots, and Clothes, and Linen, and Bands and Cuffs of our English Fashion, which are all of them very much different from their Fashions and Habits, and yet make them all exceeding neatly. (374–78)

Extract V

Of our safe and secure living amongst the Natives there, if we do not provoke them. of their faithfulness unto those that entertain them as Servants: For how little they serve, and yet how diligent they are &c.

I travelled from *Surat* with four English-men more, and about twenty of the Natives in our Company, we beginning our journey the first of *Jan.* towards *Sir Thomas Row,* at the Mogol's Court, then above four hundred miles distant from Surat.[10] We had six Wagons drawn with Oxen in our Company, laden with rich English Goods (the principal part whereof was English broad Cloth) assign'd to an English Merchant at the Court. . . .

In our journey towards the Court (after we had been in our way about seven dayes from Surat) we rested in a place called *Ditat,* where many of the Inhabitants offered to guard us and our goods, though we (observing there no danger) desired it not; but they would do it, and in the Morning expected and asked something of us, by way of recompence. One of our Company (who had been in East-India a year or two before) told them, that what they had done they did without our desire, and therefore they should have nothing from us, but some ill Language which he then gave them. We set forward in the Morning according to our wonted custom, they followed after us, to the number at the least of three hundred Men, (for the place was great and populous) and when we were gone about a Mile from that Town, stopped our carriages; he of our company who told them that they should have no recompence, was presently ready to shoot at them with his Musket; which made them all to bend their Bows at us: but I happily and suddenly stepping in, prevented his firing at them, and their shooting at us; which if I had not by Gods good Providence done, but we had madly engaged a great multitude, there could not have been less expected in the sad issue thereof, than the loss of all our lives and goods. But having a little Parlee with them, for the value of three shillings of English money given amongst them, they were all quieted and contented, and immediately left us, wishing us a good journey.

After this, when we had gone forward about twenty days journey, (which daily Remoovs were but short, by reason of our heavy carriages, and the heat of the weather) it hapned, that another of our Company, a young Gentleman about twenty years old, the Brother of a Baron of *England,* behaved himself so ill, as that we feared it would have brought very much mischief on us.

This young man being very unruly at home, and so many others that have been well born, when their friends knew not what to do with them, have been sent to *East-India,* that so they might make their own Graves in the Sea, in their passage thither; or else have Graves made for them on the Indian shore, when they come there. A very cleanly conveyance (but how just and honest, I leave to others) for Parents to be rid of their unruly Children; . . .

For the young Gentleman I spake of, his imployment was to wait upon our Chief Commander in his Cabin, who very courteously, when he came to Sea, turn'd him before the mast amongst the common Saylors, . . . but for all this he out-liv'd that harsh usage and came safely to *East-India* , and my Lord Ambassadour hearing of him, and being well acquainted with his great kindred, sent for him up to Court, and there entertain'd him as a Companion for a year; then giving him all fit accommodations, sent him home again as a passenger for *England,* where after he safely arrived.

But in our way towards that Court, it thus happened, that this hot-brains being a little behind us, commanded him [then near him] who was the Princes servant [before spoken of] to hold his horse; the man replied, that he was none of his servant and would not do it. Upon which this most intemperate mad youth . . . first he beat that stranger, for refusing to hold his horse, with his horse-whip, which, I must tell you, that people cannot endure, as if those whips stung worse than Scorpions . . . this stranger (being whipt as before) came up and complained to me;[but] . . . that frantick young man (mad with rage, and he knew not wherefore) presently followed him, and being come up close to him, discharg'd his Pistol laden with a brace of bullets directly at his body, which bullets, by the special guidance of the hand of God, so flew, that they did the poor man no great hurt; only one of them first tearing his coat, bruised all the knuckles of his left hand, and the other brake his bow which he carried in the same hand. We presently disarmed our young Bedlam, till he might return again to his wits. But our greatest business, was how to pacifie the other man, whom he had thus injured: I presently gave him a Roopee, in our money two shillings and nine pence; he thanked me for it . . . . Then we did shew (as we had cause) all the dislike we could against that desperate act of him, from whom he received his hurt . . . . He told us, that we were good men, and had done him no wrong, and that he would till then rest contented; but he did not so, for about two hours after we met with a great man of that Country . . . He presently went towards him, that to him he might make his complaint . . . shewing him his hand and his other breaches. The great man replied, that it was not well done of us, but he had nothing to do with it; and so departed on his way. That night, after we

Figure 7. The lively portrait of the Great Mogul, Jehangir, from *A Voyage to East-India* (1665 edition).

came to a strong large Town . . . he did what he could (as we imagined) to raise up that people against us, some of them coming about us to view us, . . . standing then upon our guard . . . by the good providence of God, who kept us in all our journey, we here felt none of that mischief we feared; but early in the morning quietly departed without the least molestation. After which, with a little money, and a great many good words, we so quieted this man, that we never after heard any more complaining from him. (391-96)

Extract VI

Of their King the Great Mogol, his discent, &c.

Now those Mahometans and Gentiles I have named, live under the subjection of the Great *Mogol,* which Name, or rather Title, (if my Information abuse me not) signifies Circumcised, as himself, and the Mahometans are; and therefore for his most general Title he is called the Great *Mogol,* as the chief of the circumcised . . .

He is lineally descended from that most famous Conquerour, called in our Stories *Tamberlain,* concerning whose Birth and original Histories much differ, and therefore I cannot determine it; but, in this, all that write of him agree, that he having got together very many huge multitudes of Men, made very great Conquests in the South-East parts of the World, not onely on *Bajazet* the Emperour of the *Turks,* but also in *East-India,* and else-where; . . . The *Mogol* feeds and feasts himself with [the] conceit, that he is *Conquerour of the World,* and therefore (I conceive) that he was troubled upon a time, when my Lord Ambassador, having business with him (and upon those terms, there is no coming unto that King empty-handed without some Present . . .), and having at that time nothing left, which he thought fit to give him, presented him with *Mercators* great Book of *Cosmography* (which the Ambassador had brought thither for his own use) telling the *Mogol,* that that Book described the four parts of the World, and all several Countries in them contained. The *Mogol* at the first seem'd to be much taken with it, desiring presently to see his own Territories, which were immediately shewen unto him; he asked which were those Countries about them, he was told *Tartaria* and *Persia,* as the names of the rest which confine with him; and then causing the Book to be turn'd all over, and finding no more to fall to his share, but what at first he saw, and he calling himself, the *Conquerour of the World,* and having no greater share in it, seemed to be a little troubled; yet civilly told the Ambassador, that neither himself, nor any of his People did understand the Language in which that Book was written; and because so, he further told him, that he would not rob him of such a Jewel, and therefore returned it unto him again.

And the Truth is, that the Great *Mogol* might very well bring his Action against *Mercator* and others who describe the World, but streighten him very much in their Maps; not allowing him to be Lord and Commander of those Provinces, which properly belong unto him.

But it is true likewise that he, who hath the greatest share on the face of the Earth, if it be compared with the whole World, appears not great. . . . The *Mogol's* Territories are more apparent, large, and visible, as one may take notice, who strictly views this affixed Map, which is a true representation of that great Empire in its large dimensions. So that although the *Mogol* be not Master of the whole World, yet hath he a great share in it, if we consider his very large Territories, and his abundant riches, as will after more appear, whose wealth and strength makes him so potent, as that he is able, whensoever he pleaseth to make inroades upon, and to do much mischief unto any of his Neighbours. . . . (446-49)

Extract VII

Of the Mogol's Policy in his Government, exercised by himself and Substitutes.

The *Mogol* sometimes by his *Firmauns,* or Letters Patents, will grant some particular things unto single, or divers persons, and presently after will contradict those Grants by other Letters, excusing himself thus, That he is a great, and an absolute King; and therefore must not be tied unto any thing, which if he were, he said he was a slave, and not a free-man: Yet what he promised was usually enjoyed, although he would not be tied to a certain performance of his promise. Therefore there can be no dealing with this King upon very sure terms, who will say and unsay, promise and deny. Yet we Englishmen did not at all suffer by that inconstancy of his, but there found a free Trade, a peaceable residence, and a very good esteem with that King and People; and much the better (as I conceive) by reason of the prudence of my Lord Embassadour, who was there (in some sense) like *Joseph* in the Court of *Pharoah;* for whose sake all his Nation there, seemed to fare the better. (455)

Extract VIII

The Merchants and Mariners Preservation and Thanksgiving . . .

Preface

You that are the Representatives, and into whose hands and trust, the managing of that great businesse of Trade is put, by that most Worthy Company of Merchants, trading for the Easterne India; you are heere met this day to offer up a voluntary, and a willing sacrifice of Praise and Thanksgiving unto Almighty God, for a great, and an unexpected mercy, in safely returning, and that of late, seaven of your Ships together, from that long, and tedious, and hazardous Voyage. A greater returne at once, for number of Ships, then ever you had, since you looked that way, since you knew that Trade. Now, as I cannot but presume you have already, more than once, sent up your private and particular acknowledgments to God for this great mercy : So you doe well, very well, now in Publick to give him thankes for it, *in the great Congregation.* (7)

Sermon

But to come up particularly to you, who are Merchants and Mariners, your dangers and deliverances are layd downe at large in this Psalme [107], and consequently your feares and joyes. You, I say, who are Merchants, and by being so, hold a correspondency by Traffique, with all places that are fam'd for Trade the world over: You by your Adventures can bring India, and Turky, and Egypt, nay Europe, Asia, Affrica, America, I meane all parts and places the World over, that know Commerce, in their rich and usefull Commodities home unto us. Certainly, as your calling is honourable, so 'tis very profitable and usefull to all

Kingdomes and Commonwealths; a calling very lawfull, while lawfully used.

The state of the World cannot stand without buying and selling, Traffique and Transportation; . . . No Countrey in the World yields in sufficiency all kindes of Commodities, and therefore there must be a path from one Kingdome to another, as there was from Egypt to Assyria, and from Assyria to Egypt back againe, to make a mutuall supply of their severall wants. (14–15)

. . . Now that which I would advise you too in the first place, that God may blesse you in your Factories abroad, & in your returnes home (which for my part I shall ever wish and pray for) is, as much as in you lies carefully to take heede that you imploy such Presidents, Ministers of the Word, Factors, and other servants, residing in all your remote places of Trade, as may take speciall care to keep God in your Families there: for let me tell you, that it is a miserable thing for such as professe themselves Christians in places where Christ is not knowne, or if heard of, not regarded . . . to play the Heathens, nay to doe worse, and that under the names of Christians . . . to shame Christianity by professing of it, by whose miscarriages, the Gospel, Christianity it selfe suffers

. . . So for a Mahumetan, or an Heathen in India, observing the very loose lives of many of the English there, the very foule misdemeanors of those that professe themselves Christians: to say of Christianity (as I have sometimes heard) *Christian Religion, Divel Religion, Christian much drunke, much Rogue, much naught, very much naught.* I speak this in their language, that is in that broken English those Indians speake, who live in those places who most converse with the English: And truly tis sad to behold there, a drunken Christian and a sober Indian; an Indian to be eminent for devotion in his seducing way, and a Christian to be remisse in that duty; for an Indian to be excellent in many moralities, and a Christian not so, for one who professeth himselfe a Christin, without which profession there is no salvation to come short of them, which come short of Heaven, what can be more sad than this? (29–30)[11]

I know how that you who are Merchants love to heare of places that are most advantagious for Trade, and I can tell you that there are richer places to be found then both the Indies, better Ports than *Surat* or *Bantam,* or any beside that can be thought on in the World . . . In the Land of the living there is durable riches to be found, which no violence can plunder, *nor Rust nor Moth,* nor fire, nor time can *consume* . . . And as those places that afford the richest Mines and Mineralls are most barren; so are those hearts that most affect them, that most seeke after them. Those Treasures which are concealed in the bowells of the Earth, are there layd up secretly, and basely; basely that wee might not overvalue them, and secretly that we should not spend too much time in the search after them. (32–33)

. . . And further, many of us heere in this Congregation may consider that time hath Snowed upon our haires, and the end of our journey cannot be farre off, and therefore it is very seasonable, now after many travells and troubles to thinke of, and prepare for our rest. . . .

And that wee may be the better prepared for that great businesse, wee must labour to be like Shippes abroad, farre from their home, that are well furnished and fitted, and richly laden, in readinesse to returne unto their Countrey, and want nothing but a winde to carry them thither. (35)

Notes

1. This information is taken (almost in the exact form of the original) from William Foster's introduction to the abridged version of *A Voyage to East-India*, reprinted in *Early Travels to India 1583-1619*, ed. William Foster (New Delhi: S. Chand & Co., reprinted, 1968), pp. 288-90. For a useful account of the historical background to Terry's travels from an Indian perspective, see Ram Chandra Prasad, *Early English Travellers in India: A Study in the Travel Literature of the Elizabethan and Jacobean Periods with Particular Reference to India* (New Delhi: Moti Lal Banarsi Das, 1965), pp. 276-322. Prasad also gives additional details about the publishing history of Terry's travel writings, pp. 281-82.
2. Deu and Daman are on the West coast of India and a part of Portuguese territory, known as Goa.
3. The Bay of Cambaia and Swally are on the northwest coast of India.
4. Terry spells *Row* rather than *Roe* throughout.
5. The Mogul Court was situated in Agra, in north-central India.
6. Indostan is a variant of the term "Hindustan," the title given to India in Urdu.
7. The term for Muslims (also spelled Mohammedan) at the time.
8. Mogul is also spelled as "Mughal" and "Mogol." Terry generally uses the term "Mogol," while referring to Jehangir the King of the Mogul dynasty.
9. Throughout Terry's narrative, he refers to Sir Thomas Roe as "My Lord Embassador."
10. Surat, on the west coast of India, was an important port of entry for the English merchants. Terry was probably traveling from the northwest to Mandu, where the Mogul King, Jehangir, moved his court temporarily. According to Foster, Terry "joined the ambassador near Ujjain towards the end of February 1617, and accompanied him to Mandu (called *Mandoa)*, where the Emperor had fixed his court until October of that year, when he removed to Ahmedabad (on the northwestern coast)"(288). As I stated in the biographical sketch, Terry never visited the main courts at Agra and Fatehpur Sikri and only saw a small area of western and central India.
11. This passage reappears in all the versions of Terry's *A Voyage to East-India*. In the 1665 version, it appears on pp. 418-19.

Special Note:

There are conflicting views of the year Edward Terry landed in India. In the 1665 version of *A Voyage to East-India* he notes it as 1615 and the same date is mentioned in the version reprinted by Purchas. However, Thomas Roe and later historians cite 1616 as the correct year. I assume the latter date to be the correct one.

HISTORY OR COLONIAL ETHNOGRAPHY? THE IDEOLOGICAL FORMATION OF EDWARD TERRY'S *A VOYAGE TO EAST-INDIA* (1655 & 1665) AND *THE MERCHANTS AND MARINERS PRESERVATION AND THANKSGIVING* (1649)

Jyotsna G. Singh

I

Faithful Observations

> I have nothing to plead for this presumption but the Novelty of my Subject, in which I confesse some few have prevented me, who by traveling *India* in *England,* or *Europe,* have written somewhat of those remotest parts but like unto poor Tradesmen who take up wares in trust, have been deceived themselves, and do deceive of others.
>
> For myself I was an eyewitness of much here related, living more than two years at the Court of the mighty Monarch, the great *Mogol*[1] (who prides himself very much in his most famous Ancestor, *Tamburlane*) in the description of whose Empire, your Highnesse may meet with large Territories, a numerous Court, most populous, pleasant, and rich Provinces, but when all these shall be laid in the balance against his miserable blindness, your Highnesse shall have more cause to pity, than envy his greatnesse.
>
> —Edward Terry, "Epistle to the King," Preface to *A Voyage to East-India* (1655), A2.[2]

How do we imagine Edward Terry's voyage and visit to India more than 400 years ago? Reopening the records of past travelers from our perspective of a new millennium, we are also led to make a postcolonial reevaluation of the beginnings of European colonialism. Records such as Terry's claim experiential veracity based on eyewitness observation, while they also discursively map the new, yet over determined, discursive terrain. Thus, ironically, it is not surprising that their claims of eyewitness authenticity are as much in keeping with generic expectations as they are in offering factual accounts. However, while each narrative stresses its unique perspective, when we consider how the English travelers' journeys overlap or follow one upon the other, we can also note how their accounts frequently corroborate and/or draw from one another. Terry's narrative about East-India, for instance, offers an interesting interplay of connections with Sir Thomas Roe's *Journal* of his experiences as the Ambassador and Thomas Coryate's *Letters* from the Mogul Court, among others.

Throughout his narrative, Edward Terry assumes the role of an eyewitness, recording what he sees. Asserting his veracity in his "Epistle to the King" (epigraph), Terry straddles both public and private worlds: as a creative ethnographer driven by the "Novelty" of his "Subject," and, it seems, as a trusted witness to the historical event of King James's Embassy to India through his representative, Sir Thomas Roe. The compiler of the 1665 edition, for instance, prefaces Terry's text by describing it as follows: "one of the Exactest Relations of the Eastern parts of the World that hitherto hath been publish'd by any Writer, either Domestick or Forreign; having been penn'd by one that attended Sir Thomas Roe in his Embassy to the Great Mogol."[3] The idea of the European as an eyewitness frequently emerges in many travel narratives. As Samuel Purchas notes in his preface "To the Reader" in his monumental anthology, *Purchas, His Pilgrimes:* "What a World of Travelers have by their owne eyes observed in their kinde is here delivered . . . by each Traveler relating what he hath seene" (xl). In travel writing as a genre, he explains, "remarkeable varieties of Men and humane Affaires are by Eye-witnesses related more amply and certainly than any collector." [4]The centrality of the firsthand observation to the travel genre noted by Purchas continues to be a topic of interest in contemporary studies of early modern travel writing. As one critic observes, the act of witnessing serves as a "foundational rhetorical device which fabricates and accredits the travel text as a recorded understanding of otherness."[5] If witnessing was the primal act in the discourse of travel, it meant that travelers/writers could then attempt to record, objectify, and claim the world they encountered. However, given the narrators' repeated claims to veracity based on *seeing,* they seem to be preempting the question as to why they should trust an eye-witness account by falling back onto a familiar generic motif.[6] Thus, while assuming transparency, the act of witnessing complicates the task of the traveler/writer.

We can note some of these complexities throughout Edward Terry's narrative. His role as an eyewitness, offering an empirical rendition of unfolding events, often converges with his role as a writer shaping his materials. His descriptive taxonomies of the landscape, the people, and their customs, among other things, illustrate the discursive shaping power of travel writing, even while the narrator stresses the authenticity of his experience. Terry is conscious of his "Reader," whom he frequently addresses, yet he also seems to accept an ideology of language, claiming transparency of representation and immediacy of experience. He reveals his own methodology, in the following passage, when he moves from one subject to another:

> And so having observed what is Truth, and what is enough to be said of the Inconveniences and Annoyances, as well as of the Commodities . . . which are to be found in those parts, I come now to speak of the People that inhabit there. And because many particulars will necessarily fall within the compass of this part of my Observations, which would more weary my Reader if they should be presented unto him in one continued Discourse, I shall therefore (as I have begun) break this into Sections, and proceed to speak. (374)

Recent literary scholars and historians suggest that while much of what Terry describes about India had already appeared in print in Europe, many of his observations were fresh and detailed. Typically, they recognize that the "vast mass of information" reflected the multiplicity of Terry's interests. As one critic observes:

> His [Terry's] eyes have moved freely from the exalted palaces of the Great Mughal [the emperor Jehangir] to the base cottages of the poor, enabling him to furnish in . . . his *Voyage* a valuable repertory of descriptions of all kinds. . . . [And] accounts which add little or nothing to what we already know from Sir Thomas Roe are outnumbered by . . . passages which contain invaluable details that no other traveller furnishes. *The Voyage to East-India* affords, on the whole, an excellent picture of Jehangir's India, drawn by one of the most acute observers of the early seventeenth century.[7]

Terry begins *A Voyage to East-India* with a catalogue of territories governed by the Great Mogul. He finds that the Mogul's dominion, the "most spacious and fertile Monarchy (called by the Inhabitants *Indostan*), so much abounds in all the necessaries for the use and service of man, to feed, and cloath, and enrich him, as that it is able to subsist and flourish of it self, without the least help from any Neighbour-Prince or Nation" (358). The landscape, in the Englishman's eyes, is a veritable cornucopia of natural resources; it is a "Countrey . . . beautified with many Woods and Groves of Trees . . . [and watered by] "many goodly Rivers" (Extract II). It also does not escape Terry's attention that nature's gifts, such as indigo, silk, and cotton wool, and "good Mineralls of Lead, Iron, Copper, Brass," among others, are useful commodities for the English merchants (Extract III). Overall, the Chaplain's evocations of the fruitful land are accompanied by reminders of "Annoyances" such as preying beasts, scorpions, mosquitoes, gnats, and rats, also a part of the fertile land. And in a typical moralizing gesture Terry frames his observations with a biblical homily: "*The Garden of Eden had a Serpent in it*" (373). Thus, he views the seemingly paradise-like landscapes of East-India as a fallen Eden, despite many "things to content and please the enjoyers of them" (371).

In describing the inhabitants of India, Terry offers painstaking details of diverse races and groups: "Hindoes, belonging to that very great number of those which are called *Heathens,* which take up almost two thirds of the. . . . People who inhabit the face of the whole Earth" mixed with "*Mahumetans,*" some of whom are "*Persians* and *Tartars,*" and *Armenians,* and "some Abissins amongst them . . . [and] some Jews [whom the] Almighty God . . . scattered among Nations of the World"(374-75) (extract IV). Protestant Christianity remains the normative marker in his approach to these cultural others. While he notes both positive and negative aspects of the major religious communities, he repeatedly views all non-Christians as "ignorant of God" (438). It is notable that this religious ethnocentrism or intolerance extends to Jews whom Terry encounters in East-India, bringing to the fore the characteristically European anti-Semitism of the period (Extract IV). However, despite his moralizing intrusions, Terry places these diverse peoples in a rich, bustling world, describing their markets, mosques, burials, feasts, books, eunuchs, marriages, among a variety of subjects. Finally, and perhaps most significantly, the Chaplain's account throws light on the general administration of the empire of Jehangir, who called himself "*Conquerour of the World*" (449) and who, in his view, was a very impressive, though absolutist, monarch (Extract VI).

Should we view Edward Terry as a historian or a quasi-ethnographer? While ethnography did not exist as a scientific discipline and social science, the Chaplain's role in his own narrative frequently resembles that of a participant observer—though not rooted in a single bounded site observing the natives—but more uncharacteristically, recording a string of episodes and observations while traveling thorough western India.[8] Following Terry's detailed typologies of nature, peoples, customs, and religions, one is struck by the interplay of allegiances and oppositions within the sway of an inclusive imagination, though given to judgmental observations. For instance, he looks for both positive and negative aspects of eastern religions, while considering Protestant Christianity the only true religion. Not surprisingly, then, Terry's factual account is permeated by images from the scriptures.

Terry covers a wide range of subjects, attempting a fairly comprehensive picture of the Mogul Empire; his empirical "witnessing," nonetheless, often, though not uniformly, takes the form of a colonizing moral imagination, mapping and cataloguing the alien lands and people through the prism of a supposedly normative English Protestantism.[9] English merchants in the early and mid-seventeenth centuries did not equally partake of an ideological consensus about the necessity or inevitability of English commercial and political domination. This consensus gathers force later on, in the late eighteenth and early nineteenth centuries.[10] However, while colonial imperatives are not yet evident, the accounts of Thomas Roe, Edward Terry, and numerous other travelers/writers, are nonetheless implicated in the social, cultural, and religious codes of the time—particularly, in a belief in the natural superiority of "Christendom" over non-Christian heathens.

II. In the "footsteps of the Almighty"

> . . . in our passage to East-India [we] may observe very large footsteps of the Almighty in works of Creation and Providence.[11]

On the opening page of *A Voyage to East-India,* Edward Terry sees the hand of God in his impending journey: It was on the "ninth of March, on which day it pleased God to send us, what we much desired, a North East Wind, which made us . . . set sail for East-India"(325). It is ironic that in part Terry envisages his sojourn in India as a discovery of another Eden, while seeing his experiences in terms of a divine destiny. For instance (as noted earlier), as a diligent geographer, Terry represents the territories of the Great Mogul as a veritable cornucopia of the gifts of nature: this "most spacious and fertile Monarchy," he states, "so much abounds in all necessaries for the use and service of man [that it] brings forth [food] in abundance"(358) (extract II). The seasonal rains seem particularly refreshing: within a few a days after "those fat enriching showers begin to fall, the face of the Earth there (as it were by a new Resurrection) is so revived, and throughout so renewed, as that it is presently covered all over with a pure green Mantle" (362). Such tropes of plenitude cast an Edenic glow on his picture of "East-India," as Terry frequently transposes the image of the primal garden (in the shadow of the fall) onto the profusion and fecundity of the tropical landscapes. How can he accommodate the image of a regained Paradise with its heathen inhabitants? Terry does so via an allegory of God's purpose in his Preface to the 1655 edition. Here he reminds his readers that they will find God's "footsteps" among "his works of Creation and Providence" in the East Indies. Thus, he suggests that divine meaning can be read in everything he records, as, for instance, in the "riches and splendor" of the Court one can "consider of the glorie of Heaven." And asserting the divine purpose of a recognizably Christian God, Terry exhorts his readers to question that if "Almighty God hath given such sweet places of abode here on the earth to very many whom he owns not; how transcendently glorious is that place which he hath prepared for them that love him" (A 6). Overall, Terry sees God's will in the Englishmen's journey to East-India not in terms of a universal, providential history for all time, but within a more local context, looking for the Christian God's presence in an alien land rich in natural resources.

Thus, while the profusion of religious references coupled with moral homilies in *A Voyage to East-India* may seem like digressions, they, in effect, underpin the commercial and political imperatives of the English merchants, as well as of Thomas Roe, the representative both of the East-India Company and King James I. If East-India, in Terry's view, is a paradise of nature's plenty, reminiscent of the biblical Eden, its natural treasures are also the valuable commodities for which different European nations compete (Extract III). Early modern Europeans generally represented their "discoveries" and trade in the East Indies (and the Americas) in providential terms, and Terry is no exception in making such claims. He does so in *A Voyage to East-India* and more explicitly in his sermon delivered to the East-India Company in 1649, and printed in the same year under the title of *The Merchants and Mariners Preservation and Thanksgiving. . . .* Called to preach before the Company on the occasion of the return of seven ships from the East Indies, Edward Terry offered the merchants a religious validation of trade. In fact, the ideological function of Christianity in justifying England's expanding commercial and colonial role becomes apparent in this religious tract in which Terry elides profitable and providential trade.

In an uncanny echo of the twenty-first-century rhetoric of globalization, the sermon represents English (and implicitly European) trading ventures as naturally beneficial to all nations and regions: "You [merchants] by your Adventures can bring India, and Turky, and Egypt, nay Europe, Asia, Affrica, America, I meane all parts and places the World over, that know Commerce, in their rich and usefull Commodities home unto us. Certainely, as your calling is honourable, so 'tis very profitable and usefull to all Kingdomes and Commonwealths" (14) (Extract VIII). While reminding the representatives of the East-India Company of "richer places" in the afterlife, and cautioning them not to "overvalue" earth's "richest Mines and Mineralls"(32-33), the Chaplain nonetheless naturalizes the pervasive mid-seventeenth-century perspective on the inevitable—and necessary—spread of trade. He sums up his attitude when he declares, the "state of the World cannot stand without buying and selling, Traffique and Transportation" (14) (Extract VIII).

Overall, Terry's narrative (in both texts) gives his readers some glimpses of the implicit religious Eurocentrism accompanying mercantile expansion to non-European and non-Christian lands. While blessing the merchants' enterprises—"God may blesse you in your Factories abroad, & in your returnes home"—Terry also advises them to "imploy such Presidents, Ministers of the Word, Factors, and other servants, residing in all your remote places of Trade, as may take speciall care to keep God in your Families there"(29) (Extract VIII). Terry's idealization of Protestant Christianity leads him to call for appropriate behavior of individual Englishmen, who must set a moral example as proponents of a superior religion. The Chaplain's anxieties about unworthy Christians (as compared to the Indian members of inferior religions) appear in a curious anecdote in his sermon (recapitulated in *A Voyage to East-India*) in which Indian voices offer their critique of the English travelers: Terry considers it unfortunate "for a Mahumetan, or an Heathen [here meaning Hindu], observing the very loose lives of many of the English there, the very foule misdemeanours of those that professe themselves Christians: to say of Christianity (as I have sometime heard) *Christian Religion, Divel Religion, Christian much drunke, much Rogue, much naught, very much naught*" [italics Terry's] (30) (Extract VIII).[12]

In both texts, the Sermon and the travelogue, Edward Terry repeatedly strikes a cautionary note about unworthy Christians who make "Christianity it self evil spoken of [in East-India], as a Religion that deserves more to be abhorred, than imbraced" (418). Perhaps unlike many other travel narratives of the period, the Protestant Christian world forms the Chaplain's overarching frame of reference. In this world view his authorial voice sets up standards of integrity, even though he takes great pains to register the otherness of the natives—Hindus, Mahumetans, Armenians, Jews, among others—via a diverse range of cultures, creeds, and representations. Yet, in doing so, he ironically seems to prefigure nineteenth-century Victorian liberalism, which recognized the diversity of native cultures within the British Empire, while holding in place the dichotomy—and hierarchy—between "civilization" and "barbarism." Such desired hierarchies and divisions are not always held firm by the English moralizing and colonizing impulses, as Terry concludes his narrative with a continued lament about debased Christians. Among the reasons he gives that "hinder the settlement and growth of Christianity in those parts" are the "most debauch'd lives of many [Europeans] coming thither, or living amongst them who profess themselves Christians." By their failings, he believes, "the gospel of Jesus Christ is scandalized, and exceedingly suffers" (480).

III. Colonial Knowledge

It is evident that the representational practices of English travelers/writers in the sixteenth and seventeenth centuries were shaped by the ideological codes of the time. Terry's investment in Christianity is one such example. However, it is also suggested that these sixteenth- and seventeenth-century narratives of "discovery" and trade were strategic in detailing the initial commercial penetration into

Mogul territories—information that was useful to later imperial designs, as Bernard Cohn explains:

> The records of the seventeenth and eighteenth centuries reflect the [East-India] Company's central concerns with trade and commerce. In it one finds long lists of products, prices, trade routes, descriptions of inland and coastal and inland marts, and "political" information about the Mughal Empire. . . .[13]

It would be obvious that accounts such as Terry's possibly served as repositories of information not only about trade, but also about the geographical features and social and religious customs of the land. Terry himself confirms, (as does Roe in his *Journal*) that he witnessed the establishment of trade in East-India: "we Englishmen . . . there found a free Trade, a peaceable residence, and a very good esteem with that King and People" (455) (Extract VII). Since travel narratives were mostly written by those who represented commercial or political interests (Terry was somewhat of an exception in being a Chaplain, but he was an employee of the King and Company), it seems obvious that later travelers and administrators drew on the existing archive of travel narratives. For instance, we know that Hakluyt was a consultant and advisor for the East-India Company.

Those of us studying the early modern travel archive face the problematic task of locating an originary moment of colonialism. Critics ranging from Mary Louise Pratt to David Spurr historicize travel writing and consider it a form of colonial discourse. Spurr, for instance, defines it as "a form of self-inscription onto the lives of people who are conceived as an extension of the landscape. Thus, for the colonizer, as for the writer, it becomes a question of establishing authority through the demarcation of identity and difference."[14] It is important to note, however, that these postcolonial theories focus on travel writing as a colonial/imperial genre in the mid-eighteenth century and the later period of high imperialism and, hence, are not clearly applicable to the earlier narratives of exploration. Thus, we should be sensitive to yoking disparate journeys and encounters within an ideological straitjacket of colonial intent. In such a proleptic reading, the traveler becomes merely a colonist and usurper.

Furthermore, I want to include another approach to travel writing—drawn from psychoanalysis and structuralism—that enables us to contrast the more concrete and empirical moments in travel writing with experiences of fantasy and frustration where the "dream of possession and occupying the other" eludes the traveler/writer.[15] Complicating the typical postcolonial perspective on cross-cultural encounters with this model affords us a way to consider the more contingent aspects of the travel experience/narrative, and in doing so, to bracket those moments when the European desire to *know, possess,* and *master* the other remains unfulfilled.

IV. Narratives of Encounter

While the narratives of the English travelers/writers such as Edward Terry reveal shared cultural/religious assumptions of the period, the Englishmen do not emerge as static allegorical entities, as agents of an incipient colonial ideology; but rather they frequently appear as complex historical subjects struggling to interpret a different culture that challenges the stable categories and assumptions of English cultural and (especially in Terry's case) religious identity. Of particular interest, therefore, in any study of travel writing as a literary/political genre is the *narrative of the encounter.* Given the hindsight of our postcolonial moment, it is easy to reduce the cross-cultural encounter to relations of domination and subordination—to trace a trajectory between the early modern proto-colonial moment to nineteenth-century imperialism. Instead, I want to bracket Terry's narratives of encounter and distance, though not separate them from any direct imperial trajectory; this is not to depoliticize their implicit collusion with an incipient colonial ideology, but rather to reveal a complex interplay of cultural signification in Terry's recounting of such "living" encounters.

Most of Edward Terry's descriptions of the East Indies in *Voyage to East- India,* and (in a less detailed form) in his earlier *Sermon,* consist of detailed classifications, vivid typologies, and religious homilies—as well as details of the types of transportation, housing, attire, weapons, and animals, among numerous other aspects of Indian life. However, within this general descriptive and moralizing frame, Terry brings to life numerous interactions between the English (including himself as either a witness or participant) and Indians. Ironically, these "real" encounters also create an imaginary space upon which the reader can participate in the playful dynamic of cross-cultural meetings that elude the narrator's classifying grasp.

Imagine Terry traveling from Surat with "four Englishmen, and about twenty of the natives" toward the Mogul court "four hundred miles from Surat" (see Extract V for the entire incident). Resting on the way in a place called "*Ditat,*" where many of the inhabitants offered to guard them, and although the Englishmen "desired it not," the natives "asked something of us by way of recompence." The English refusal led to a minor incident: followed by "at least of three hundred Men," a young, hot-headed (aristocratic) young man, shot his musket and stirred the crowd of the Indian men. Had Terry not intervened, he fears, the conflict may have resulted in "the loss of . . . [their] lives and goods" (394-95). The English negotiations of offering three shillings brought temporary peace that was again disrupted by the "unruly" "young Gentleman," the "hot-brains," when he whipped a stranger for "refusing to hold his horse." This incident seems to have an uncanny resonance: a scene of Englishmen surrounded by natives with some misplaced expectations on both sides. But what did the native Indians think of the English? Why did such large numbers follow the English party? Though Terry indicates brief exchanges, we do not know whether there were some effective translators in their midst, since at other times, the Chaplain mentions Indians speaking in a "broken English." It is telling that Terry records the voices of the Indian natives, especially commoners, but almost like a ventriloquist, since he does not offer us much depth of interpretation of their motives, feelings, reactions. In fact, despite the textual density of the European records and the sporadic voices of the Indian natives emerging from the descriptive narrative, we get little "sense of an indigenous actuality, except as mediated by English drives and desires."[16] Terry's moral disapproval of the young Gentleman offers an important insight into what became a historical trend: the younger, sometimes undisciplined sons of aristocratic families being sent to the colonies. This young man remains nameless, though Terry reveals that the ambassador was "acquainted with his kindred" and "entertained him as a companion for a year" (395).

While we can contrast the improvisational nature of this episode with the more formal interactions between the English Ambassador's party and the Mogul King, Jehangir, even in these latter narratives of encounter we can find dissonant strains in Terry's claim to knowledge and moral superiority. The Chaplain's frequently awe-filled accounts of the Mogul resonate with mythic/historical associations of the times. The figure of "Tamburlain" as the ancestor of the Moguls obviously had apocryphal associations for Renaissance Europeans facing the Islamic other in the form of the Ottoman Empire on its borders. In this vein, Terry begins by recapitulating Tamburlain's legendary expeditions, "having got together very many huge multitudes of Men, made very great Conquests in the South-East parts of the World" (446) (Extract VI). That the Mogul Court, like the Ottoman Court, stirred awe and wonder in the European consciousness is also apparent throughout the account. Terry goes into painstaking detail regarding the Mogul King's displays of wealth as well as his imperious, though politically manipulable, modes of governance. While Sir Thomas Roe attends to the Mogul King, Terry proves to be a valuable witness, leaving a record that frequently validates Roe's own observations.

The Mogul King figures prominently in several episodes of live interactions recorded by Edward Terry. In all these encounters between the Englishmen and the Mogul Court (frequently in the person of the King), Terry's classificatory, moralizing, Christian modes of description are often replaced by an account of different economies of exchange—political, cultural, commercial—in which the Englishmen are compelled to participate. Gift-giving, premised on a different system of reciprocity, seems an important aspect of their relations with the Mogul ruler as Terry observes, "there is no coming unto that King empty-handed without some Present" (448). In some of these exchanges, we can hear Jehangir's voice (even though from a second hand source), often giving us glimpses of how the Mogol Court interpreted the early modern European forms of knowledge with their attendant claims to power.

A curiously telling anecdote is one in which the English ambassador presents the King with Mercator's *Book of Cosmography,* "telling the Mogol, that that Book described the four parts of the World, and all severall Countries in them contained" (448) (see Extract VI for the full episode). The King rejects the gift of the book on the pretext that neither "himself, nor any of his People did understand the Language in which that Book was written," but Terry also notes that the ruler who called himself the "*Conquerour of the World*" seemed to resent that he did not have a greater share in its representation of the world's territories. The European mapmakers, in this instance, Terry acknowledges, "streighten him [the Mogul] very much in their Maps; not allowing him to be . . . Commander of those Provinces, which rightly belong unto him" (449). That Sir Thomas Roe had brought this book of maps for "his own use" would suggest that books such as Mercator's were important European sources of knowledge of the world they were "discovering"; but Renaissance maps, as we also know today, both literally and ideologically often simply substantiated and embellished the landscapes of the colonizing European imagination.[17] This narrative of encounter reveals a power struggle on multiple levels: between the ambassador and the Mogul King, who constantly demands gifts as a condition of his patronage, and between competing modes of and understandings of knowledge.[18] Thus in this scene, as in other such moments scattered through *A Voyage to East-India,* the Chaplain plays the role of a "participant observer," while suspending his classificatory drives and following the contingent turn of events and interactions with the cultural others he encounters.

To conclude, as Edward Terry takes his readers on a journey through seventeenth-century "East-India," he assumes many different roles: geographer, moralist, historian, and quasi-ethnographer. Like other European travel writings of the period, *A Voyage to East-India* marks the early stages of colonial domination, as it maps the geographical, commercial, and imaginative space of Mogul India. Yet, the "dream of possession" remains unrealized by these English travelers/writers, as they struggle to interpret the literal and imagined landscapes in which they find themselves. Terry sums up the elusiveness of the English claims to power in India when he describes their relations with the Mogul King as follows: ". . . what he [the King] promised was usually enjoyed, although he would not be tied to a certain performance of his promise. Therefore, there can be no dealing with this King upon very sure terms, who will say and unsay, promise and deny" (455) (Extract VII).

Notes

1. Note the spelling of *Mogol* in Terry's account here (and in the primary texts) differs from *Mogul* (my spelling) and *Mughal* as spelt by Prasad and other critics. Such variations in the English spellings of Indian terms and names are common.
2. This excerpt is taken from the Preface "To the Reader" in the 1655 edition of *A Voyage to East-India.* When the work was republished in the folio volume in 1665, the Preface was cut.
3. The compiler of the 1665 edition is one G. Havers who dedicates the folio collection to the Earl of Orrery (A3). The primary extracts accompanying this essay, and from which I cite here, are taken from the 1665 edition.
4. See Samuel Purchas, *Hakluytus Posthumous or Purchas His Pilgrimes,(1625)* vol. I, (Glasgow: James MacLehose and Sons, 1905), pp. xxxix-xlviii, for a detailed account of the methods and cultural assumptions of travelers/writers in the period.
5. This quotation is taken from Brian Mugrove's essay, "Travel and Unsettlement: Freud on Vacation," *Travel Writing and Empire: Postcolonial Theory in Transit* (London: Zed Books, 1999), p. 40. It refers to Stephen Greenblatt and de Certeau's readings of the "act of witnessing" in European travel writing, especially in the early modern period.
6. Stephen Greenblatt examines the structure of witnessing in early modern travel narratives in some detail in *Marvelous Possessions: The Wonder of the New World* (Chicago: University of Chicago Press, 1991), p. 122. I draw on his formulations in my discussion of witnessing.
7. Ram Chandra Prasad, *Early English Travellers in India: A Study in the Travel Literature of the Elizabethan and Jacobean Periods with Particular Reference to India.* (Delhi: Motilal Banarsi Dass, 1965), p. 293. Prasad gives many examples to show both how Terry's accounts are often authenticated by other travelers/writers of the period and also reflect the religious and cultural biases of his contemporaries. For instance, according to Prasad, Terry's descriptions of the grandeur of the Mogol's Court and his particular love of watching elephant fights are verified by references in the historical chronicles of the period as well as in modern studies, 309-13. Sometimes Terry's

information overlaps with Roe's *Journal,* as, for instance, his "list of provinces under the Mughal rule is almost a reproduction of Roe's geographical account of the Mughal territories"(311). In frequently looking upon Indian religions with disfavor, however, Terry reflected the prevailing Christian ideologies. Like other European travelers, he simply described what he saw, but without investigating the origins of unfamiliar practices he observed (297). Overall, Prasad, echoing other critics, views Terry's narrative as an important supplement to Thomas Roe's *Journal.*

8. In viewing Terry as a quasi-ethnographer, here and elsewhere in the essay, I am indebted to James Clifford's formulations of reading and writing culture in historical relations of travel. See *Routes: Travel and Translation in the Late Twentieth Century* (Cambridge, MA: Harvard University Press, 1997), pp. 1-39.
9. Here I draw on David Spurr's concept of a colonizing imagination, which assumes a "metaphorical relation between colonizing and writing." Thus, following this premise, one can approach colonial interactions as an effect of power relations inscribed within cultural and linguistic forms. See Spurr, *The Rhetoric of Empire: Colonial Discourse in Journalism, Travel Writing, and Imperial Administration* (Durham, N.C.: Duke University Press, 1993), especially pp. 6-7. His assumptions about the language of colonization, however, apply more explicitly to a later period of colonial rule, beginning in the eighteenth century.
10. For a fuller account of the gradual emergence of the English colonial presence, see Jyotsna G. Singh, *Colonial Narratives/Cultural Dialogues: "Discoveries" of India in the Language of Colonialism* (London: Routledge, 1996), especially pp.1-50.
11. In his Preface "To the Reader" in the 1655 edition of *A Voyage to East-India,* Terry offers a moral, providential justification for the "discovery" of India, especially see A6 and A2-A4.
12. This anecdote of Terry giving voice to native speech in English can be found in a slightly different wording in the 1665 reprint of *A Voyage to East-India,* p. 418
13. Bernard Cohn, "The Command of Language and the Language of Command." In *Subaltern Studies: Writings on South Asian History and Society,* vol. IV, ed. Ranajit Guha (Oxford: Oxford University Press, 1985), p. 276.
14. This quote is from Spurr, *The Rhetoric of Empire.* Mary Louise Pratt historicizes European travel writing from 1750 to 1980 in *Imperial Eyes: Travel Writing and Transculturation* (London: Routledge, 1992).
15. I am indebted for this model to Brian Musgrove's essay, "Travel and Unsettlement: Freud on Vacation," and especially to his reading of Meghan Morris's essay, "Panorama: The Live, the Dead, and the Living," pp. 40-41.
16. See Singh, *Colonial Narratives/Cultural Dialogues,* p. 28.
17. Much has been written about the formation of cartography as a cultural formation in early modern Europe and Anthony Grafton's *New Worlds, Ancient Texts* (Cambridge: Belknap Press, 1992) is one such study. It points to the importance of cartography in European discourses of "discovery."
18. See Bernard Cohn's essay, "The Command of Language and the Language of Command," for a full account of the Indians' understanding of power and rights—an understanding that, unlike Europeans, was not premised on the authority of texts and language, but on the literal being and person of the king. Cohn also discusses the translation problems faced by Roe and his companions, who had to rely on seemingly unreliable interpreters.

PART III:

Travel to Africa

SECTION EIGHT

THE RED DRAGON IN SIERRA LEONE

The following extracts are both taken from the only surviving journals kept on board the *Red Dragon* in 1607, en route from England to India and the Spice Islands on the Third Voyage of the East India Company.

The manuscript journal kept by John Hearne and William Finch, two English merchants, is now in the British Library (L/MAR/A/v); it has never been published in its entirety. Finch had previously been a "servant to Master Johnson in Cheapside," and later in the voyage was left in charge of some goods at Surat. Finch also compiled a retrospective of "Remembrances touching Sierra Leone, in August 1607," which was published by Samuel Purchas in *Hakluytus Posthumus; or, Purchas his Pilgrims* (1625). I have quoted relevant extracts from these "Remembrances" in some of the notes below.

Sir William Keeling (1578-1620), author of the second journal, was in command of the entire expedition; extracts from his journal were published by Purchas, who noted that "This journal of Captain Keeling's . . . written at sea leisure, very voluminous in a hundred sheets of paper, I have been bold to so shorten as to express only the most necessary observations for sea or land affairs." The manuscript is now lost, but it survived among the papers of the Company until at least the middle of the nineteenth century; some additional extracts were published in *The European Magazine* (1825/26) and in *Narratives of Voyages towards the North-West,* edited by Thomas Rundall (1849).

—Gary Taylor

Journal of Finch and Hearne

August 6 . . . Within half an hour after we came to an anchor,[1] there came some of the people to the waterside, waving with a white flag to have us come ashore. So our General[2] caused our pinnace to be manned, which rowed ashore unto them. But none of our people could understand them, only but by signs. So the pinnace returned aboard, bringing four negroes in her, leaving two of our men in gage[3] for them; which, after kind usage, and making signs unto them for fresh victuals, and giving them many odd trifles, they were set ashore, and our men returned aboard for that night.

This night John Pawling, a Norfolk man, one of the Hector's[4] company, died within one hour after we came to an anchor.

7 In the morning divers[5] negroes came aboard, bringing with them a small quantity of limes and hens,[6] for which we gave them beads and knives, our General giving them the best content that possibly he could, the better to procure some refreshing of them, making signs unto them for beeves, goats, sheep or other victuals.[7] They made signs that none was to be had about that place, but all the cattle were far up in the country. These negroes were very important[8] to have somebody sent unto *Capitan*[9] Pinto, their commander;[10] and by their great persuasion our General sent Edmund Buckbury with them, retaining for hostage aboard three negroes, and there was sent divers trifles for a present, which were thankfully received. And in the evening Buckbury returned, bringing from Pinto a small ring of gold for a present to our General. This their commander dwelleth near unto the point of *Sierra Leona*[11] and hath his name Pinto (quasi *de Punto*),[12] because his command lieth thereabout, and dwelleth so near the point.

8 The negroes were set ashore, and with our seine we caught good store of fish, and refreshed ourselves with some limes.

9 We had very foggy weather and much rain.[13]

10 We filled some water and caught great store of fish, and bought off the people some limes and a few hens for our sick men.[14]

11 We filled about eight tons of water and caught much good fish. This day we weighed,[15] and sailed farther within the next bay, which is about the third point from *Sierra Leona* point, and is the fourth cove or bay[16] on the south side of this harbor, where we came to an anchor near unto the watering place in eleven and a half fathom water. This day we also bought store[17] of limes and some hens.

12 We got aboard about twelve tons of water and caught some good fish. All night and all the forenoon fell very much rain.

13 We drew our seine and killed good store of fish. This day certain negroes came aboard in a *canoa*,[18] bringing with them an elephant's tooth,[19] which we bought off them for some calico and iron. All this day fell very much rain.

14 Fell very much rain both day and night, yet we filled some ten tons of water and caught great store of fish, as also bought many lemons.

15 We got aboard some wood and filled five ton of water. This day in less than three hours [time] we got with our seine above six thousand fishes (all of a fin,[20] being small fishes), for which God be blessed.

16 Our General with Captain Hawkins, accompanied with a good number of men, marched up into the country unto one of their towns,[21] where we filled eight biscuit bags with limes; and having recreated ourselves with walking, in the evening returned aboard.

17 We filled more water and caught some fish. This day fell much rain.

18 Fell much rain, yet we got aboard some wood. This day by the persuasions of some of the people our General sent John Rogers unto their great commander called Captain Boree, retaining three negroes for a hostage; sending unto him for a present some calico, with a bottle of wine and some other trifles. This afternoon we caught much fish.

19 Fell very much rain both day and night, that we could not stir with our boats.

20 Fell very much rain all day, and in the afternoon returned John Rogers in a *canoa* from Borea, having been by him very well entertained. He brought one gold ring and some fruits for a present to our General. John Rogers met with a negro called Lucas Fernandez, which spoke the Portingal[22] language very well and was interpreter to the king, by whom we learned many things.

21 Fell some rain, yet we filled some water and got some wood aboard.

22 Fell very much rain, yet this day we bought good store of lemons and made seven *barricas*[23] full of lemon water for both ships, bringing also store of lemons aboard with us.

23 We marched into the country to refresh ourselves, and in the evening returned aboard, having bought about 20,000 lemons. This day our General sent some of our sick men ashore into certain Portingal houses which stand by the waterside, being now empty, where they remained, the sooner to recover their healths.

24 Fell very much rain with much thunder, and in the afternoon we went ashore and bought some limes.

25 Our General with Captain Hawkins went to Captain Pinto's dwelling to seek for some fresh victuals, but could find none; so in the evening, having caught good store of fish, returned to our ships.

26 Fell some rain, yet we got aboard some wood and bought 20,000 limes.

27 In the morning Captain Hawkins, with his and the Dragon's pinnaces being well manned, went to the eastward into the bay about three leagues from our ships, unto the place where the principal commander or king doth keep, called Captain Borea,[24] to confer with him, and if it were possible to procure him to be a means to help us to some cattle; at which place we found Lucas Fernandez the negro, who told Captain Hawkins that there were neither cattle, sheep nor goats to be had; but for hens and plantains he would procure all that he could. Also he could help us with some elephants' teeth, but no quantity worth the speaking of, and so dear[25] that no profit could arise thereby—he rating the quintal[26] of teeth at twelve bars of iron, and if in linen cloth (as Rouen)[27] thirty-six *varas*[28] of Rouen for every quintal, he accounting every bar of iron at three *varas* of Rouen; so that in linen cloth the quintal of teeth will cost 54 shillings, and in iron 48 shillings, at the rate the commodities cost in England.

Some gold in small rings is to be had, and it may be some in bars, if they will bring it down. Their gold is likewise dear. He doth rate the gold at three shillings; I say three *varas* of Rouen, an "ochava,"[29] as he [called] it, which I take to be the one-eighth part of an ounce (for the negro did not know the weight when we showed it him, only the name "ochava"), which is four shillings three pence the "ochava", or the one-eighth part of an ounce. These prices I do nominate for if any man coming hither to trade, and doth deal for any quantity, these rates are to pay. But to buy here and there a ring of gold of some negro, you may have it very cheap; as also elephants' teeth, to buy them of some negro which do sell them by stealth, you may buy a great tooth, which may weigh one-half hundredweight, for eight *varas* of Rouen.

All sorts of linen cloth, both white and brown, are good commodity; small brass kettles or basins of brass or tin, and knives both great and small of eighteen pence per dozen, are also good commodities. You may buy for a two-penny knife upwards of 300 lemons, or a hen, or a great bunch of plantains. Beads white, yellow, blue and black of all sizes are good also; for twenty beads you shall buy 300 lemons and more.

28 Fell very much rain, and about one o'clock we descried a small sail coming in with us right before the wind; but having no sooner made[30] our ships, as we judged, but she tacked about and stood to the northward,[31] and in the evening we discerned her running alongst the north shore into the bay, having her pinnace sounding[32] before her.

29 Fell very much rain, and in the morning we perceived the small ship working into the bay. Our General, being desirous to speak with her, sent off his pinnace well manned, wherein went Mr. Hippon; who, when he fetched her up,[33] haled her and aske[d] of whence she was. And they answered that she was a Portingal barque, belonging to the isles of Cape Verde, of the burden of forty tons, laden with rice and salt, coming from Santiago;[34] the master of her was Bartolmeu Andrea. This Portingal doth use continual trade in this place, and

may go on land as freely as the country people. He hath in sundry places built him houses for himself and all the elephants' teeth and gold they gather together, against[35] his coming, for him. He brings them rice, salt, cloth made of cotton, wool whereof they make their apparel, and also linen cloth. Whilst he is there, they will trade with none other.

Mr. Hippon would gladly have gone aboard him[36] to have talked with him about conveyance of letters for England,[37] but the Portingal desired him not to come aboard him until he was come to his houses within the bay, and then he should be welcome. Mr. Hippon would gladly have gone aboard him, but for two reasons did not: the one, because we knew where the Portingal houses are, nothing doubting but he should find him there the next day; secondly, for that he had no order from our General to enter him perforce.[38] So he returned again to the ships.

30 By reason of a tumult raised ashore amongst the negroes, our General (being careful of the safety of his sick and impotent[39] men which lay ashore) commanded forthwith[40] our boats to be well manned; he himself with Captain Hawkins going also ashore, where we found some twelve negroes, which presently[41] increased to the number of forty or upwards, all being well appointed with their weapons, which caused us greatly to admire.[42] They urging our General with much and very earnest talk, which we could by no means understand, they speaking only their natural[43] language; they pointing very earnestly towards their town; so our General with a good guard with him marched towards their town, where to our great admiration[44] perceived it to be abandoned of the people, having carried all things out of their houses, leaving nothing but the bare walls. Mistrusting[45] that the Portingal had incensed them against us, not knowing of any offence offered unto them; but at the length our General was certified that one of the Hector's company had been missing all night. Partly conjecturing him to have given the occasion of some misdemeanor, at last gathering by one of the negro's speeches who spoke broken Portuguese that he was kept prisoner some five or six mile up in the country, where he with others had been straggling,[46] being kept in the stocks, as he at his return certified to the General, being indeed taken[47] instead of others of the Hector's company, which at that instant had stolen from the people six latten[48] basins and some cotton cloth, in all eleven parcels. Which fellow (after much arguing) was brought down unto our General, his jailer accompanying him, who had lost the aforesaid goods. Which fellow after examination confessed who they were that had done this thing; but he himself was clear[49] of it. Our General with Captain Hawkins, taking three negroes with them and leaving hostages for them, went aboard the Hector where they soon found out the matter, part of the things being aboard, and the rest hidden ashore in the woods. So our General caused all, being eleven parcels, to be brought together, and to be restored to their right owner, and the offender to be ducked[50] at the yard arm, the negroes beholding the same; which being done, one of them kneeled on his knees and held up his hands to heaven, approving the justice which our General had done them, and in their manner dutifully thanked him for the same. And being farther contented

with knives which our General bestowed on them, carried them ashore and fetched our hostages; so that I doubt not but[51] all of our nation that shall hereafter come hither will be the better used,[52] and may the more boldly go about their business on shore, they not abusing themselves. This brabbling[53] matter being so well ended, in the ev'ning we all returned aboard, giving thanks unto God for the protection of our sick men ashore, whom He had so graciously preserved from the fury of this heathenish nation.

31 Fair weather. This day we went ashore and were very lovingly entertained of the negroes, all things being forgotten, and their women and children returned to the town again.

September 1607

1 Captain Hawkins in the Hector's pinnace and Mr. Hippon in the Dragon's pinnace, being both well manned, went to see if they could come to the speech of Bartolmeo Andrea, and if it were possible to send letters by him to be conveyed into England. But we could not find him where he went in, the place being so full of small islands that it was a thing impossible to find him out, he being gone into one place or other to hide himself. Upon one of the islands as we passed along we espied houses, where we went ashore, and went into the town; but the people were all fled, only[54] one old cripple to whom we gave some beads. Here we espied three goats which were very tame, feeding about the houses, but left all things as we found them and retired[55] to the waterside, and at length the captain of that place, called Captain Beleyn, came down to us; whom Captain Hawkins used kindly, and would gladly have bought the goats off him. But he would not sell them, not having any more upon that island. But he brought down some hens and a gold ring, which we bought off him for knives. So we departed thence, not offering them any injury at all, and put over to the southward to the town of Captain Borea. And at our coming ashore Lucas Fernandez met us at the waterside, promising the Captain to come over to our ships and either to bring the Portingal with him, or a letter from him whereby we should know what to trust unto. So we returned to our ships. This negro Lucas is a Christian and can argue well of his faith—only he is led by the delusions of the friars, according to the popish[56] religion.

2 Fell very much rain all day and all night.

3 Very fair and temperate weather.

4 Being very fair weather, and towards evening Lucas Fernandez came aboard with three negroes with him. He brought with him a letter from Bartolmeo Andrea unto our General. He with the rest had very kind entertainment aboard. This night fell much rain.

5 Fair weather. After dinner our General and Captain Hawkins went ashore, where we understood by the negroes of an elephant which was not far

off. So our General caused four good shot[57] with their muskets to go along with the negro to see if they could shoot him; which they did, all four at once being close by him. Yet he made way from them so violently that they were not able to follow him; but they espied that he bled very much all the way as he went. But they could do no good upon him, so they returned.

6 Being fair weather, our General dispatched his letters to the right worshipful Company,[58] certifying them therein the estate of our voyage to this day, and sent John Rogers (one of the Dragon's company who speaketh the Portingal language very well) alongst with Lucas to deliver the letters unto Bartolmeo Andreas' own hands, with a piece of calico for a present to him. So towards night we returned aboard.

7 Fell some vehement showers of rain.

8 Fair weather. This day we caught good store of fish.

9 Fair weather. This day likewise we caught good store of fish.

10 Very fair weather. Also this day we caught much fish with our seine. In the afternoon our sick men were fetched aboard, who were (God be praised) somewhat amended.[59] And in the evening our General willed all our men to be called before him, where [he] examined divers about stealing of shirts and other things which were missing, amongst whom was George King, who on the 22 of May last was punished for a most wicked crime then proved against him, as on that day at large doth appear. Then our General caused every man to deliver up the keys of his chest unto the Master, which was done. And in the night the said King (being as it seemeth guilty of much villainy, and fearing some severe punishment), no man distrusting him, stole privately into the pinnace with his fardel[60] of clothes, and loosed her from the stern, departed in her all alone, no man seeing him, the night being dark. But by the will of God she was presently[61] missed, and our longboat being presently manned, by our Lord's direction, themselves not seeing, following her directly, and fetched her up, being driven almost as far as the point where we came in. King no sooner perceiving the boat approach him but leaped overboard, thinking to drown himself. Our men, much admiring to see a man in the water (supposing that the pinnace had broke loose), made haste to save the man, and haling him into the boat found him to be that wild wretch, and bringing him aboard [he] remained in the bilbos until the morning.

11 At two o'clock in the morning our General, admiring at John Rogers so long stay at Borea's in delivering the letters to the Portingal, sent the Dragon's pinnace to fetch him.

In the morning all our men were called again before our General, at which time King was had in examination about his wicked pretence, and whether he had any consorts. But of himself haled up the longboat, which he thought also to have loosed, and out of her he went into the pinnace. Then being examined about the shirts that were stolen, he would neither by fair persuasions nor by punishments inflicted upon him confess anything, but with a bold face denied all that was laid to his charge. (For what will not such a wicked wretch do?) So our General caused him to be loosed and put into the bilbos, and going down whilst they were fetching the bilbos he desired he might go forward; his arms being bound, leaped overboard and drowned, so that we could not see him rise above water. Notwithstanding, with the pinnace our men sought very diligently for him. Thus was the wretched end of such a wicked person, and I pray God that all offenders may take warning by his most miserable end and become new men.

This afternoon our pinnace returned with John Rogers, who brought all the letters back again, with the piece of calico which he should have delivered to Bartolomeo Andrea, not having come to the speech of him all the time of his being there. He brought with him a few hens and plantains, which he bought off the people.

This day we made an end of wooding, and watering.

12 Lucas Fernandez came aboard again, and brought a letter from Padre Bartolmeo Barrera,[62] a friar; wherein he did proffer, if so be it did please our General, to see his letters conveyed into some part of Portingal or Spain, for that he was to go for the islands of Cape Verde on that barque. And for that our General should know that he had means for conveyance of his letters, he writ him that he had often advice out of Portingal how matters passed, and amongst others he had advice of the great plot of treason[63] which should lately have been committed against our King's majesty. Also required[64] to have two copies to send by several conveyances, that if one should miscarry,[65] yet the other might come to hand; which was also a sign that he was experienced in such business. So our General delivered the letters to Lucas Fernandes, with a present with them to defray this charge of their conveyance, desiring him to deliver them to the friar. Yet fearing that neither of his letters should come to hand, he left his and Captain Hawkins' names, with both the ship's names, graven[66] upon a great stone by the watering place (where we found Sir Francis Drake's name anno 1580, and Captain Ca'ndish his name anno 1586).

This friar with two more doe keep here about *Sierra Leona* among the Portingals to say mass; also to procure some of the black people to become Christians, they having drawn some few already to be christened, as Captain Boree and others.

In the time of our being here in this harbor we traded for above 100 [thousand] lemons, a small quantity of hens and plantains, and two large elephants' teeth. In this place also is great store of oranges, very good and large. Here we made good store of lemon water. Also in both ships we filled above a hundred tons of water, and wooded so much as was needful, for in this harbor is as good wooding as can be desired.

These people are very lusty[67] men, strong and well-limbed, and a good people and true.[68] They will not steal, as others of their color will do in other places; for many of our men lost many things ashore, and they that found them brought them and restored them to the right owner. And in all the time of our being here we had no injury offered to any of our people; but all the kindness that might be expected at the hands of such a black heathen nation.

This 12 day in the evening we had much lightning at the northeast, and about midnight exceeding much thunder and rain, with a gale at northeast.

13 Our ship loose and under sail. In the morning we did set Lucas Fernandez the negro ashore (the wind blowing a fresh gale at east and fair weather, and having tide of ebb[69] underfoot), where we met with other negroes which were come down to the waterside, expecting his coming on land. They all wept, and in outward appearance they seemed to be very sorrowful for our departure. So this morning about seven o'clock we set sail from this part of *Sierra Leona,* commending ourselves and our voyage to the protection of the Almighty . . .

Journal of William Keeling (abridged)

[August 6] This afternoon, being anchored, we espied men to wave us ashore. I sent my boat, which leaving two hostages, brought four *negroes,* who promised refreshing. . . .

The seventh, there came *negroes* of better semblance aboard with my boat (for whom, as for all other, we were fain to leave one of my men, for two of them in hostage), who made signs that I should send some of my men up into the country, and that they would stay aboard in hostage. I sent Edward Buckbury and my servant William Cotterell with a present, viz. one coarse shirt, three foot of a bar of iron, a few glass beads and two knives. They returned towards night, and brought me from the said[70] Captain[71] one small earring of gold, valued at seven, eight, or nine shillings sterling. And because it was late, the hostages would not go ashore, but lay aboard all night, without pawn[72] for them.

[The tenth,][73] I sent my boat and fetched five tons of fresh water, both very good and easy to come by.

The eleventh, I went ashore a-fishing, where the people brought their women unto us, but feared we would carry them away.[74] I gave some trifles; we bought good store of lemons, two hundred for a penny knife. Wind at east.

The twelfth, I went but took little fish. Wind from northwest to south, rainy weather.

The thirteenth, it rained without intermission. We got fish enough for a meal. I bought an elephant's tooth of sixty-three pound English, for five yards blue callico, and seven or eight pound of iron in bar.

The fourteenth, I kept aboard, all day rainy.

The fifteenth, I went and took within one hour and a half, six thousand small and good fish *cavallos.*[75] After noon, with Captain Hawkins and a convenient guard, I went ashore and to the village, where we bought two or three thousand lemons. We esteem it a fair day wherein we have three hours dry overhead.

The sixteenth, I licensed our weekly workers to recreate themselves with me ashore, where in our large walk we found not past four or five acres of ground sowed with rice. The superficies of the ground is generally an hard rock. This only day, hitherto, we had fair weather.

The seventeenth, it was all day fair weather. I appointed making of lemon-water.

The twentieth, John Rogers returned and brought me a present of a piece of gold, in form of an half moon, valued at five, six, or seven shillings sterling. He reporteth the people to be peaceable, the chief without state,[76] the landing two leagues up, and the chief village eight miles from the landing.

The two and twentieth, we went ashore, where we made six or seven *barricos* full of lemon-water. I opened the Company's firkin[77] of knifes to buy limes withal.

September 4.[78] Towards night, the king's interpreter came, and brought me a letter from the Portugal, wherein (like the faction)[79] he offered me all kindly services. The bearer is a man of marvelous ready wit, and speaks in eloquent Portuguese. He laid aboard me.

September 5. I sent the interpreter, according to his desire, aboard the *Hector,* where he broke fast, and after came aboard me, where we gave[80] the tragedy of Hamlet;[81] and in the afternoon we went all together ashore, to see if we could shoot an elephant;[82] we shot seven or eight bullets into him, and made him bleed exceedingly, as appeared by his track, but being near night we were constrained aboard, without effecting our purposes on him. . .

September 29. Captain Hawkins dined with me, when my company acted King Richard the Second. . .

March 31 [1608]. I invited Captain Hawkins to a fish dinner, and had Hamlet acted aboard me; which I permit, to keep my people from idleness and unlawful games, or sleep.

Notes

1. (off what is now Whiteman's Bay).
2. person in overall command of the fleet (Keeling).
3. as guarantees, hostages.
4. other ship in the fleet (commanded by Captain John Hawkins).
5. several.
6. guinea hens.
7. (The English had been at sea since April, blown off course and caught in the doldrums, and running short of supplies of food and fresh water.)
8. importunate.
9. captain (Portuguese), general term for "leader."
10. "a wretched old man" (Finch).
11. "lion range" (referring to the highlands, visible from the sea); originally a name only for the bay.
12. "as in the Portuguese word point," referring to the tip of the peninsula. But a village with this name was recorded here as early as 1500, so the name probably originated in an African language.
13. (The English arrived in the middle of the rainy season.)
14. (One-third of the crew was sick, mostly with scurvy, a disease cured by the vitamin C in limes and lemons.)
15. raised anchor.
16. (modern Kroo Bay, seven kilometers upriver, with an important freshwater spring).
17. plenty.
18. canoe.
19. (valued as ivory).

20. i.e., the same species.
21. "Their towns consist of thirty or forty houses all clustered together (yet each hath his own), covered with reed and enclosed with mud walls, like our hovels or hogsties in England, having at the entrance a mat instead of a door locked and bolted, not fearing robbery . . ." (Finch).
22. Portuguese. (Portuguese contact with this region began in 1462.)
23. cask keg.
24. Borea "hath power to sell his people for slaves (which he proferred unto us) . . . The King, with some about him, are decently clothed in jackets and breeches, and some with hats, but the common sort go naked, save that with a cotton girdle about their waist they cover their privities; the women cover theirs with a cotton cloth, tacked about their middles and hanging to the knees, wrapped round about them; the children go stark naked. They are all, both men and woman, raced and pinked on all parts of their bodies very curiously, having their teeth also filed betwixt, and made very sharp. They pull off all the hair growing on their eyelids. Their beards are short, crisp, black, and the hair of their heads they cut into alleys and cross-paths; others wear it jagged in tufts, others in other foolish forms; but the women shave all close to the flesh" (Finch).
25. expensive.
26. hundredweight.
27. cloth from Rouen (France).
28. *vara* (Portuguese) = 1.1 meters.
29. Portuguese "oitavo" (one-eighth).
30. sighted.
31. (using the less navigable north channel of the estuary, in order to avoid the English in the south channel).
32. measuring the depth of the water.
33. caught up with the ship.
34. (most important island in the Cape Verde group, seat of the administrative capital).
35. in anticipation of.
36. the ship.
37. (Keeling wished to inform his superiors in London of the delays in their progress, which would affect their return date.)
38. by force, against his will.
39. helpless.
40. at once.
41. immediately.
42. wonder, be astonished.
43. uncivilized.
44. astonishment.
45. suspecting.
46. wandering.
47. captured.
48. brass.
49. innocent.
50. tied up, weighted, and dropped in the water till near drowned, then pulled back up (a standard form of punishment at sea).
51. have no doubt that.
52. treated.
53. contentious.
54. except for.
55. retreated.
56. Roman Catholic. (The English were Protestants.)
57. riflemen.
58. East India Company.
59. improved.
60. bundle.
61. immediately.
62. Jesuit missionary who had arrived in Sierra Leone in 1605 and had just returned from a trip upcountry.
63. Catholic plot to blow up Parliament (1605).

64. requested.
65. go astray.
66. carved.
67. healthy, energetic.
68. honest.
69. ebb tide (going out to sea).
70. aforementioned (evidence of abridgement, since the published extracts do not mention any "captain" until this point).
71. Pinto, leader of the nearby village.
72. hostages.
73. (Purchas does not give a date, thereby implying that this happened on the seventh; but all the other journals make it clear that the fresh water was only discovered on the tenth.)
74. (Previous English and European visitors had taken slaves.)
75. (compared by Finch to bleaks, a small, pale, quick-moving, soft-fleshed and wholesome river-fish common in Europe).
76. regal magnificence.
77. small cask or barrel.
78. (The four remaining entries, extracted here, were omitted by Purchas.)
79. (The English were suspicious of the Portuguese and Spanish—their commercial, political, and religious competitors.)
80. (The other printed version of this extract reads "had" instead of "gave".)
81. Shakespeare's *Tragedy of Hamlet, Prince of Denmark* was published in 1603 and again in 1604/5; an earlier play on the subject was never printed, so presumably Shakespeare's play is meant.
82. (Purchas places the elephant hunt on the seventh, but other accounts confirm that it took place the afternoon of the fifth.)

HAMLET IN AFRICA 1607

Gary Taylor

To Stanley Wells

"Methinks I see my father."
"Where, my lord?"
"In my mind's eye, Horatio."[1]

In your mind's eye, I would like to conjure up a company of British seamen, far from home. These men will spend the afternoon on shore, sweating, shooting an elephant.[2] But in the cool of the morning they gather on board ship for a different kind of sport. Within sight of conspiratorial packs of long-tailed monkeys on the rocks, within earshot of the estuary's cranes and pelicans, a sailor steps onto the deck.[3] He holds a weapon that combines a spear with a hatchet. He points this weapon in the direction of another man, and says, "Who's there?"—The first words of Shakespeare spoken outside of Europe.[4]

Who *was* there? Another British sailor, playing the role of Francisco, soon followed by others to play Horatio and the Ghost and Hamlet and all the rest. A unique company, they played to a unique audience: a boatload of 150 men on the first British diplomatic mission to India—and in their midst, four guests with filed teeth, plucked eyelashes, rings on their fingers and in their ears and noses, braided hair shaved into "elegant patterns" forming "various shapes, some oval, others like half an orange," and scarified black bodies, artistically tattooed "by means of hot irons" with "pictures of lizards, fish, gazelle, monkeys, elephants and all other kinds of animals, insects and birds."[5] We even know the name of one of those four African spectators. "Who's there?"—Lucas Fernandez, the first identifiable black man (or woman) whose life intersects with the works of our most canonical dead white male, William Shakespeare. But when Lucas Fernandez encountered *Hamlet,* the Great White Bard was still alive, and *The Tragical History of Hamlet, Prince of Denmark* was a new play. The performance of *Hamlet* I have been asking you to imagine took place on September 5, 1607—almost a century and a half before the first known performance of *Hamlet* anywhere in the New World.[6]

In the four centuries since this African performance of *Hamlet,* Shakespeare himself has acquired an unrivalled global reputation, and *Hamlet* in particular has become his most international play. By 1626 a touring troupe of English actors had performed *Tragoedia von Hamlet einen Printzen in Dennemarck* in Dresden;[7] in the later eighteenth century, it was translated into both French and German, and successfully adapted in theatres in both countries.[8] By the nineteenth century the character had become central to German national identity: "Deutschland ist Hamlet." He was, or became, equally important in Poland and Russia.[9] Even farther East, *Hamlet* was performed in Japan, in English, as early as 1891, and by 1902 it had been adapted into Japanese for "shinpa" actors.[10] It had been translated into Arabic by 1901, Afrikaans by 1945, and Zulu by 1954.[11] By 1998 Harold Bloom could claim that "There are many signs that global self-consciousness increasingly identifies with Hamlet, Asia and Africa included."[12]

But even if that is true now—and I doubt it—it was not true in 1607. For English and American readers, as Terence Hawkes says, *Hamlet* has "always already begun";[13] it is everywhere, has been everywhere for a long time, and so in a sense it can never startle us. If we want to experience that sense of astonishment, I think we need to see *Hamlet* performed in Sierra Leone. But the performance we need to witness is only available in "the mind's eye." The bay where *Hamlet* was performed, in 1607, is now one of the world's largest deep-water seaports, adjoining the sprawling city of Freetown, built by an English colonial administration in the nineteenth century and now the capital of one of the world's five poorest countries, devastated by years of brutal civil war; the elephants, crocodiles, and hippopotami noted by early European observers have long since disappeared. The Sierra Leone I want you to imagine is as dead as Hamlet's father. But you can picture it: a drowned estuary sheltered by a forested hilly peninsula, where a river then called "Mitombo" descends from the Futa Jallon highlands and empties into the Atlantic Ocean. The first Europeans there—comparing the mountains to the mane of a lion, the frequent summer thunderstorms to the roaring of a lion, the wild rough countryside to an untamed lion—called it Sera Lyoa ("Lion Range").[14]

In 1923 Professor F. S. Boas celebrated the 1607 performance of *Hamlet* in this extraordinary setting as a prototype of English cultural imperialism: "At a time when our mercantile marine has been covering itself with glory on every sea, it is an act of *pietas* to reclaim for it the proud distinction of having been the pioneer in carrying Shakespearean drama into the uttermost ends of the earth."[15] Such attitudes have not disappeared; even in the 1990s, Shakespeare scholars who mention this incident treat Africa not as a subject of interest in itself, but as a temporary way-station on the high road to something more familiar and important—the British Empire, or the cultural politics of subversion and containment, or the canonization of Shakespeare as a global cultural icon. Even for postcolonial critics like Ania Loomba, these performances represent "the moment when Shakespeare first traveled to India"—although there is no evidence, on this voyage, of a performance of Shakespeare anywhere near India.[16] Even for a respected theatre historian like Dennis Kennedy, the significance of this African premiere is so self-evident, so simple, that *Othello* is inadvertently substituted for *Hamlet*.[17] Shakespearians all know what "Africa" means, right?

But the sailors who performed *Hamlet* on September 5, 1607, stayed in the Sera Lyoa estuary for 38 days; surely we can spare that many minutes to ponder one of the most extraordinary encounters in the long history of Shakespeare's reputation and the even longer history of Euro-African interaction. Who are these people? What brought them together, here? Why are they performing (or watching) *Hamlet?* How is *Hamlet* being performed? What did *Hamlet* mean to them? And what does that African performance of *Hamlet,* late in the summer of 1607, mean to us in the twenty-first century?

Who are these people?

On September 4, 1607, the English merchant ship *Red Dragon,* commanded by Captain William Keeling, was anchored just off the coast of what is now Sierra

Leone. With the *Red Dragon* were two other ships: the *Hector* (another English merchantman, commanded by Captain William Hawkins) and a small Portuguese craft. As Captain Keeling recorded in his diary,

> Towards night, the King's interpreter came, and brought me a letter from the Portugal . . . The bearer is a man of marvelous ready wit, and speaks in eloquent Portuguese. He laid aboard me.[18]

"The King's interpreter," who spent the night aboard Keeling's ship, was not sent by the king of Portugal. The "King" he represented was the local African sovereign. That is why Keeling was surprised, and thought it worth recording in his diary, that an ambassador could speak good Portuguese and possess a "marvelous ready wit": he did not expect eloquence or intelligence from a mere negro. In fact, all the English were impressed by the interpreter's acuity, and virtually every European visitor to the area in the late sixteenth and early seventeenth centuries remarked upon the "keen-witted . . . intelligence" of the locals, and their "gift of artistic imagination."[19] The next day, Keeling recorded,

> I sent the interpreter, according to his desire, aboard the *Hector,* where he broke fast, and after came aboard me, where we gave *The Tragedy of Hamlet.*

William Keeling, to whom we owe our knowledge of this performance, was only 29 years old, but he had already distinguished himself as one "of our principle merchants" by 1604, and from 1604 to 1606 had commanded one of the three ships on the Second Voyage of the English East India Company to the Spice Islands.[20] Now, Keeling was what contemporaries called "the General," in overall command of all three ships of the Third Voyage of the East India Company. In 1614 the Company described the ideal "General" as "partly a navigator, partly a merchant (to have knowledge to lade a ship), and partly a man of fashion and good respect";[21] at about that time, they chose Keeling as "General" of yet another voyage, commissioning him to remain commander-in-chief of their Asian operations for five years. A Groom of the Chamber to King James, his epitaph described Keeling as "a merchant fortunate, a captain bold, a courtier gracious."[22] The distinguished Jacobean diplomat Sir Thomas Roe, who shipped with Keeling in 1615, wrote that he "did use his authority with more moderation and better judgment and integrity than most men would, and will not be easily matched for sufficiency every way, and did as well deserve the trust as any, I believe, [the Company] can ever employ."[23] Modern historians agree, calling him an "extremely able and efficient administrator," whose organizational skills were instrumental in creating an "interregional and integrated trading system."[24]

The English East India Company, which Keeling represented, had been chartered in 1600; it would become, by the eighteenth century, the world's most powerful company, indeed the first truly global multinational corporation, shaping the daily lives and political destinies of millions; the tea dumped in Boston Harbor, at the beginning of the American Revolution, was part of their government-supported commodity empire. In 1607, though, it was, like Shakespeare, not very important.[25] Although Sierra Leone would later become the first English colony in Africa, England was the European Johnny-come-lately. Before 1600, the frequency and range of English activities was dwarfed by the Portuguese. Indeed, the map of Africa used by English navigators on this voyage was a Portuguese map; they had nothing of their own as reliable.[26] Even the French began visiting the Guinea coast in strength in the 1530s, decades before the English made any impact.[27] Shakespeare's own superpower status tends to make us forget that in his lifetime Britain was not the international power it eventually became. Indeed, in 1608 the Portuguese commander at Surat described his royal majesty King James, monarch of England, Scotland, and Wales, as a "king of fishermen and of an island of no importance."[28]

So much for the English *dramatis personae;* what about their African counterparts? On August 18, Captain Keeling sent one of his men with presents to the principal native sovereign, "their great commander called Captain Borea." Farim Buré governed the southern and eastern shores of the estuary, and he was still ruling Sera Lyoa in 1625, when he was 80 years old.[29] The English emissary returned to Keeling's ship two days later, "having been by [Borea] very well entertained" and having "met with a negro called Lucas Fernandez which spake the Portugal language very well, and was interpreter to the King, by whom he learned many things."[30] Putting together a variety of scattered documents, we can "learn many things" about this black man called Lucas Fernandez. In 1615 a Dutch captain noted that "The Interpreter spake all kind of languages, one with another."[31] In 1607 an eyewitness from the *Hector* recorded that "Lucas the Interpreter . . . seemed very sensible, and plentiful in Spanish complements, both in speech and action, and very humane in his carriage, whose sister was wife to Borea . . . This Lucas Fernandus sat with him [King Boreah] at meat, though his own brother was not permitted so to do."[32] A merchant on board Captain Keeling's ship recorded that "This negro Lucas is a Christian and can argue well of his faith, only he is led by the delusions of the friars according to the popish religion."[33] The English were all Protestants; Fernandez was Catholic. Finally, Fernandez was "a negere . . . who had lived at Santiago in former time."[34] Santiago was the most important of the Cape Verde Islands, off the coast of northwest Africa. During the African wars in the middle of the sixteenth century, some local Temne inhabitants of the Sera Lyoa estuary fled to Cape Verde, where they lived as refugees, and were educated in local Portuguese Catholic schools.[35] Lucas Fernandez—an African who was also a Catholic and who spoke exceptionally good Portuguese—was probably one of them, which means he was probably also able to read and write European languages. Moreover, we also know that King Buré's "chief wife" was "a Christian, and had been brought up among Portuguese."[36] Since Lucas had a sister who was married to Buré, it would not be surprising if his sister had also been raised in Cape Verde; if his sister were the king's chief wife, that would help explain his particularly high status with the king. To sum up: Lucas was a multilingual, literate, Catholic African, a brother-in-law and favorite of his king. A worthy match for Sir William Keeling. But

What brought these people together?

The English were not explorers or ethnographers, and not really interested in Africa at all.[37] Keeling and his men were on their way to the Spice Islands in the eastern Indian Ocean, because that was where they could make money. The profit on the first two East India voyages had been 98 percent; the profits on this third voyage would, in the end, be even greater, an amazing 234 percent return on its investment.[38] But the expedition began badly. It got a late start, and one of Keeling's three ships became permanently separated from the *Red Dragon* and the *Hector* almost immediately. Those two ships got blown off course toward Brazil, then caught in the doldrums for months; with their supplies running low, they considered returning to England. According to Keeling's journal, at a shipboard council meeting "we had some speech of Sierra Leona. I, having formerly read well of the place, sent for the book." The book was "Master Hackluyt's book of Voyages," which, as Samuel Purchas observed in 1625, proved to be "of great profit." After examining it, Keeling and his council decided to sail for Sierra Leone. "This saved the Company," according to Sir Thomas Smith (effectively their first CEO), "20,000 pounds, which they had been endamaged if they had returned home, which necessity had constrained, if that book had not given light."[39] The English showed up in the Sera Lyoa estuary as a last-ditch effort to save their investment.

By the time they arrived, a third of the *Dragon*'s crew was sick with scurvy and the flux. As one anonymous journal records, "For the purpose of refilling the water casks, the General put into Sierra Leone."[40] But a hundred tons of fresh water was not all they needed. On August 15 they "took within one hour and a half, six thousand small and good fish"; that same afternoon, they "went ashore and to the village, where we bought two or three thousand lemons."[41] Indeed, one Englishman claimed that, over the course of a month, they bought more than 100,000 lemons.[42] The fish could be salted and stored; the lemons cured scurvy.

On another English visit to shore, Lucas Fernandez promised "that he could help us with some elephants' teeth".[43] Captain Keeling personally acquired some ivory "for five yards blue calico, and seven or eight pound of iron in bar."[44] Ivory, of course, unlike water and food, was not essential for life on board ship; but it was a profitable commodity, one of the few products of the region with resale value. In fact, Keeling's commission from the East India Company had specified "it shall not be amiss if you can have them, to buy some elephants' teeth . . . that commodity is exceedingly well requested in Cambaya."[45] In Sera Lyoa, Keeling traded "blue calico" cloth, which had been acquired on previous voyages to India, in exchange for African ivory, ivory for resale in India, in exchange for goods that could in turn be traded for spices in the Indonesia archipelago, spices that could then be sold in England, partly for the domestic English market but partly for re-export to various European markets. Keeling and Fernandez, in August 1607, were already participating in a fully global economic system.

And that system had advantages for the local inhabitants, as well as the visitors. As Jean Boulègue and Jean Suret-Canale have pointed out in the *History of*

West Africa, the arrival of the Portuguese, in the middle of the fifteenth century, had upset the economic equilibrium of the region "by re-routing the commercial currents towards the Atlantic coast, where henceforth the Europeans supplied, at much lower cost, products traditionally imported" from the interior.[46] The key phrase here is "at much lower cost"; economically, the coastal Africans got a better deal from the Europeans than from other Africans.

It is hard for postcolonial readers to imagine these precolonial advantages because we have become painfully aware of European exploitation of Africans. The first known English visitor to Sera Lyoa was also the first English slave-trader, Captain John Hawkins, who in the 1560s had taken hundreds of slaves from this very estuary. In 1586 two separate English fleets had burned down a village, within weeks of each other, in a kind of attack apparently unprecedented in the history of European relations with West Africa.[47]

But Captain Keeling enslaved no one. Indeed, he insisted that his men treat the local people with respect. In retaliation for thefts committed by some of the crew, the Africans detained one man a prisoner at their village; when the affair was reported to Keeling, he caused the English offenders to be punished in the presence of the African owners of the stolen goods, to whom also he restored their property.[48] Keeling was here enforcing company policy; his commission instructed him "that at every place where you shall water and refresh your men, you call the companies together, giving them severe warning to behave themselves peaceably and civilly towards the people of that place (if any be there), the better to procure their friendship, towards the supply of your wants, and the like in every place where you come, lest the loss of your lives and overthrow of our voyage pay for your disorders."[49] For their part, the Temne were also remarkably "strict about theft. Goods may be left in the streets and outside houses, and no one will take them. Nothing goes missing from the churches." Convicted thieves were impaled at a crossroads.[50] Keeling and Buré shared a commitment, common in trading cultures, to protecting the rights of strangers.

Within that trading culture, the expansion of European economic ambition enabled coastal Africans to play one set of foreigners off against another. For instance, the English were not the only Europeans in the bay of Sera Lyoa that month. Bartolomeu André, frequently mentioned in the English journals, was Portuguese, but he had lived for years in Sera Lyoa, was probably married to an African woman, and certainly had adult sons. He was recognized as *capitao* of the local Portuguese traders, and fiercely defended their interests; he had once massacred a boatload of Dutch traders, and on another occasion had persuaded a local African ruler to murder another group of Dutchmen, living further south.[51] Indeed, in the decades on either side of 1600, Portugal attempted to establish a colonial settlement in Sera Lyoa.[52] A Portuguese Jesuit, Father Baltasar Barreira, had arrived in Sera Lyoa in September 1605, and was still there; another Jesuit, Manuel Álvares, had landed in May 1607. Although Portuguese ships had reached this part of the African coast by 1462, and although the spread of Christianity was part of the justification for their empire, these were the first Christian missionaries sent to Sera Lyoa. They did indeed have some initial success converting the local inhabitants, beginning with Buré himself. But they also wanted to secure a Portuguese monopoly on trade, which the Africans resisted: Buré himself traded with Keeling in 1607. In 1608 Father Barreira left; Father Álvares remained, but in 1615 he was imprisoned by Buré for attempting to persuade the locals not to trade with a visiting English vessel. By 1617 the Jesuit mission had completely collapsed, and was not replaced.[53] (Islamic missionaries eventually succeeded where the Jesuits had failed.)

As the 1607 journals make clear, the English did not trust the Portuguese, who in turn did not trust them. After Vasco da Gama circumnavigated Africa in 1497/98, the Portuguese maintained a virtual monopoly on European sea-trade to Asia for almost a century. But eventually that monopoly began to be contested by the Dutch and the English. In the 1560s, English trading voyages to the West African coast had almost led to war between England and Portugal, and early in the seventeenth century the Portuguese belligerently resisted English access to the Indian Ocean.[54] But on this occasion both the English ships were heavily armed, and Bartolomeu André's 40-ton barque would have been badly outclassed by the 700-ton *Dragon* and the 500-ton *Hector.* So he avoided them—which forced him to rely on Lucas Fernandez to convey excuses and seek information.

The rivalries between the Europeans empowered the Africans. Lucas Fernandez, negotiating for Farim Buré with the English, also acted as an intermediary between the English and the Portuguese businessman André, and between the English and the Portuguese missionary Barreira; through Fernandez, Buré controlled the Europeans' interactions with each other. And the Africans had their own entrepreneurial agendas. In order to get ivory, Keeling had to barter, not only calico, but also "seven or eight pound of iron"—iron being a notional currency crucial in intra-African commerce. This pattern of African control is typical of European trade with West Africa in this period.[55]

What brought these people together?—Mutual economic advantage. Of course, those advantages were not evenly distributed, on either side. The pay of an ordinary English sailor was only about £5 a year; by contrast, a ship's captain made twice that much each month, and he could supplement his salary with private entrepreneurial activity.[56] Keeling, for instance, had a hundred pounds of his own merchandize on board, which (by permission of the Company) he could sell or trade for personal profit; it is not clear whether he bought that ivory as an employee or an entrepreneur. On the other side of the shoreline, the rulers of Sera Lyoa belonged to an ethnically and linguistically distinct clan, which had invaded the region from the south at some time between 1545 and 1560; Farim Buré was apparently the son of Farim Sheri (or Xeri), one of the original conquerors;[57] like his father, Buré was the Mani overlord of a population of Temne peoples, whose fathers and grandfathers had been violently subjugated. (Those who resisted had been enslaved or slaughtered; their leaders were eaten.)[58] The Jesuit missionary Álvares concluded that the Mani kings and their relatives—who had "no right or title here other than by conquest"—were "to blame for the land being impoverished," because they were interested only in enriching themselves.[59]

So the main motives of our *dramatis personae,* European and African, are fairly clear: they all want to get rich, and the leaders on both sides expect to get much richer than their followers. But if that is why our cast is gathered on board the *Red Dragon* on September 5,

Why this performance of *Hamlet*?

Obviously someone on board the *Red Dragon* loved theatre. After they had left Sera Lyoa, Captain Keeling's journal records that on September 29, becalmed along the western coast of Africa, "Captain Hawkins dined with me, where my company acted King Richard the Second." Six months later the expedition was sailing north up the eastern coast of Africa; on March 31, 1608, becalmed near the equator, Keeling notes that he "invited Captain Hawkins to a fish dinner, and had Hamlet acted aboard me: which I permit to keep my people from idleness and unlawful games, or sleep."[60] (Is that why millions of students have been required to read *Hamlet?* To keep them from idleness, video games, or sleep?)

Actually Keeling's explanation quotes from the commission he had been given by the East India Company: "ITEM that no blaspheming of God, swearing, theft, drunkenness, or otherlike disorders be used, but that the same be severely punished, and that no dicing or other unlawful games be admitted, for that most commonly the same is the beginning of quarreling, and many times murder." The ban on "unlawful games" has an entirely pragmatic justification: the recognition that certain behavior will "most commonly" divide the crew, and that in the confined and vulnerable community of a ship at sea on a long voyage such "quarreling" may be disastrous for everyone. The success of the East India Company, the very lives of Keeling and his men, depended upon collective effort, which in turn depended upon the maintenance of social unity. The first item in the commission, after the appointment of the expedition's leaders, was the instruction that, because "religious government and exercise doth best bind men to perform their duties, it is principally to be cared for that prayers be said morning and evening in every ship, and the whole company called thereunto, with diligent eyes that none be wanting."[61] This is not an environment that encourages dissidents or (what so many recent critics claim to find in Shakespeare) the subversion of authority. Louis Montrose cites Keeling's defense of these shipboard performances as evidence that Elizabethan theatre did not threaten the status quo in England.[62] I don't know that Keeling's journal entry permits us to generalize about all Renaissance plays, but if *Hamlet* in particular had seemed subversive, Keeling would never have permitted it to be performed once, let alone twice.

Nevertheless, Keeling felt it necessary, in his journal, to justify the fact that he "permitted" such performances. He knew that they were unorthodox, and that the management back home might not approve. The East India Company did recognize that something had to be done to keep sailors from getting into trouble during idle time—inevitable, especially, on the outbound voyage. Historian K. N. Chaudhuri has called attention to the fact that "The total number of men in the outward bound ships was often more than what was actually required to man them; for the company shipped extra deckhands, fearing that the ships on their way home might be weakly manned." Even on "successful" voyages, mortality could be as high as 85 percent. Keeling himself would later abandon and burn the *Hector,* because he only had enough surviving crew to man a single ship safely for the return voyage to England; on his next voyage, in 1617, he lost half his crew.[63] Sailing around Africa, Keeling had too many men, with too little to do, and the London headquarters would understand his need to keep them occupied. But the kind of diversion the Company had in mind is suggested by their commission to a later voyage, in 1611: "ITEM for the better comfort and recreation of such of our factors as are residing in the Indies, we have sent the *Works* of that worthy servant of Christ Master William Perkins to instruct their minds and feed their souls, with that heavenly food of the knowledge of the truth of God's word, and *The Book of Martyrs* in two volumes, as also Master Hackluyt's *Voyages* to recreate their spirits with variety of history."[64] *Hamlet* and *Richard II* provide "variety of history"—but not the variety the London office had in mind.

Naval historian N. A. M. Rodger notes that reading was a common recreation at sea, and as early as the 1550s English seamen voyaging to Africa carried books with them.[65] Reading the sermons of William Perkins or the printed text of *Hamlet,* silently to yourself, in your cabin or your bunk (or your condo), is a solitary, individuating activity: it creates, and makes you conscious of, a difference between yourself and others. But performing *Hamlet,* or watching a live performance of *Hamlet,* has a conspicuously different effect. Like the sailing of a 600-ton wooden ship, theatrical performances depend upon communal effort; they require interaction and mutual trust; they will succeed only if each participant pulls his own weight—and each participant is motivated in part by a fear of disapproval or even ostracism, if he lets his companions down. This communal imperative is heightened when the performers belong to the same small community as their audience. Members of any audience of course remain individuals, but they become individuals-suspended-in-solution, laughing or crying or clapping or booing together, for the most part, collectively focussed upon the same stimulus, and reciprocally influenced by each other's responses, in a massive feedback loop, which can itself create a powerful sense of community.[66] And when every performer knows every spectator, when cast and audience have already been confined in each other's company for months, the pressure on each individual to do his best, not to let his fellow actors down, not to embarrass himself in front of his work-mates, can become extraordinarily compelling.

Performing plays did not just give Keeling's "company" something to do; it gave them something to do that they might have enjoyed rather more than the compulsory praying sessions morning and evening. In Sera Lyoa he recorded in his journal not only the performance of *Hamlet,* but also, on August 16, "I licensed our weekly workers to recreate themselves with me ashore"—a pleasure obviously appreciated by the "good number of men" who joined him in "our large walks."[67] Keeling was chosen, in 1614, to command the Company's operations in Asia for five years because of his skill at "merchandizing," but also because of his "good command . . . over his men abroad (whom they loved and respected for his kind usage of them)."[68] Keeling's "kindly treatment" of his men—what one modern study describes as his "concern for the minutiae of life on board"—seems to have included permitting his men to perform plays.[69] Unlike gambling, the production of plays required collaboration and created community—a community like that of an acting company. Shakespeare himself, after all, belonged to such a community. Although "Shakespeare in Business" would not attract as many moviegoers as "Shakespeare in Love," the life of England's favorite playwright is more conspicuous for loyalty to a company than for romantic passion. Shakespeare was a founding member of the King's Men, a long-lived and cohesive group of coworkers, founded in the reign of Queen Elizabeth and organized as a joint-stock company. The East India Company, also founded in the reign of Queen Elizabeth, was also a joint-stock company. These two enterprises were arguably the two most successful businesses in British history. Certainly our global village is still living with the legacy of both.

But the performance of *Hamlet* in the estuary of Sera Lyoa on September 5, 1607, suggests that the rehearsing and performing of plays on board the *Red Dragon* had at least one other motive. When Lucas Fernandez and his entourage came on board, "he with the rest" was given "very kind entertainment."[70] He was treated, in other words, as a visiting dignitary. In England it was entirely normal for such dignitaries, including ambassadors, to be entertained by a command performance, at court, of a play or masque.[71] Indeed, as many critics have remarked, masques differed from plays primarily because some of the court "audience" also became performers, thereby breaking down the barriers between illusion and life, in a collectively willed image of communal sociability; on board ship, although the formal and expensive conventions of the court masque could not be reproduced, the performance of a play by members of the crew before members of the crew might have had some of the same effect.

Of course, Keeling had not originally intended to make landfall in Sera Lyoa, or to entertain an African dignitary. But he might easily have anticipated that it would be useful, on this voyage, to be prepared to give impressive theatrical performances. Most Anglo-Americans associate the colonization of the New World with the Puritans of New England, who were of course programmatically hostile to all forms of theatre. But other European colonists enthusiastically imported plays as part of the culture they wanted to display to indigenous peoples. Spanish-American theatrical performances took place as early as 1530, in what is now Mexico City, only five years after Cortez overthrew the Aztec empire. In the Great Lakes in the seventeenth century, French Jesuit missionaries organized a performance of Corneille's *Le Cid*. And in between these Spanish and French performances, in 1583, an English colonial expedition to Newfoundland, led by Sir Humphrey Gilbert, found space for such "toys" as "Morris dancers, hoby horse and May-like conceits" and "music in good variety," which was brought along for the "solace of our people, and allurement of the savages."[72]

If North America provides evidence of how natural it seemed to European voyagers and colonists to take their drama with them, recent English experiences in Asia gave them particular and potent motives for doing the same. On the second East India voyage, the English (including Keeling) had witnessed a spectacular theatrical festival of "pageants" and "triumphs" and "shows," which went on "every day for a month" at the royal court in Bantam. These are described in detail in a book written by a member of that earlier expedition, Edmund Scott's *An Exact Discourse of the Subtleties, Fashions, Policies, Religion, and Ceremonies of the East Indians, as well Chineses as Javans, there abiding and dwelling.* One day, for instance, "a crew attired like masquers . . . before the King did dance, vault, and show many strange kinds of tumbling tricks." On another occasion, "amongst some of these shows there came in junks sailing, artificially made, being loaden with cashews and rice. Also in these were significations of historical matters of former times . . . of chronicle matters of the country and kings of Java."[73] The Europeans at court in Bantam were expected to participate in this festival honoring the local king, and the Dutch and Portuguese spectacularly obliged. The English were not so well prepared or equipped.[74] The English offering included an apologetic speech, declaring that they "would have presented his Majesty with a far better show," if they had been able.[75]

As the events in Java demonstrate, when alien cultures encounter each other, both sides almost inevitably resort to theatre—a complex semiotic performance that can, without language, communicate power. For their return voyage, the English were better prepared to engage in such competitive theatrical displays. Keeling's journal records that on September 13, 1609, while he was anchored in the bay at Bantam "upon the King's request, I sent five-and-twenty armed men to make him pastime, which he willed in honour of his having the last night made conquest of his wife's virginity."[76] We don't know what kind of "pastime" the English troupe provided, but it was certainly something spectacular and performative. And it was not only in Java that such skills might be needed. Keeling carried with him a letter from King James to "the king of Surat"; this was the first English diplomatic mission to the Indian subcontinent.[77] One of the Company's objectives was to impress India's rulers, and thereby persuade them to grant the English commercial access to their ports, which had been until this time a Portuguese monopoly. In the event, this task fell to Captain Hawkins and the *Hector,* but what Keeling might have planned in 1607 is suggested by his voyage of 1615, when he transported Sir Thomas Roe as an ambassador from King James to the Mogul Emperor Jahangir, who ruled two-thirds of the subcontinent. When Roe landed at Surat, by order of Keeling "the ships in their best equipage" fired "their ordnance as I passed; with his trumpets and music ahead my boat in the best manner." A shipboard observer recorded "48 pieces great ordnance discharged from our fleet; this day our ships were all handsomely fitted with their waistclothes, ensigns, flags, pendants and streamers." Sir William Keeling appreciated the political usefulness of theatrical spectacle. And Sir Thomas Roe, Keeling's partner in this enterprise, also displayed English artistic skill as an index of the splendor of the civilization he represented. As a gift from King James, Roe gave the governor of Surat "a very large fair map of the world."[78] When he reached the emperor, Roe's gifts included an English cornet player and an English painting.[79] Under Keeling's command, the English sailed to Asia armed with their art. Sera Lyoa gave them, unexpectedly, their first opportunity to show off.

Of course, by the standards of modern criticism the shipboard performance of *Hamlet* on September 5, 1607, may not have been entirely satisfactory. Indeed, in 1950 one English critic, Sydney Race, derided the idea that "a crew of rude sailors" could have performed the play at all. "What unnamed man of genius played the part of Hamlet, and who was the young sailor who took the part of Ophelia? Was the play performed with or without costumes, scenery and properties? The Dragon had indeed an unusually docile crew if in the heat of Sierra Leone they were prepared to listen to Hamlet."[80] Race was obviously a sarcastic snob, and his hysterical attack on the authenticity of the records of this performance has been magisterially refuted by some of the most respected (and cautious) scholars of the twentieth century.[81] But Race does raise an interesting question:

How was *Hamlet* performed on the *Red Dragon* in the bay of Sera Lyoa?

We don't know who played Hamlet, but the *Dragon* had a crew of 150 men; depending on whether they used the short text of the 1603 "bad" quarto or the long text of the 1605 "good" quarto, and on what they omitted, *Hamlet* could have been performed easily by 17 or 18 actors.[82] Although by the social criteria of the day most of the crew were indeed "rude mechanicals," the status of actors in 1607 was hardly higher than that of sailors. Moreover, the ship's company included well-educated merchants and officers of the East India Company. Although they may not have been "the best actors in the world, either for tragedy, comedy, history, pastoral," any Englishman with a grammar school education would have "taken part in the school plays which were so prominent a feature of Tudor life."[83] Indeed, in 1582, during the last extended stay of a group of English ships in this bay, a ship's chaplain had compared various members of the expedition to the *dramatis personae* of "an elegant and witty comedy."[84] In 1631, aboard another East India Company ship, Walter Mountfort actually wrote a play, *The Launching of the Mary,* which was performed in London when he returned.[85] In the centuries since 1607, *Hamlet* has often been performed by amateur companies, and the crew of the *Red Dragon* would have had the advantage of not being inhibited by the belief that their script was a sacred classic, far beyond mere mortal comprehension. For them, it was just a popular new play.

It was also a play designed for playing conditions surprisingly similar to those on board ship.[86] Indeed, the two great joint stock companies of early modern England, the King's Men and the East India Company, both did their business in large, multistory, rope-worked, hollow, resonating wooden structures, designed and built by London carpenters. A wooden stage is indistinguishable from a wooden deck, a trap door resembles a ship's hatch, a tiring house facade is remarkably similar to a forecastle, a theatre's "cellerage" is structurally parallel to below-decks. The stage, like the deck, is open to the weather, and dependent on sunlight for illumination; the company at the Globe performed *Hamlet* in the afternoon, but Sera Lyoa is hotter in the afternoon than London, so the company on the *Red Dragon* performed *Hamlet* in the morning, while it was still cool. The company at the Globe, being dependent on ticket revenue, performed rain or shine; the company on the *Dragon* often faced terrible weather, but September 5 was (as we know from the shipboard journals) a fine clear day, perfect for an outdoor theatrical performance.[87] When Hamlet said "I am too much i'th' sun," the meteorological pun would have been unavoidable; no doubt, whatever Osric said, the audience could feel that the air was indeed "very sultry"; the actor could literally point to "yonder cloud," shaped "very like a whale," and gesture toward "this most excellent canopy the air, look you, this brave o'erhanging firmanent, this majestical roof fretted with golden fire, why it appeareth nothing to me but a foul and pestilent congregation of vapors." Moreover, the *Red Dragon,* being an unusually large ship for the time—600 tons according to some accounts, 700 tons according to others—provided a playing space comparable in size to those in early modern theatres.[88] On the outbound voyages to India, merchant ships were not heavily laden, and the deck would have been uncluttered, resembling the "bare stage" of Renaissance playhouses.[89] Admittedly, the *Red Dragon* did not have a proscenium arch—but then neither did the Globe.

Actually I should specify which "Globe" I mean, because early modern London carpenters produced at least two structures with that name: the East India shipyards built a 350-ton vessel called the *Globe,* which left London for Asia on January 5, 1611.[90] That was about the time the King's Men began performing, at *their* wooden structure called the Globe, a play called *The Tempest,* which begins with a scene in which the audience has to imagine that the stage is a ship at sea. Indeed, while Keeling's men were performing *Hamlet* in the course of circumnavigating Africa, the King's Men were performing *Pericles,* a play that repeatedly depends upon the structural similarity between stage and ship.[91]

Sydney Race is right to suspect that a shipboard performance of *Hamlet* would have had to do "without . . . scenery"—but there wasn't painted scenery at the Globe, either. The King's Men's costumes almost certainly made no attempt at historical or Danish accuracy; their chief function would have been to indicate variations of social rank, and such variations were clearly signaled by the clothing worn on board. The *Red Dragon* was armed for combat at sea; it could easily provide armor for Hamlet's ghost, military accessories for Fortinbras's army and Laertes's uprising, artillery sounds for the first and last act, swords for the duel, lanterns to indicate a night watch. On the *Red Dragon* as at the Globe, flour was on hand to produce a ghostly pallor, boys were available to play Ophelia and Gertrude and the Player Queen, hoboys and trumpets stood ready to create impressive sound effects.[92]

Keeling's men performed *Hamlet* in conditions that the King's Men would have recognized, and approved. But how did Keeling's men interpret the play?

What did *Hamlet* mean to the English?

The crew of the *Red Dragon* performed two plays, *Hamlet* and *Richard II;* both plays are preoccupied with sovereignty, political legitimacy, and the administration of justice. Captain Keeling carried with him a commission from King James himself, which did "hereby straitly charge and command all and every person and persons employed, used or shipped in this voyage in any of the said ships to give all due obedience and respect unto you during the said voyage." Keeling was authorized "to chastise, correct and punish all offenders and transgressors in that behalf according to the quality of their offences . . . And for capital offences, as for willful murder (which is hateful in the sight of God) or mutiny (which is one offence that may tend to the overthrow of the said voyage) . . . We do hereby give unto you the said William Keeling during all the time of the said voyage or during so long time as you shall live in the same, full power and authority to use and put in execution our law called martial law in that behalf."[93] A ship at sea is a little kingdom, and Keeling was its king; indeed, the leader of a previous English voyage to Sera Lyoa, in 1582, was repeatedly (sarcastically) called "our little king."[94] Keeling's commission from the East India Company expected him, "having procured him sufficient authority from our sovereign lord the king's majesty for that purpose," to "so behave himself, as he may be both feared and loved."[95] Fear and love are, of course, the two prerequisites of political authority, recognized by Machiavelli among many others.

The maintenance of authority was, for Keeling and his fellow officers and merchants, not simply a theoretical problem, or a legal fiction; it was a practical reality of long voyages, including this one. On October 28, Captain Hawkins recorded that Keeling came to speak with him: "the reason was for that we were then drawing into a colder climate, which would prejudice [the health of our men, if provision of warm clothes were not made for] them, whereof some of his men had already complained." On December 17, Hawkins recorded the men's "discontent for not putting in there [at Saldania Bay], and by the scanting of our allowance which our small spare of water would enforce us to be so cast down, as it might work the utter overthrow of the voyage."[96] A ship at sea is, as writers from Herman Melville to Joseph Conrad and William Golding have recognized, a microcosm of human society; a play, too, is a microcosm of human society, and the two microcosms intersected, perhaps very powerfully, in what we might call the *Red Dragon Hamlet*.

After all, *Hamlet* has sailors in its *dramatis personae* and a sea voyage in its plot—indeed, several sea voyages. The Danish ambassadors to Norway go and return by sea, Laertes travels by ship to Paris, the Norwegian army led by Fortinbras must cross the Baltic Sea to attack Poland, Hamlet travels by sea with pirates, while Rosencrantz and Guildenstern continue by sea to England, and the English ambassadors return by sea to Denmark; at the end of the play, the stage is dominated by Fortinbras, his soldiers, and the ambassadors from England, who have all traveled here by sea. When someone says "Hamlet" in the twenty-first century, sea voyages are probably not the elements of the play that immediately come to mind. But they are obviously elements of the play that would have seemed particularly interesting to the crew and cast of 1607. And I can't say that the cast and crew of 1607 was wrong. The play's language—including some of its most famous lines—repeatedly encourages such preoccupations. Not just "to take arms against a sea of troubles" or "That undiscovered country, from whose bourn No traveler returns," but also the "impress of shipwrights" and "the moist star, Upon whose influence Neptune's empire stands." And on and on: "My necessaries are inbarqued . . . as the winds give benefit and convoy is assistant" . . . "the wind sits in the shoulder of your sail" . . . a cliff "looks so many fathoms to the sea and hears it roar beneath" . . . "the fat weed That roots itself in ease on Lethe wharf" . . . "your bait of falsehood take this carp of truth" . . . "with windlasses . . . by indirections find directions out" . . . "I'll board him presently" . . . "I am but mad Northnorthwest; when the wind is southerly, I know a hawk from a handsaw" . . . "enterprises of great pitch and moment, with this regard their currents turn awry" . . . "Haply the seas, and countries different, with variable objects" . . . "in the very torrent, tempest, and as I may say, whirlwind of your passion" . . . "Neptune's salt wash" . . . "If thou canst mutine in a matron's bones" . . . "I lay worse than the mutines in the bilbo" . . . "mad as the sea and wind when both contend which is the mightier" . . . "but yaw neither in respect of his quick sail." Ophelia, of course, dies by drowning, after floating awhile "Mermaid-like," and the gravediggers argue "here lies the water, good, here stands the man, good,"—surely you can imagine a sailor/actor on a deck/stage, in 1607, speaking these lines—"if the man go to this water and drown himself, it is willy nilly, he goes, mark you that, but if the water come to him, and drown him, he drowns not himself." One of their shipmates had, only a few days before, done exactly that: gone to this water and drowned himself.

I don't mean to imply that Shakespeare wrote this play with a shipboard performance on his mind. But it is worth reminding ourselves that London was, in Shakespeare's lifetime, a major international port city, and that ocean-going ships sailed up the Thames to within sight of the Globe theatre. Britain until recently could only be reached by sea, and could only exert an influence on the rest of the world by sea. The summer of 1607, when *Hamlet* was performed in Africa, is also the summer when the English founded, at Jamestown, their first permanent colony in America. In Shakespeare's lifetime, English national identity was being defined by the circumnavigations of Drake and Cavendish (who both stopped in Sera Lyoa), by the defeat of the Spanish Armada, by the voyages of the East India Company.

It was also being defined by an increasingly complicated relationship to Europe, a relationship always mediated by the sea. In the sixteenth century, English political and religious life was dominated by a struggle between competing European influences (Catholic Rome, Protestant Germany). Hamlet and Horatio were students together at Wittenberg, a university most famously associated with Martin Luther. Indeed, *Hamlet* is Shakespeare's most European play; it is saturated with references to "this side of our known world." Set in Denmark, it is based upon a Viking saga, written down in the twelfth century by Saxo Grammaticus, first published in Paris in 1514; that Latin original was translated into French by François de Belleforest in 1570; Shakespeare's play derives, directly or indirectly, from that French translation. And Shakespeare's Denmark is conspicuously allied to France: Laertes begs permission to "return to France," Polonius sends a servant named "Reynaldo" to visit Laertes in Paris, Claudius describes the visit to his court of "a gentleman of Normandy," the king bets "six French rapiers" that Hamlet will win the wager, and Hamlet imagines himself wearing shoes adorned with "Provençal roses."

When Keeling's expedition carried *Hamlet* in its cargo, it was not bringing along a play by Shakespeare, or even an English play; it was bringing along a

European play, a play that, literally and symbolically, represents Europe. In the first scene, Denmark and Norway stand on the brink of war; later, Denmark aids Norway in its attack on Poland; Hamlet is shipped to England, which owes tribute to Denmark. Claudius is guarded by "Switzers," and "The Mousetrap" is "the image of a murder done in Vienna"; "the story is extant, and written in very choice Italian" (and probably based on the murder of the Duke of Urbino in 1538). The characters' names mix Danish (Rosencrantz, Guildenstern), Latin (Claudius, Horatio, Marcellus), Italian (Barnardo, Francisco), and Greek (Laertes, Ophelia). They speak of Julius Caesar and Aeneas and "when Roscius was an actor in Rome," of Mount Olympus, Alexander the Great, the siege of Troy. They drink wine from Germany and swear "by St. Patrick" of Ireland.

The Europe that *Hamlet* represents was racially prejudiced.[97] Hamlet expects the devil to "wear black," and Laertes invokes "the blackest devil." Hamlet describes the intentions of a murderer as "thoughts black," and imagines the soul of Claudius "as damned and black as hell"; Claudius agrees, describing his own sinful heart as a "bosom black as death." Hamlet says that his own evil "imaginations are as foul as Vulcan's stithy"—that is, they resemble a blacksmith whose face has been darkened by soot and smoke. (Blacksmiths were, by contrast, among the most revered figures in West African society.)[98] Contrasting his mother's first husband with her second, Hamlet describes the detestable second husband as a "Moore," which at the very least puns on the word for Muslim, blackamore, negro. Most importantly, the violent murderer Pyrrhus, in the player's speech, is as black as the Africans in that 1607 audience: his "sable arms, Black as his purpose, did the night resemble." You may think this simply means he wore a suit of black armor, but Hamlet specifies that Pyrrhus has a "black complexion" which has been "smeared" with blood. Whether or not Homer testifies to the presence of African warriors at the siege of Troy, Shakespeare imagined one. And the one he imagined is a figure of inhuman savagery.

Shakespeare scholars until recently have paid almost no attention to the prejudices implicit in such language. As a teacher at the University of Alabama, where one-sixth of our undergraduates are African-American, I cannot help but be embarrassed by such images, because I read them now through the eyes of my black students. And I read them, too, through the ears of those four Africans attending *Hamlet* in 1607. Of course, none of them spoke English; they heard the play's language mediated through a running translation into Portuguese, and perhaps another running translation into Temne, the local African language.[99] This was, indeed, probably the first translation of the play into any language, and certainly the first into Portuguese (by more than two centuries).[100] One can only hope that the translator discretely omitted the more offensive phrases in Shakespeare's text.[101]

But the issue of color raises a much more general issue:

What did *Hamlet* mean to its African audience?

In a famous essay called "Shakespeare in the Bush," published in 1966, the American anthropologist Laura Bohannan described her attempt to tell the story of *Hamlet* to a group of Tiv tribesmen in what is now Nigeria. They repeatedly objected to her account, certain that she must have got the story wrong, because some of the details simply did not make sense the way she told it, and the details that *did* make sense to them, made a sense that would have been incomprehensible to an early modern Englishmen—for the Tiv interpreted the story in terms of their own very unEuropean customs, which naturally they believed to be universally valid, because "people are the same everywhere."[102]

There was no anthropologist on board the *Red Dragon* to record the reactions of that first African audience, and we cannot simply project the beliefs of the Tiv elders in 1966 onto Temne spectators in 1607: the cultures of Africa are multiplied across space and time. Nevertheless, we do know something from early accounts of the people of Sera Lyoa about their customs and beliefs. They would have understood king-poisoning, a king's fondness for wine, a son carrying an image of his dead father and extravagantly mourning him, the importance of a king's counsellors, the outrage over scanted funeral ceremonies for a nobleman, kneeling to show deference to a sovereign, and the complex relationships of war, alliance, and tribute among neighboring states.[103] Spectators with a history of ritual political cannibalism no doubt relished the wit of Hamlet's famous line about the whereabouts of the dead Polonius: "not where he eats, but where he is eaten."

But other local customs fundamentally departed from the customs prevailing in Shakespeare's play. Most significantly, the African spectators would have found the position of Claudius, at the outset of the play, perfectly comprehensible: "Sons do not inherit but the brother, and from the brother the nephew, the son of the original owner of the inheritance."[104] Indeed, this had just happened in the spectators' own kingdom, which had passed from Farim Xere to his brother Sacena, then to Xere's son (Sacena's nephew), Farim Buré.[105] In Sera Lyoa, Claudius would have been the proper legal heir after the death of his brother Hamlet Senior; Hamlet Junior would have had no reason to complain that Claudius had "popped in between th'election and my hopes," or to describe him as a "cutpurse of the empire and the realm." All Hamlet Junior had to do was wait patiently for Claudius to die, for he would automatically inherit the kingdom from his paternal uncle. Unlike innumerable English critics, the Temne would not have been puzzled by Hamlet's "delay," but by his senseless impatience.

Other aspects of the play would have been equally confusing. The Temne believed in an afterlife, but not in hell; they ritually secluded young women that European observers often called "nuns," but "get thee to a nunnery" would have meant "go to a place where you will be initiated into womanhood by being taught various sexual practices, and undergoing clitorectomy"; what Hamlet expresses by wearing black they would have conveyed by shaving their heads; they would have found nothing incongruous about Claudius drinking rowdily so soon after his brother's death, because they commemorated their own dead with "great parties" featuring "eating and drinking excessively," making so much noise that "for the space of eight days no one can sleep by night or day."[106]

And what would they have made of English sexual customs? "The men do not worry much if their wives go to others for pleasure, for when they are found to have committed adultery they have to pay with money."[107] So a Portuguese observer recorded sometime between 1506 and 1510; in 1582 these attitudes were still in place, according to an English report: "If a man be taken in adultery, having a wife, he shall fine [=pay a fine] for it; but she shall go free, because he (said they) might have had enough of fleshly appetite with his wife, but the woman not of one man."[108] In 1616 Álvares reported that Temne men regularly acted as panders for their own wives and daughters.[109] So much for Hamlet Senior's obsession with Gertrude's adultery, and Hamlet Junior's disgust with female sexuality generally: the Temne did not expect sexual fidelity from women and would have found the outrage of Shakespeare's men strange, or perhaps even ridiculous. But although Gertrude's adultery should not have surprised them, her presence would have been puzzling for other reasons: in their culture, "when the king dieth, his concubines or wives shall be put to death with him."[110] Gertrude, the dead king's wife, should not be alive to remarry.

Hamlet's experience is only universal if you make the assumptions that Shakespeare made. And Shakespeare's African spectators, in 1607, would not necessarily have granted all his assumptions.

Actually, it occurs to me, there *was* an anthropologist on board the *Red Dragon,* that September 5. But the anthropologist was not white. These English seem to have learned, or cared, little about Sera Lyoa customs and beliefs. After all, Shakespeare treats Danes as though they were just Englishmen with odd names; he does not recognize the existence of a Danish culture significantly and legitimately different from English culture. ("All the world's a stage, And all the men and women . . . Englishmen.") No, the astute and respectful observer, comparing an alien culture to the multiplicity of possible human worlds—Mandi, Temne, Cape Verde, Portuguese, Dutch, French, English—was not a Westerner voyaging to the uttermost ends of the earth: the ethnographer was "the negro, Lucas Fernandez."

A cynic might object that my respect for Fernandez, like the respect he earned from his English visitors, simply and chauvinistically reflects the fact that he spoke a European language, and therefore had become Europeanized, "one of us," the victim of "linguistic imperialism"—as though a language were some sort of cookie cutter, imposing its fixed shape upon an unyielding mass of dough. But languages do not teach themselves, and those who learn them are agents, individuals intelligently mastering obstacles in order to increase their own linguistic capability and authority, and in the process reshaping and adapting the discourse they appropriate.[111] By learning Portuguese, or other languages, he was not replacing his Temne mother tongue, but supplementing it; like Englishmen who knew both English and Latin, Fernandez knew both Temne and Portuguese (and probably other languages); he was bilingual, and he helped contribute to the development of the modern Luso-African Crioulo language born of both tongues. Bilingualism gave Fernandez and his people an advantage over their European trading partners—whom they regarded not as oppressors, but suckers. "The natives . . . regularly and commonly say that the whites are like flies: despite the danger of falling in, they are always attracted by honey."[112]

In the end, we can only conjecture what that performance of *Hamlet,* in the summer of 1607 on a ship anchored in a bay off the coast of Africa, meant to its audience. But we can answer, for ourselves,

What does that long-ago performance of *Hamlet* mean to us now? In 1607 there were no European forts in this part of Africa; Sera Lyoa was not a colony; it was not suffering from an economic monopoly imposed by military force. As modern historians have observed, in this part of Western Africa in "the seventeenth century . . . commerce was relatively free and differentiated."[113] That balance, as we know, would not last. Portuguese racial tolerance was replaced by Spanish color prejudice, after the union of the two Iberian monarchies in 1580; after 1630 a new long cycle of drought debilitated indigenous agriculture; increasingly African elites paid for prestige commodities by selling their own people into New World slavery.[114] But all that was in the future, a future made by people other than William Keeling and Lucas Fernandez. Keeling and Fernandez did business together. Both peoples profited from the encounter.

And some of them learned from each other. Not all of them, of course. One striking feature of this encounter is the fact that we can witness it from so many individual perspectives. The slave-merchant Captain John Hawkins, after his own experience of Sierra Leone, comprehensively reviled "the Negro (in which nation is seldom or never found truth."[115] One of the merchants on Keeling's ship recorded in his journal that "This people are very lusty men, strong and well-limbed, and a good people and true; they will not steal as others of their color will do in other places." Prejudice here runs up against experience, and though the prejudice does not entirely yield, it admits exceptions. Indeed, the English journals express, if anything, more prejudice against Catholic Europeans than against the African infidels. "And in all that time of our being here we had no injury offered to any of our people: but all the kindness that might be expected"—I wish he had stopped here, but he didn't: "all the kindness that might be expected at the hands of such a black heathen nation."[116] But even that last belated reflex spasm of bigotry is itself a testimony to the overcoming of prejudice: the point of the sentence, a point which its distorted syntax cannot conceal, is that the Englishman did not expect any kindness from "such a black heathen nation," and what he got instead was "all the kindness that might be expected" of any civilized people.

And the English reciprocated. Whether or not the African spectators understood *Hamlet* in the way that Shakespeare or Keeling's actors intended, they surely understood one thing: that the English were inviting them to join in that community formed by actors and their audience. "We *gave* the tragedy of Hamlet." The English were offering *Hamlet* as a gift to their guests. And their guests, in turn, were "giving an audience," giving the English an opportunity and excuse to perform. Noticeably, the three recorded performances of plays on this voyage all took place when there were guests on board: someone from shore or someone from another ship. The purpose of such visits, as Keeling recorded in another journal, was "to increase affection."[117] So both sides, African and European, black and white, were getting something, and giving something. Something voluntary and superfluous, something that did no harm, something meant to cause pleasure, something translated from one people to another. "Over all this the Master of the *Dragon* presides."[118]

What better epitome for this moment of civilized multicultural exchange than *Hamlet?* And what did these civilized men do after they had celebrated and internalized Shakespeare's most famous play, after their self-consciousness had, as Harold Bloom declares, come to "identify with Hamlet"? They went out in the woods "all together" and bonded, as men will, by brutally and mortally wounding, without provocation, a magnificent, intelligent, wild animal. What better after-piece for a play that ends, "Go, bid the soldiers shoot"?

Notes

1. In my mind's eye, I remember with gratitude my tireless research assistant, Isaac Taylor, and the helpful responses from audiences at Morehouse, Spelman, the University of New Mexico, and the Shakespeare Institute, which contributed so much to this essay.
2. The most comprehensive edition of the surviving documents is P. E. H. Hair's *Sierra Leone and the English in 1607: Extracts from the Unpublished Journals of the Keeling Voyage to the East Indies,* Occasional Paper No. 4 (Freetown: Institute of African Studies, University of Sierra Leone, 1981); although Hair's extracts omit some passages and contain some errors, they are generally accurate and relatively complete for the days in Sierra Leone. In citing the manuscript journals (from the originals), I supplement the manuscript folios with page numbers from Hair. All the extant manuscripts are now in the British Library: William Hawkins, Egerton MS 2100; Anthony Marlowe, Cotton MS Titus B VIII; John Hearne and William Finch (aboard the Dragon; see extracts), India Office MS L/MAR/A/v; Unidentified, India Office MS L/MAR/A/iv. This fourth manuscript breaks off abruptly at the foot of a page, after entry for 30 August, then begins again 18 February, then ends mid-entry 12 March.
3. The most thorough early eyewitness account of flora and fauna in the estuary is André Donelha, *An Account of Sierra Leone and the Rivers of Guinea of Cape Verde,* ed. Avelino Teixero da Mota, trans. P. E. H. Hair (Lisbon: Junta de Investigaçoes Cientificas do Ultramar, 1977).
4. In quoting *Hamlet* throughout this essay, I cite the text of the Second Quarto, with modernized spelling and punctuation, occasionally emended. For what seem to me necessary emendations of the Second Quarto (as opposed to authorial variants), see the textual notes to *Hamlet* in Stanley Wells, Gary Taylor et al., *William Shakespeare: A Textual Companion* (Oxford: Clarendon Press, 1987), 396-420. My conclusions would not be significantly altered if Keeling's crew used the 1603 edition.
5. Manuel Álvares, *Ethiopia Minor and a Geographical Account of the Province of Sierra Leone* (ca. 1615), trans. P. E. H. Hair (Liverpool: privately published, 1990), f. 62-62v (ch. 4, p. 4), f. 77 (ch. 10, p. 3). For ear and nose rings in particular, see André Alvares de Almada, *Brief Treatise on the Rivers of Guinea* (ca. 1594), trans. P. E. H. Hair (Liverpool: privately published, 1984), ch. 15, p. 5 (and commentary note). Hair's "interim translations" have no through pagination; Álvares is cited by chapter and page, Almada by chapter and paragraph.
6. Arnold Aronson, "Shakespeare in Virginia, 1751-1863," in *Shakespeare in the South: Essays on Performance,* ed. Philip C. Kolin (Jackson: University Press of Mississippi, 1983), 25; Charles H. Shattuck, *Shakespeare on the American Stage from the Hallams to Edwin Booth* (Washington, D.C.: Folger Shakespeare Library, 1976).
7. For early German performances, see Simon Williams, *Shakespeare on the German Stage, Volume I: 1586-1914* (Cambridge: University Press, 1990), 27-45; for the relationship between Shakespeare's play and the German adaptation, *Bestrafte Brudermord,* see Harold Jenkins, ed., *Hamlet,* Arden Shakespeare (London: Methuen, 1982), 118-22.
8. Helen Phelps Bailey, *Hamlet in France from Voltaire to Laforgue* (Geneva: Librairie Droz, 1964), 1-23; Williams, *Shakespeare on the German Stage,* 1-26, 46-87.
9. Gary Taylor, *Reinventing Shakespeare: A Cultural History from the Restoration to the Present* (New York: Weidenfeld & Nicolson, 1989), 122-24, 167-68, 317. There are more than 20 verse trans-

lations of the play into Polish. For *Hamlet* in Russia, see Laurence Senelick, *Gordon Craig's Moscow "Hamlet": A Reconstruction* (Westport, Conn.: Greenwood Press, 1982); Grigori Kozintsev, *Shakespeare: Time and Conscience,* trans. Joyce Vining (New York: Hill and Wang, 1966); and Eleanor Rowe, *Hamlet: A Window on Russia* (New York: New York University Press, 1976).

10. Takeshi Murakami, "Shakespeare and *Hamlet* in Japan: A Chronological Overview," in *"Hamlet" and Japan,* ed. Yoshiko Uéno (New York: AMS Press, 1995), 250, 252.
11. Barry Gaines, "Shakespeare Translations in Former British Colonies of Africa," *International Encyclopedia of Translation Studies,* ed. Armin Paul Frank, Norbert Greiner et al. (Berlin: Walter de Gruyter, 2001), article 374.
12. Harold Bloom, *Shakespeare: The Invention of the Human* (New York: Riverhead Books, 1998), 420.
13. Terence Hawkes, *That Shakespeherian Rag: Essays on a Critical Process* (London: Methuen, 1986), 94.
14. P. E. H. Hair, "The Spelling and Connotation of the Toponym 'Sierra Leone' since 1461," *Sierra Leone Studies,* 18 (1966): 43-58.
15. Frederick S. Boas, *Shakespeare and the Universities and Other Studies in Elizabethan Drama* (New York: Appleton, 1923), 95.
16. Ania Loomba, "Shakespearian transformations," in *Shakespeare and National Culture,* ed. John J. Joughlin (Manchester: University Press, 1997), 111. Loomba incorrectly states that "two performances of Shakespeare took place aboard the *Hector* while the two ships were anchored in Sierra Leone"; the single performance in Sierra Leone clearly took place aboard the *Red Dragon.*
17. Dennis Kennedy, "Introduction: Shakespeare without His Language," in *Foreign Shakespeare: Contemporary Performance,* ed. Dennis Kennedy (Cambridge, UK: Cambridge University Press, 1993), 2.
18. See extracts. For two independent nineteenth-century transcriptions of these additional entries from Keeling's (now lost) manuscript journal, see *Narratives of Voyages towards the North-west,* ed. Thomas Rundall (London: Hakluyt Society, 1849), 231, and G. Blakemore Evans, "The Authenticity of Keeling's Journal Entries on 'Hamlet' and 'Richard II,'" *Notes and Queries,* 196 (1951), 313-15; 197 (1952), 127-28. At the beginning of the text of Keeling's journal printed by Samuel Purchas in *Haklytus Posthumus, or Purchas His Pilgrims* (London, 1625), Purchas notes that "This journal of Captain Keeling's . . . I have been bold to so shorten as to express only the most necessary observations for sea or land affairs" (Part I, book iii, 188); the printed text skips from August 22 (189) to September 7 (190).
19. Álvares, *Ethiopia Minor,* f. 55-55v (ch. 2, pp. 5-6). See also Donelha, *Sierra Leone,* f. 13 (p. 111); Almada, *Brief Treatise,* ch. 15, par. 17, etc. Sera Lyoan woven mats and carved ivories were prized by West Africans and Europeans alike.
20. "Master William Keeling, another of our principal merchants," *The Voyage of Sir Henry Middleton to the Moluccas 1604-1606,* ed. Sir William Foster (London: Hakluyt Society, 1943), lxxxviii (quoting Middleton's commission from the Company).
21. *The Journal of John Jourdain, 1608-1617,* ed. William Foster (Cambridge, UK: Hakluyt Society, 1935), xxix.
22. *The Embassy of Sir Thomas Roe to the Court of the Great Mogul, 1615-1619, as narrated in his journal and correspondence,* ed. William Foster (London: Hakluyt Society, 1899), 18.
23. Roe to Pepwell, December 30, 1616 (quoted ibid).
24. Philip Larson, *The East India Company: A History* (London: Longman, 1993), 25-26.
25. It was founded with an initial capital investment of only £30,000; by contrast, the Dutch East India company, founded in 1602, began with £540,000 (18 times as much). In the first decade of the seventeenth century, the English sent a mere 17 ships around Africa to the East Indies; the Dutch sent 60. See Brian Gardner, *The East India Company: A History* (New York: McCall, 1972), 32.
26. Captain John Hawkins on March 9, 1608 noted "our master's plot being a Portingale plot": see "A Journal kept by m[e William Hawkins in] my voyage to the East I[ndies, beginning the 28 of] March a0 1607 . . .", in *The Hawkins' Voyages during the reigns of Henry VIII, Queen Elizabeth, and James I,* ed. Clement R. Markham (London: Hakluyt Society, 1878), 379.
27. P. E. H. Hair, "Guinea," in *The Hakluyt Handbook,* ed. D. B. Quinn, 2 vols (London: Hakluyt Society, 1974), 204.
28. Gardner, *East India Company,* 32.
29. Adam Jones, "Sources on Early Sierra Leone (22): The Visit of a Dutch Fleet in 1625," *Africana Research Bulletin,* 15 (1986): 51, 61.

30. Hearne and Finch, 20 August, f. 7v (Hair, *English,* 26).
31. William Scouten, "The Sixth Circum-Navigation," in Purchas, *Pilgrims,* II, i, 88-9. For another account of this same voyage, see P. E. H. Hair, "Sources on Early Sierra Leone (10): Schouten and Le Maire, 1615," *Africana Research Bulletin,* 7 (1977): 36-75; it specifies "a brother-in-law of the king" (66). Hair also identifies the source of the text printed by Purchas (who edited, abbreviated, and sometimes misunderstood it, sometimes conflating material from different days, or misdating entries); this is significant in relation to his abbreviation of Keeling's journal. Purchas, for instance, describes the interpreter as a "Moore," where his source reads "black"—an important distinction, because Lucas Fernandez was a Christian.
32. Unidentified, 27 August, f. 15 (Hair, *English,* 22).
33. Hearne and Finch, 1 September, f. 9 (Hair, *English,* 33).
34. Unidentified, joint entry for 18, 19, 20 August, f. 14v (Hair, *English,* 21).
35. P. E. H. Hair, "Hamlet in an Afro-Portuguese Setting: New Perspectives on Sierra Leone in 1607," *History in Africa,* 5 (1978): 36-37.
36. P. E. H. Hair, "Early sources on Sierra Leone: (5) Barreira (letter of 23.2.1606)," *Africana Research Bulletin,* 5 (1975): 101.
37. "In his 1598 Preface, Hakluyt distinguished between the voyages in the northern seas, where the English were explorers and innovators, and the voyages in the more southern seas, where they were only following up the Portuguese and the Spaniards. Africa, being wholly to the south, therefore received less attention than did, say, Muscovy or North America": see Hair, "Guinea," 197-207.
38. John Keay, *The Honourable Company: A History of the English East India Company* (New York: Harper Collins, 1991), 39.
39. Purchas, *Pilgrims,* I, 3, 188 (July 30, 1607).
40. *The Voyages of Sir James Lancaster, Knight, to the East Indies,* ed. Clements R. Markham (London: Hakluyt Society, 1877), 113, 111.
41. Keeling, August 15, *Purchas,* 189 (Hair, *English,* 20); the same number of fish caught is reported in Hearne and Finch, f. 7 (Hair, *English,* 17).
42. Hearne and Finch, September 12, f. 10 (Hair, *English,* 42).
43. Hearne and Finch, August 27, f. 8 (Hair, *English,* 27).
44. Keeling, August 13, *Purchas* 189 (Hair, *English,* 20), confirmed by Hearne and Finch, f. 7v (Hair, *English,* 17). In their summary of the visit to Sierra Leone, Hearne and Finch report—September 12, f. 10 (Hair, *English,* 42)—the purchase of "2 large eliphants teeth," but they don't indicate the date or price of the second purchase.
45. *The Register of Letters etc of the Governour and Company of Merchants of London trading into the East Indies 1600-1619,* ed. Sir George Birdwood and William Foster (London: 1893; reprint Quaritch, 1966), 199 (Commission, item 11).
46. Jean Boulegue and Jean Suret-Canale, "The Western Atlantic coast," in *History of West Africa,* ed. by J. F. Ade Ajayi and Michael Crowder, 3rd ed. (Longman: New York, 1985), I, 507.
47. P. E. H. Hair, "Sources on Early Sierra Leone: (21) English Voyages of the1580s—Drake, Cavendish and Cumberland," *Africana Research Bulletin,* 13 (1982): 62-8.
48. August 30, reported by Marlowe, f. 255, Hearne and Finch, f. 8v, and Unidentified, f. 15v (Hair, *English,* 35-8).
49. *Register of Letters,* 117.
50. Álvares, *Ethiopia Minor,* ff. 57v, 58v (2,11; 3,3).
51. P. E. H. Hair, "Sources on Early Sierra Leone: (7) Barreira, Letter of 9.3.1607," *Africana Research Bulletin,* 6 (1976): 66, 70; Hair, "Sources on Early Sierra Leone: (8) Bartolomeu André's Letter, 1606," *Africana Research Bulletin,* 6 (1976): 46; Adam Jones, "The Kquoja Kingdom: A Forest State in Seventeenth Century West Africa," *Paideuma,* 29 (1983): 29.
52. P. E. H. Hair, "The Abortive Portuguese Settlement of Sierra Leone 1570-1625," in *Vice-Almirante A. Teixeira de Mota: In Memorium,* 2 vols. (Lisbon, 1987), I, 171-208.
53. For the earliest contact, see *The Voyages of Cadamosto, and other documents on Western Africa in the second half of the fifteenth century,* trans. and ed. G. R. Crone (London: Hakluyt Society, 1937), xx, xxvii, 81. On early missionary activity, see P. E. H. Hair, "Christian Influences in Sierra Leone before 1787," *Journal of Religion in Africa,* 27 (1997): 3-14.
54. John W. Blake, *West Africa: Quest for God and Gold 1454-1578* (London: Curzon Press, 1977), 188-92.
55. George E. Brooks, *Landlords and Strangers: Ecology, Society, and Trade in Western Africa, 1000-1630* (Boulder: Westview Press, 1993), 167-83.

56. K. N. Chaudhuri, *The English East India Company: The Study of an Early Joint-Stock Company* (London: F. Cass, 1965), 105-6.
57. Donelha, *Sierra Leone,* 257.
58. On oligophagy, see Almada, *Brief Treatise,* ch. 16, par. 10. All the many early accounts of the Mane invasions agree on their military and political use of cannibalism.
59. Álvares, *Ethiopia Minor,* f. 60 (ch. 3, p. 6).
60. For the Keeling entries, see Rundall, *Narratives,* 231, and Evans, "Authenticity," 314. Hearne and Finch record on September 29 that "wee were becalmed vntill 6 a clock in the eveninge" (f. 11; not in Hair, *English*); likewise, on March 31 "wee lay becalmed vntill noone" (f. 22; not in Hair, *English*). This evidence, not previously noticed, corroborates the reliability of the Keeling entries; only a ship becalmed, or a ship at anchor, is a plausible site for dramatic performances. It also, like other evidence noted by Evans, demonstrates the greater reliability of the 1824 transcription (which dates *Richard II* on September 29, where Rundall has September 30). Hawkins has no journal entry for either date, but does note, "We were becalmed," on March 29, and "The wind calm," on April 1 (*Hawkins' Voyages,* 381).
61. *Register of Letters,* 116.
62. Louis Montrose, *The Purpose of Playing: Shakespeare and the Cultural Politics of the Elizabethan Theatre* (Chicago: Chicago University Press, 1996), 102.
63. Chaudhuri, *East India Company,* 105; Michael Strachan and Boies Penrose, ed., *The East India Company Journals of Captain William Keeling and Master Thomas Bonner, 1615-1617* (Minneapolis: University of Minnesota Press, 1971), 45. This was also true of other Europeans voyaging to India: see C. R. Boxer, *From Lisbon to Goa, 1500-1750: Studies in Portuguese Maritime Enterprise* (London: Variorum Reprints, 1984), I, 58-59.
64. *Register of Letters,* 419 (Eighth Voyage, commission dated April 4, 1611),
65. N. A. M. Rodger, *The Wooden World: An Anatomy of the Georgian Navy* (London: Collins, 1986), 45; P. E. H. Hair and J. D. Alsop, *English Seamen and Traders in Guinea 1553-1565,* Studies in British History, 31 (Lewiston: Edwin Mellen, 1992): 346-7. For books aboard ship on Keeling's 1617 voyage, see Strachan and Penrose, *Keeling and Bonner,* 68, 95.
66. For this affective community, see Gary Taylor, "Feeling Bodies," in *Shakespeare and the Twentieth Century,* ed. Jonathan Bate, Jill L. Levenson, and Dieter Mehl (Newark: University of Delaware Press, 1998), 258-79.
67. Keeling, in *Purchas,* 189; Hearne and Finch, f. 7v (Hair, 18, 20).
68. Chaudhuri, *East India Company,* 46, citing Court Book, III, 211-12, September 7, 1614; *The Voyage of Thomas Best to the East Indies, 1612-14,* ed. Sir William Foster (London: Hakluyt Society, 1934), 282-83.
69. Strachan and Penrose, *Keeling and Bonner,* 3.
70. Hearne and Finch, September 4, f. 9v (Hair, 33).
71. "It was quite in the fashion of the period that the 'very kynde interteynment' of a semi-royal personage should include the performance of a play": Boas, *Shakespeare and the Universities,* 93. Boas was the first scholar to examine all the surviving manuscripts (84-95).
72. Claire Sponsler, "Medieval America: Drama and Community in the English Colonies, 1580-1610," *Journal of Medieval and Early Modern Studies,* 28 (1998): 453-78.
73. Extracts from Scott's book (printed by W. W. for Walter Burre, 1606) are reprinted in *The Voyage of Sir Henry Middleton,* ed. Foster, 152-61.
74. "Amongst all others, we were to make a show, the best we could; the which it must be understood could not be great, by reason of our small number; yet it was pretty, and such as they had not seen the like before" (156).
75. "Now we had no women to carry these things, wherefore we borrowed thirty of the prettiest boys we could get, and also two proper tall Javans to bear pikes before them. Master Towerson had a very pretty boy, a Chinee's son, whose father was a little before slain by thieves. This youth we attired as gallant as the King, whom we sent to present these things and to make a speech to him, signifying that, if our number had been equal to our good wills, we would have presented his Majesty with a far better show than we did, with many other compliments" (July 14, 157-58).
76. Keeling, in *Purchas,* 204.
77. *Register of Letters,* 105-6, 122.
78. *Embassy of Sir Thomas Roe,* 46 (September 26, 1615), 29 (Walter Peyton's journal, September 26), 63 (October 3).

79. Gardner, 36, 37.
80. Sydney Race, "J. P. Collier's fabrications," *Notes and Queries,* 195 (1950), 345-46; 196 (1951), 513-15; 197 (1952), 181-82.
81. For the authenticity of the documents see—in addition to previously cited studies by F. S. Boas, G. Blakemore Evans, and P.E.H. Hair—E. K. Chambers, *William Shakespeare: A Study of Facts and Problems,* 2 vols. (Oxford: Oxford University Press, 1930), II, 334. Keeling's surviving journal for the 1617 voyage demonstrates the normality of routines of text-production which meant some men were "continually writing" on board ship (*Keeling and Bonner,* 104, 119, 127, 165). Perhaps because he was aware of such conditions, the authenticity of the Keeling entries was defended—in *Notes and Queries,* 145 (1900), 41-42, and 195 (1950), 414-15—by William Foster, an archivist and scholar well-known to historians of the East India Company but less familiar to literary critics: for his "meticulous accuracy" and "vast knowledge of the India Office records," see C. F. Beckingham, "William Foster and the Records of the India Office," in *Compassing the Vaste Globe of the Earth: Studies in the History of the Hakluyt Society 1846-1996,* ed. R. C. Bridges and P. E. H. Hair (London: Hakluyt Society, 1996), 191-99. Foster's suspicions of Purchas as an editor/abridger were no doubt based, in part, upon his editing of *The Embassy of Sir Thomas Roe* (1899), where he was able to compare the Purchas printed abridgement with its manuscript source: "his editing of this particular journal is a very bad piece of work. That he should cut it down to a third or less . . . while leaving untouched many trivialities . . . that he should excise passages vital to the comprehension of others which were allowed to stand; that his dates should often be wrong; and that the carelessness of his copyist (or his printer) should be allowed to make nonsense of important passages, will scarcely admit of excuse" (lxiii).
82. T. J. King, *Casting Shakespeare's Plays: London Actors and Their Roles, 1590-1642* (Cambridge: Cambridge University Press, 1992), 88.
83. Boas, *Shakespeare and the Universities,* 3.
84. *An Elizabethan in 1582: The Diary of Richard Madox, Fellow of All Souls,* ed. Elizabeth Story Donno (London: Hakluyt Society, 1976), 195 (September 24).
85. W. W. Greg, *Dramatic Documents from the Elizabethan Playhouses: Commentary* (Oxford: Clarendon Press, 1931), 300-5.
86. For summaries of scholarship on early modern theatre conditions, see for instance Alan C. Dessen, "Shakespeare and the theatrical conventions of his time," *The Cambridge Companion to Shakespeare,* ed. Stanley Wells (Cambridge: Cambridge University Press, 1986), 85-100, and *Shakespeare: An Illustrated Stage History,* ed. Jonathan Bate and Russell Jackson (Oxford: Oxford University Press, 1996), 10-44.
87. Hair, "Hamlet," 33.
88. Keeling's commission from the East India Company (March 9, 1607) lists "the Red Dragon of 700 tons" (*Register of Letters,* 114), but other sources say 600.
89. C. R. Boxer, "The *Carreira da India* (Ships, men, cargoes, voyages)," in Boxer, *From Lisbon to Goa,* I, 52. By contrast, on the return voyage the ship would have been crowded, deeply laden, undermanned, and very cluttered; noticeably, no performances are recorded on the return voyage.
90. *Peter Floris, his voyage to the East Indies in the Globe, 1611-1615,* ed. W. H. Moreland (London: Hakluyt Society, 1934).
91. On the first performances of *Pericles,* see Gary Taylor, "The Canon and Chronology of Shakespeare's Plays," *Textual Companion,* 130-31.
92. *Embassy of Sir Thomas Roe,* August 24, 1615: "Hearing our hoeboyes in the Generall's [=Keeling's] boat" (34-35). The only prop required by the play that they would not normally have carried on board was a skull for the graveyard scene—but if the actors insisted on realism, they would have been able to find, or buy, bones on shore. They had, by this time, already been in Sera Lyoa for a month.
93. *Register of Letters,* 112-13.
94. Madox refers in Latin to Captain Fenton, the "general" of the expedition, as "our little king" (*Diary,* 168, 169, 172, 174).
95. *Register of Letters,* 115.
96. *Hawkins' Voyages,* 372, 374.
97. See, for instance, Eldred Jones, *Othello's Countrymen: The African in English Renaissance Drama* (London: Oxford University Press ["on behalf of . . . the University College of Sierra Leone"], 1965); Winthrop P. Jordan, *White over Black: American Attitudes toward the Negro, 1550-1812*

(Chapel Hill: University of North Carolina Press, 1968); Kim F. Hall, *Things of Darkness: Economies of Race and Gender in Early Modern England* (Ithaca: Cornell University Press, 1995).

98. Brooks, *Landlords and Strangers,* 33-46.
99. Generally, in dealing with Europeans the Africans who understood Portuguese would "then explain to the others what they have heard": see Hair, "Barreria . . . 23.2.1606," 95. However, all the African spectators may have understood Portuguese; Álvares notes that "Some of them, such as the nobles and those who have been brought up among our people, understand Portuguese, and these listen without saying anything" in order to deceive the whites, who think the Africans don't understand what they are saying (*Ethiopia Minor,* f. 56v, ch. 2, p. 8). Farim Buré, for instance, understood Portuguese, although he apparently allowed the English to assume that he did not—perhaps because local sovereigns preferred to speak through intermediaries (Almada, *Sierra Leone,* ch. 18, par. 4).
100. Otherwise, the first recognized translation of *Hamlet* into Portuguese was Oliveira Silva's text for a performance in Brazil by João Caetano in 1835; the first translation published in Portugal itself was by Dom Luís (king of Portugal) in 1877. See Eugênio Gomes, *Shakespeare no Brasil* (Rio de Janeiro: Ministry of Education and Culture, 1961). My thanks to Michael Warren for this reference.
101. Álvares (f. 54-54v) notes that "the language most commonly used today is the native Temne language," which differs from European languages in its treatment of plurals and tenses; he also complains that "native interpreters . . . never express what they are translating sufficiently exactly. Furthermore, even the best interpreters here . . . are ignorant concerning the figures of speech, especially metaphors" (ch. 2, pp. 3-4). Although Temne is as capable of metaphor as any other language, the metaphors of Christian belief (with which the missionary Álvares was chiefly concerned) would certainly have been confusing to non-Christians, and in general the metaphors that seem "natural" in one culture often seem bizarre in another.
102. Laura Bohannan, "Shakespeare in the Bush," *Natural History,* 75 (1966), 28-33.
103. Álvares, *Ethiopia Minor:* poisoning, f. 55 (ch. 2, p. 4), 56v (2,8); wine, f. 57 (2,10); images of dead parents, f. 64 (5,6); son mourning father, f. 72v (8,5); the king's counsellors, f. 58-58v (3,2); elaborate funerals, f. 68-68v, 73-73v (6,3-4; 8,7-8); kneeling, f. 61v (4,2-3). For an overview of their complex inter-state politics, see Brooks, *Landlords and Strangers,* 299-305. All these practices are amply documented, and sometimes more clearly explained, in other contemporary sources; I cite Álvares only because he was present in the estuary at the same time as the English.
104. Valentim Fernandes, *Description de la Côte Occidentale d'Afrique,* ed. Th. Monod, A. Teixeira da Mota, R. Mauny (Bissao: Centro de Estudos da Guine Portuguesa, 1951), 81-97; excerpts translated in Christopher Fyfe, *Sierra Leone Inheritance* (London: Oxford University Press, 1964), 28. This practice, recorded at the beginning of the sixteenth century, was still in place after the Mane invasions, in 1582: "if the king die, leaving his sons under years of discretion to govern, then he appointeth the eldest of his kindred to be his protector, who shall govern the kingdom; but if the king's son during this protector's life come to his years, yet he, the protector, will be king during his own life" (August 22, in Madox, *Diary,* 308). Álvares, early in the seventeenth century, makes the same observation repeatedly: ff. 58, 68v, 89v (3,1; 6,4; 15,5-6).
105. Álvares, *Ethiopia Minor,* f. 89v (ch. 15, pp. 5-6); Almada, *Brief Treatise,* 15:1 (and note), 18:12 (and note).
106. Álvares, *Ethiopia Minor:* afterlife, ff. 63, 67 (5,1; 6,1); "nuns," ff. 66, 69v-70 (5,11; 7,2-4); funerals, 73-73v (8,7-8). On funeral parties, see also Fernandes, ff. 132, 132v; Donelha, *Sierra Leone,* 118; Almada, *Brief Treatise,* 15, 6.
107. Fernandes, trans. Fyfe, 28.
108. Madox, *Diary,* 307 (transcript of John Walker's diary).
109. Álvares, *Ethiopia Minor,* ff. 62v, 71v (4,4; 8,4); see also, Almada, *Brief Treatise,* ch. 14, par. 5. Virtually all observers report the Temne practice of hosts offering male guests the use of one of their wives.
110. Madox, *Diary,* 307; Álvares, *Ethiopia Minor,* ff. 72v, 73v (8,6; 8,8-9); Donelha, *Sierra Leone,* f. 15 (p. 117). Widely reported in early sources.
111. See Janine Brutt-Griffler, *English as an International Language: Historical, Linguistic, and Pedagogical Dimensions,* Bilingual Editions and Bilingualism (Clevedon, UK: Multilingual Matters, 2001), ch. 4.
112. Álvares, *Ethiopia Minor,* f. 78v (10,8).
113. Boulegue and Suret-Canale, *History of West Africa,* I, 509.

114. Brooks, *Landlords and Strangers,* 57, 183-88, 318-19.
115. P. E. H. Hair, "Protestants as Pirates, Slavers, and Proto-missionaries: Sierra Leone 1568 and 1582," *Journal of Ecclesiastical History,* 21 (1970): 215.
116. Hearne and Finch, f. 10v (Hair, *English,* 42-3).
117. Strachan and Penrose, *Keeling and Bonner,* 59 (February 21, 1617). Keeling's 1617 journal records dozens of such social visits between ships (35, 59, 61, 69, 71, 72, 74, 77, 87, 89, 94, 104, 108, 109, 110, 116, 118, 1'33, 134, 135, 144, 150, 151, 158, 161, 165). No play performances are recorded—perhaps because the London office did not approve of Keeling's 1607 experiment.
118. John N. Morris, "*Hamlet* at Sea," in *A Schedule of Benefits* (New York: Athenaeum, 1987). My thanks to Peter Holland for calling this poem to my attention. As poets are permitted to do, Morris gets many of the facts wrong.

SECTION NINE

LEO AFRICANUS

Hasan Ibn Muhammad al-Wazzan, better known as Joannes Leo Africanus, was born in the Muslim Kingdom of Granada just a few years before it was conquered by the Christians in 1492. His family left Spain for Morocco when he was a small child; there they found a high social status and were close to the royal court of Fez. Hasan, from a very young age, had the opportunity to travel extensively, often as an ambassador to the Wattasid Sultan Muhammad. His travels led him to visit all the countries of North Africa, the Sahara desert, the sub-Saharan countries of West Africa, Egypt, Arabia, and Turkey. He was returning from a mission to Istanbul in 1518 when his ship was attacked in the Mediterranean sea by Christian pirates, who captured him and gave him to Pope Leo X. He spent one year in the Castel Sant' Angelo, a fortress just outside Vatican City, and converted to Christianity; he was baptized in Saint Peter of Rome by the Pope himself, who gave him his names, Joannes Leo de Medici, but he is generally called Leo Africanus. In Italy, he taught Arabic and wrote a number of books, many of which have not been found.[1] The most famous of these texts is a geographical opus about Africa. The manuscript was completed in March 10, 1526; an editorialized version of it was published for the first time in 1550. We know next to nothing about the rest of his life, not even if he stayed in Italy until his death, or if he went back to North Africa.

The following text is a translation of excerpts from the first book of the manuscript of Leo Africanus's most famous text (Biblioteca Nazionale in Rome, V. E. 953). This manuscript does not bear any title. I kept the title "Description of Africa" under which the text has come to be known. I stayed as close to the original as possible. I have kept the spelling of the proper names, except when indicated otherwise in the footnotes.

—Oumelbanine Zhiri

Description of Africa

Africa,[2] in the Arabic language, is called Ifrichia [Ifriqiya]. This name comes from the verb "faracha," which means the same thing as "separavuit." There are two opinions as to why it is called Ifriqiya. One is that it is separated from Europe and from a part of Asia by the Mediterranean Sea. Another is that its name comes from Ifricus King of Arabia Felix[3] who was the first who went and inhabited this part of the world; having been defeated and driven out by the King of Assiria, he could not go back to his kingdom, so he suddenly passed the river Nile with his army and went west, until he settled in the region of Cartagine.[4] Thus the Arabs do not use "Africa" except for the region of Carthagine, and the whole of Africa they call the Western Part.[5]

Of the limits of Africa

According to the African scholars and cosmographers, the eastern limit of Africa is the river Nile, which begins at the affluents of the lake of the desert of Gaugau, this in the south; and Africa ends in the north at the feet of Egypt, where the Nile enters the Mediterranean Sea. Northward, Africa begins at the delta of the Nile, and extends westward until the straits of the Columns of Hercules.[6] At the west, it extends from the straits on the coast of the Ocean until Nun, which is the ultimate part of Libia[7] on the Ocean. Southward, it begins at Nun and extends on the Ocean, which surrounds all Africa until the deserts of Gaugau.

Of the division of Africa

According to our scholars and cosmographers, Africa is divided in four parts: Barbaria, Numidia, Libya and the Black Land.[8]

[The rest of this chapter and the next three chapters provide detailed information on the delimitations of these regions, and their divisions in provinces and kingdoms.]

Of the population of Africa and the peoples who inhabit it

According to the cosmographers and the historians, Africa in Antiquity was not inhabited except for the Black Land, but Barbaria and Numidia were without inhabitants for several centuries. All its inhabitants, that is the white people, are called "el barbar,"[9] which derives, according to many scholars, from the verb "barbara" which means in Latin "murmurauit," because the African language is meaningless to the ears of the Arabs and seems, like the sounds of animals, to be just a meaningless noise. Thus the Africans were called the Berbers.[10] Others say that the word "Berber" is a duplication, because "bar" in Arabic means desert. They say that a long time ago when King Ifricus had been defeated by the Assyrians or by the Ethiopians, he fled toward Egypt and was followed by his enemies. Unable to defend himself, he consulted his people on what he should do to save them, and they all shouted "al bar bar," i. e. "to the desert, to the desert," meaning that in their view there was no other remedy than to pass the River Nile towards the desert of Africa. This explanation convenes with the view that all Africans descend from this people of Arabia Felix.

[. . .]

Of the division of the Africans, that is the Whites, in several peoples

The people inhabiting Africa are divided in 5 peoples : Sanhagia [Sanhaja], Masmuda, Zeneta [Zanata], Haoara [Hauara], Gumera [Gumara].[11]

[. . .]

Thus, anyone can see that all of the five peoples that we have talked about have ruled in these regions. It is true that Gumara and Hauara did not attain the supreme power. They have nevertheless ruled smaller regions, as it appears in the chronicles of the Africans. All these peoples ruled after they had converted to Islam. But before that, each people stayed by themselves in the country, and each people who inhabited the desert favored their own who lived in cities. The task of the ones who lived in the country was to raise the chattel, and the inhabitants of the cities were craftsmen and peasants. These five peoples are divided in 600 lineages as can be seen in the genealogical tree of the Africans, work authored by Ibnu Rachic[12] chronicler of Africa. A number of historians assert also that the present King of Tombuctu [Timbuktu] and the previous Kings of Malle [Mali] and of Agadez are of Zanaga descent, that is the ones who live in the desert.

Of the diversity and equality of the African language

All the peoples we have talked about are divided in hundreds of lineages and thousands of houses. However, they use one language which they call "agual amazig" which means the "noble language,"[13] and the Arabs of Africa call it the Berber language, and this is the native African language. It is specific and different from other languages. However it contains some Arabic words, and some believe that this proves that Africans descended from Saba, people of Arabia Felix. But others say that these Arabic words in the African language have been added after the Arab invasion and domination of Africa. These people were so ignorant and idiotic that there is not one author who could give a decisive argument to support any of these views.

There is some difference in their way of talking and in the vocabulary. The ones who live closest and have the more interaction with the Arabs use a lot of Arabic words, and almost all of the people of Gumara speak a corrupted Arabic, as well as many lineages of the people of Hauara, because they have long lived in the company of the Arabs.

As for the people of the Black Land, they talk several languages. They call one of them the language of Sungai, which is used in many regions like Gualata, Timbuktu, Ghenia, Mali and Gago. They call the other language Guber, and it is used in Guber, Chasena, Zegzeg and Guangra. Another language is spoken in the kingdom of Borno and it resembles the one that is spoken in Gago. Another one is spoken in the Kingdom of Nuba, and is mixed with Arabic, Chaldean and Egyptian.

In all the coastal cities of Africa from the Mediterranean to the Atlas mountains, the inhabitants speak a corrupted Arabic, except in the regions of the Kingdom of Marrocos [Marrakech] where the true Berber language is spoken as well as in the regions of Numidia that are neighbouring Mauritania and

Cesaria; but those who live in the vicinity of the Kingdom of Tunis and the Kingdom of Tripuli[14] all speak a corrupted Arabic.

Of the Arabs who live in the cities of Africa

There is a great number of Arabs who arrived in Africa in the army sent by Otmen, third Caliph,[15] in the year 24 of Hejira,[16] and there were around 80,000 people, noble and not. And when they had conquered many regions, all of the princes and nobles went back to Arabia with the exception of the general chief of the army, whose name was Hucba ibnu Nafih.[17] He settled there and built the city of Cairaoan [al-Qayrawan][18] because he was always afraid to be defeated or betrayed by the people of the coast of Tunis, and he thought that help could come to them from the island of Sicily or from Apulia. That is why this captain took all the treasures he had acquired, went to the desert in a place far from the coast and distant from Carthage around 120 miles, and there he built al-Qayrawan. He ordered all his lieutenants to live in castles and fortresses, and those who lived in towns that were not fortified had to fortify them. This done, the Arabs stayed secure and became citizens, mixing with the Africans who, at the time, used to speak Italian because they had been dominated by Italians for many years. That is the reason why the Arabic language had been corrupted and mixed up in every city of Africa as one can see in the modern times the Berbers mixed with the Arabs. However an Arab always notes his origin from the paternal side, and a Berber does the same, which is important because every notary and official, when he is writing a document about a person, always notes with his name the name of his origin and of his lineage, whether he is an Arab or a Berber.

[The next chapter is a long account of the history of the Fatimid dynasty,[19] who first conquered Ifriqiya and then Egypt.]

Of the division of the Arabs who went to live in Africa and are called Berberized Arabs

There are three Arab peoples who have emigrated in Africa, the first is called Chachim, the second Hilel and the third Mahchil.

[The next several chapters provide extremely detailed information about the different lineages into which these three people are divided, and the regions they inhabit.]

One must note that historiographers call the ancient Arabs who lived before the birth of Ismael "Arabi hariba," that is Arabic Arabs; as for the Arabs who are the descendents of Ismael, they are called "Arabi Muztahriba" that is Arabized Arabs, which means Arabs by accident because they are not native Arabs. The Arabs who went to live in Africa are called "Arabi Mustehgeme," which means foreignized Arabs, because they went to leave with a foreign people, and thus their language was corrupted, they changed their customs and became Berbers. All this is exposed better in the Histories of the Arabs by Ibn Calden[20] who has written a thick volume almost entirely devoted to the genealogy and the origins of the Berberized Arabs. It would be better to refer to this text if possible, because the memory of the author of the present work is feeble; he says that it has already been 10 years that he has not seen any book on the history of the Arabs. But because he has seen the peoples that he has mentioned previously, and had contact with them, he still has some information engraved in his memory.

[The next several chapters describe the customs of the different peoples of Africa.]

Of the faith of the ancient Africans

The Africans of Barbaria were quasi idolatrous like the Persians who used to worship fire or astral bodies, and had temples dedicated to them. They used to keep the fire always lit in these temples day and night, taking great care to not let it extinguish, as one can see in the chronicles of the Persians. Some of the Black Africans used to worship Guighimo which means in their language the Lord of the Heavens. They had gotten this idea by themselves, without having been informed by any prophet or man. After a long time, some of them converted to Judaism, and they stayed that way a long time. Then, some of the Black Kingdoms became Christian.

Then came the religion of Mucametto,[21] and in the year 268 of the Hegira the peoples of Libya became Muslims. They were converted by some predicators, who caused the peoples of Libya to wage big wars against the Blacks, until all the Kingdoms of the Blacks that are adjacent to Libya became Muslim. Now, there are still some Christian Kingdoms, but the Jewish ones have been destroyed by the Christians or the Muslims. As for the Kingdoms that are on the coast of the ocean, their people are all pagans and worship idols. The Portuguese know them and are in contact with them.

The people of Barbaria remained idol-worshippers for a long time until the year 250 before the birth of Muhammad. Then all the people on the coast of Barbaria became Christians because the region of Tunis and its people were dominated by some Apuglessi and Christian lords, while the coast of Cesaria and Mauritania was dominated by the Goths. At the same time, several Christian lords of Italy fled from the Goths toward Carthage, and created a new power there. But the Christians of Barbaria did not follow the observance and order of the Roman Christians, but the rules and faith of the Arians, among whose was Saint Augustine.[22]

When the Arabs invaded Barbaria, they found the Christian lords of this region and waged war against them, until their final victory. The Arians left for Italy, and some of them to Spain. Around ten years after the death of Muhammad, almost all of Barbaria became Muslim. But its people many times rebelled and rejected the faith of Muhammad and killed their governors and priests. But each time they heard of that happening, the pontifes[23] sent immediately their armies against the Berbers until they remained with the schismatics who fled from the pontifes of Bagdad.[24] Then the faith of Muhammad began to settle in Barbaria but gave rise to many heresies and conflicts among them. There will be a more complete view of those in the book of the faith and law of Muhammad, that is the most important aspects of it that are common between Africans and Asians if time does not impede the writing of the present work.[25]

[The next chapters are devoted to the writing of Africa, the agricultural products of the regions, their climate, the health and diseases of Africans.]

Of the virtues and laudable qualities of the Africans

The Africans who live in the cities of Barbaria, specially those on the Mediterranean coast, devote themselves with great care to studies and sciences, specially in humanities and in their theology and in matters of faith and law. They used to study the sciences of the Ancients, mathematics and astrology and philosophy. But it has been 400 years that many sciences have been prohibited by the doctors and by the princes, such as philosophy and judicial astrology. The inhabitants of the cities of Africa are also devout in their faith and they obey theirs doctors and priests. They take care of learning the things that are necessary to their faith. They go to their temples to do the ordinary prayers, and for that they wash their whole body, as will be seen in the second book of the faith and law of Muhammad.

Moreover, the inhabitants of the cities of Barbaria and Africa are ingenious and elegant in their arts and crafts, and they build beautiful edifices and works, which demonstrates their great mind. They are honest, without a lot of malice. They are virile men who will confirm in your presence what they had said in your absence. They have courage and spirit, specially the mountainers and the countrymen. They keep their word when they have promised something. They are jealous and despise their own life when they have been dishonored, specially when it concerns their wives. They have a great desire to obtain riches and honors, they travel the whole world for commerce, and as professors and masters in the sciences. Africans can be found in Egypt, Ethiopia, Arabia, Persia, India, Turkey, and wherever they go, they are well received and honored because each one of them is very competent in his art or craft. The inhabitants of the cities of Barbaria are also honest and proper men, they never say anything inappropriate in public, the younger respects the older, specially in discussions. The son will never mention love or women, or sing a love song, in the presence of his father or his uncle. If somebody talks about women or love, all the children leave the room. These are the good customs of the inhabitants of the cities of Barbaria.

As for the ones who inhabit tents, that is the Arabs and the pastoral tribes, they are generous, courageous, pious, patient, open, familiar, honest, obedient, reliable, agreeable and merry; they don't care much about trade. Mountainers are generous and courageous, proper and honest in their common living. The inhabitants of Numidia are more ingenious and educated than the mountainers because the Numids devote themselves to virtue and study law and faith, but they are not very versed in natural sciences. They are well trained in arms, very spirited and they are good companions. The inhabitants of Libya, both the Africans and the Arabs, are generous and agreeable. They're very good friends and hospitable to foreigners. They have spirit, purity, integrity and frankness. The inhabitants of the Land of the Black People have a lot of integrity and good faith. They are very nice to foreigners. They all are very merry. They dance and give banquets all the time. They show a lot of respect to scholars and religious men. Of all the Africans, they are the ones who lead the happiest life, and even more than anybody that the author says to have seen in the world.

Of the vices and flaws of the Africans

The inhabitants of the cities of Barbaria are like the fleas who stay in filth, poor, brave and terrible. They are contemptuous. When somebody displeases them in the least, they hold an eternal grudge against them. They are also unpleasant, and it is impossible for a foreigner to become their friend. Outside their area of competence, they are very credulous and simple, they believe any impossibility. The commoners are very idiotic and ignorant of the natural things, thus they believe that all the natural movements are divine acts. They are irregular in the way they live, trade and reason. They always talk in a loud voice and in every street one can always witness a fight at knife point. They are vile and despised by their lords. A lord has more esteem for an animal than for a citizen. They have no principals, captains, magistrates, or consuls who could rule them or advise them in how to govern themselves. They are also very ignorant and gross when it comes to trade. They have no banks of change nor a delegate who could send their merchandise from one town to another. Every merchant wants to stay close to his merchandise, and where the merchandise goes, his owner goes. They are extremely avaricious. There are among them numerous people who have never received in their home strangers nor out of courtesy neither for the love of God. Few among them reward the favors they have received from others. There are also very melancholy men who have no pleasure in life, as they are always busy surviving, because of the great poverty that exists in these regions, and the small gain they make.

As for the inhabitants of the mountains and the country, that is the pastoral tribes and a part of the Arabs, they all live a difficult life, working all the time, but they stay in misery and necessity. They are very beastly and they are thieves, very greedy and they tend to not pay for they have bought with credit. They are ignorant of letters. They are as cuckold as chicken. Each one of them picks a boyfriend for his daughter and gives a banquet and he joins the hand of the boyfriend with the hand of the daughter or the sister and they become friends. But when the girl gets married, the boyfriend does not see her anymore, and flees from the ill will of the husband. Most of them are not real Muslims, neither Jews nor Christians. They know no God nor saints, they don't pray, they don't know how, they have no temples, no priests, no doctors. The few among them who are devout are quite lost because they have no laws and no rules, but live with no knowledge of the limits.

As for the inhabitants of Numidia, they have no knowledge of things, and they are very remote from any natural way of life. They are traitors, murderers, and thieves without any conscience. They are very vile, they go in Barbaria and take all kinds of vile jobs, like stablemen and cooks; they would do anything for money.

The inhabitants of Libya are beastly, thieves, and ignorant in any science.

They steal and murder, and live like wild animals. They have no faith and no rule. They live and will always live in misery. They are traitors to the supreme degree, specially when it comes to merchandise. They have very long horns and waste their whole life in vanity, either hunting or waging war among themselves, and, in the meantime, they go nude and barefoot and lead their beasts in the desert.

The inhabitants of the Black Land are very beastly and have no reason; they are irrational. They have no mind and no experience, they have no good knowledge of anything, they live like animals without law and without rules. Among them there are a lot of prostitutes and a lot of cuckolds. One has to except some of them, who live in the big cities and who have a little more rationality and human decency.

The author does not feel a little shame and confusion in telling and uncovering the vices and flaws of Africa, where he has been nurtured and raised, and since he is known as a good man. But it is necessary to one who wants to write and to describe things as they are and finds himself in the same situation as it is said in the book of the *Cento Novelle.* In one of the stories, briefly, one man was sentenced to be whipped, and he was a very good friend of the executioner. When he arrived to the place where he was to be punished, he saw his friend and began to cry, in which he found comfort and relief, thinking that since the executioner was his friend he will show him compassion because of their friendship. And the first hit that the executioner gave him was very hard and cruel and the accused shouted loudly and told the executioner:

"O my friend, you are treating me very badly even though I am your friend."

The executioner said back: "Be patient, my friend, I must accomplish my duty as it must be done."

Because it would be more scandalous for me if I was accused for good reason, and thought of as a man who has a lot of the vices and flaws that I have talked about, but utterly deprived of any virtue and good quality of the Africans, I am going to do like the bird whose story can also be found in the *Cento Novelle.* This bird could live underwater as well as on the earth. When the King of the Birds asked them to pay their taxes, the bird fled underwater and told the fish: "You know that I have always been one of yours, and this cowardly King of the Birds asked me to pay the taxes." And the fish answered: "He has some nerve to demand that you pay his taxes! Let him come to us and we will show him what taxes he is going to get from you." And the bird stayed very comfortable and protected. One year passed, and the King of the Fish came to ask for his taxes. When the other fish payed their taxes, the bird left the water and fled to the birds telling them the same excuse.

What the author means by that is that where man sees his advantage, there he goes. For example, if people disparage the Africans, he will have the clear excuse that he was not born in Africa but in Granada. And if the Granadins are disparaged, he will have another excuse, that he has not been raised in Granada, and he does not even have a memory of it. So, in order to tell the truth, it is necessary to ignore who nurtured you. Besides, he will not say anything except what is publicly and generally known, which he could not hide even if he wanted to.

End of the first part of the book
of the cosmographia of Africa according
to the author mister Joanni Affricano.[26]

Notes

1. Besides the *Description of Africa*, two of these works have been published: "Libellus de Viris Quibusdam Illustribus apud Arabes," in J. H. Hottinger, *Bibliothecarius Quadripartitus*, 1664; "De Arte Metrica Liber," in A. Codazzi, "Il trattato dell'Arte Metrica di Giovanni Leone Africano," *Studi Orientalistici in Onore di Giorgio Levi Della Vida*, vol. 1 (Rome: Instituto per l'Oriente, 1956).
2. Throughout the text, Africa and its derivatives are spelled with two "fs." The rest of the translation conforms to modern usage.
3. Yemen.
4. Carthage.
5. "Western part" is a literal translation of the word "Maghreb," which designates the northern region of Africa, with the exception of Egypt.
6. The straits of Gibraltar.
7. Leo calls the Sahara "Libia." In the rest of the translation, I use the modern spelling, "Libya."
8. On these denominations, see later in the article. For the first two names, the manuscript offers different spellings (Barbaria, Barberia, Berberia for the first, and Numidia, Numedia, for the second). I adopt the following spellings in this translation: Barbaria, Numidia. "Black Land" is a translation of "Terra Negresca."
9. Modern "Berber"; the Berbers are the native inhabitants of North Africa. I use the modern spelling in all other occurrences.
10. "Barbari" in the manuscript.
11. The names of these five people are spelled in various ways throughout the manuscript. As in other names, the variations concern chiefly the vowels. I indicate in parentheses the standard transcription, which I use in the rest of the translation.
12. Ibn al-Raqiq, historian who died in the eleventh century.
13. In Berber.
14. Tripoli, in modern Libya.
15. Uthman was the third Caliph, that is the third ruler to succeed the Prophet Muhammad. He was Caliph from 644 to 655.
16. The Islamic era.
17. Uqba ibn al-Nafi was one of the Arab conquerors of North Africa.
18. Al-Qayrawan, in modern Tunisia. I use this spelling in all occurrences.
19. The Fatimids were a Shiite dynasty who first established themselves in North Africa where they ruled from 910 to 969, then in Egypt, from 969 to 1171.
20. Ibn Khaldun (1332-1406) is one of the most famous Arab historians. He is the author of the *Kitab al-Ibar*, a universal history. Its most celebrated part is the introduction, the *Muqqadima*, often published on its own; it contains an extremely novel view of history, for the first time seen as an autonomous science.
21. The religion of the Prophet Muhammad, Islam. I use the spelling "Muhammad" in all other occurrences.
22. This mistake shows Leo's lack of familiarity with Christian history. Saint Augustine (354-430) was on the contrary the most famous adversary of the heretic Arrianists.
23. Leo uses many times the word "pontif" as a translation of the Arabic "caliph."
24. The schismatics are the Fatimids, and the pontifes of Bagdad are the Abassid dynasty.
25. Leo had conceived and written a book on certain aspects of the Islamic religion. He mentions this text a few times in the *Description of Africa*. Unfortunately, it has not been found.
26. When referring to himself in the manuscript, Leo uses any of his two Christian names (Joannes and Leo).

LEO AFRICANUS'S *DESCRIPTION OF AFRICA*

Oumelbanine Zhiri

At the end of the first book of Leo Africanus's description of Africa, the author reflects, through two stories, on his extraordinary situation. An author of Arab origins, he is writing in Europe, in faulty Italian, a text about Africa, and addressing European readers at the time when they had very little reliable information about this continent. This set of circumstances was unique for an early sixteenth-century writer,[1] and a great deal about the text and its impact on European representations of Africa can only be understood in light of this fact. As we will see, one of the particularities of Leo Africanus's influence is that his biography seems as compelling as his work; it has attracted almost as much interest through the centuries, and it informs his text in important ways.

From the time that his text was first known to our present day, Leo's work has been considered a first-rate source on the history and geography of North and West Africa in the Early Modern period. However, from the beginning it has also attracted attention beyond its purely informative content, thanks to its multilayered and multicultural richness. Nowadays in particular, Leo Africanus and his work have clearly ceased to be confined solely to the field of historical studies. This is due in large part to Leo's unusual life story, and to the fact that he conceived and wrote his work while in exile. There is a growing interest among historians and literary scholars in authors whose work crosses the boundaries of civilizations and who belong at the same time to different cultural traditions. It is now widely, and rightly, thought that texts written under such circumstances can inform us acutely about the processes of interaction between the cultures, and that these interactions themselves also play a crucial role in the evolution of civilizations. While scholarship has been long constrained by rigid national traditions, it is now recognized that the people who have associations with different cultures in their lives, and whose works make manifest those associations, play as vital a role in culture as works by individuals rooted in a single tradition.

This essay aims to show how Leo Africanus and his work may inform us about the interrelation between two major Mediterranean cultures: the Arab and Islamic, and the European and Christian. Moreover, it means to expose that this multicultural perspective on Leo and his text is probably what makes them still vital to the modern reader, and what allows us to reflect on the import of cultural hybridity. Finally, the purpose of this essay is to show how much this hybridity not only is a useful frame to understand his work but also that the work itself presents a reflection on the issues raised by this hybridity—a reflection by an author for whom this hybridity was far from merely an academic concern but which constituted the horizon of his own literary production as well as of his own subject position.

From the sixteenth to the twenty-first century, the adventurous life of Leo Africanus has attracted the attention of many authors. Better to understand the appeal of his life and the perils of life in the Mediterranean in general, we need to look at the larger historical context. When Leo was writing, Africa had already been the object of European imperialist attempts for some time.[2] The first successful Portuguese attack on the Moroccan coast occurred in 1415 and was the prelude to a long series of assaults on the African-Atlantic coast. The aim of the Portuguese was to have direct access by sea to the gold mines of sub-Saharan Africa, and not to have to depend on North-African countries for this commerce. Their purpose was to control the Atlantic coast of the continent, which they accomplished by taking parts of the coast and building fortresses. That enabled them to launch exploratory expeditions around Africa, which culminated in the circumnavigation of the continent by Vasco da Gama in 1498. They left the Mediterranean coast to the Spaniards, who despite undeniable successes finally abandoned Africa to focus on the conquest of America. They were rivaled by the Ottoman Turks, who after conquering Egypt began to establish their influence in North Africa. The Turks prevailed in this struggle and conquered almost all of North Africa. Only Morocco escaped Ottoman domination, largely due to the new Saadi[3] dynasty that emerged during that time, which established its power over the whole territory by the middle of the sixteenth century. The Saadis set as one of their main goals the liberation of the coastal cities taken by Europeans. They had great success against the Portuguese, who in an attempt dubbed by historians as the "last crusade" tried to invade Morocco in 1578 and were thoroughly defeated by the Moroccan armies, in what became known as the Battle of Three Kings. The impact of this victory[4] was to shield Morocco against further attacks from Europeans as well as from the Turks, until the nineteenth century.[5]

Many events in Leo Africanus's life are directly linked to this history of conflict between Europe and North Africa, most importantly his capture by Italian pirates in 1518. That capture ultimately led him to write a text that simultaneously belongs to the history of the interaction of these two cultures and is also a result of that history. The essential hybridity of this work is present at all levels, the most important being the text itself. Unfortunately this does not appear as clearly as it should in the numerous editions of his text. This is an issue of paramount importance, and it is necessary to explain how it impacts on the full understanding of the mixed nature of Leo's work.

In 1550 one of the most important collections of travel accounts of the Renaissance, *Delle Navigazioni e Viaggi,*[6] coordinated by Gian Battista Ramusio, was published in Venice. The first geographical text of the first volume was Leo's geographical opus on Africa, in a version that was considered for centuries to be faithful to the original. In 1931 the Biblioteca Nazionale of Rome acquired an untitled manuscript, identified and described by Angela Codazzi[7] as a copy of the original manuscript used by Ramusio. This allows us to assess Ramusio's considerable editorial work. Indeed, the manuscript was barely publishable as it was. It is not punctuated and is written in extremely flawed Italian. Most importantly, Ramusio did not content himself with correcting the grammatical mistakes and making the text easier to read. He also gave it its title, *Descrittione dell'Africa* (*Description of Africa*), structured it differently, changed the spelling of many proper names, omitted passages, even added some remarks of his own, and significantly altered the meaning of many sentences and even entire pages. This heavy editorializing cannot be overlooked. Briefly, Ramusio aimed to make the text more easily readable for Europeans, and in fact Europeanized it significantly, adding references to the history and culture of Europe, and omitting passages that evoked in some detail events and works belonging to the history and culture of Africa. In other words, it tampered with the very quality that makes the *Description* so interesting, the mixing of cultural traditions that it attempts. Later in this essay, we will scrutinize a few examples of these alterations. The Ramusio version deserves to be considered a far from reliable "*translation*" rather than merely a grammatically more correct text. It is nevertheless the one that has been republished several times up to the twentieth century, and is the basis of all subsequent translations.

These translations have considerably helped the impact of Leo's work on European representations of Africa. Too often, they are still the texts used by modern readers, despite their lack of accuracy. Two were published in 1556, one in French, the other in Latin. The more important one historically is the latter, written in a language accessible to all scholars in Europe.[8] It is also, unfortunately, the less reliable: it is flawed and contains an incredible number of mistakes.[9] Yet it served as the basis for subsequent translations, the English translation by John Pory in 1600 prominent among them. This unreliable translation of the faulty Latin translation of the already problematic Italian version, is unfortunately to this day the only one available in English.

None of the published editions of the *Description* takes fully into account the discovery of the manuscript, despite the fact that the interest in Leo's work has greatly heightened since the nineteenth century. The European investment in the exploration and colonization of Africa has provoked a flourish of new texts, travel accounts, and historical works, bringing new information on what was considered the "dark continent." During that time, Leo's work was used to map new explorations, was reassessed in light of new information, and was unanimously considered the most important text on Africa published in Europe before the nineteenth century. This renewed interest is reflected in a steady flow of new publications of the old Italian, French, and English versions during the nineteenth and twentieth centuries.[10]

A most influential new French translation was given by Alexis Epaulard.[11] His edition, through a great number of historical notes, gives the reader an invaluable background to the work of Leo Africanus. But the problems concerning the text itself are still present. In the preface, Epaulard asserts that he has collated the text published by Ramusio with the manuscript identified by A. Codazzi. This assertion elevated Epaulard's to the status of standard text and was used mainly by historians of Africa who use Leo's text as an important source, but also by a new wave of translators who translate the Epaulard French version rather than go back to the manuscript.[12] Unfortunately, Epaulard's contention should not be accepted at face value for two important reasons. First, he has not, as he states, corrected all the discrepancies between the manuscript and the text—far from it. Secondly, in some cases he has actually added his own mistakes to the ones made by Ramusio. The manuscript is the closer text to the original, to the one that Leo conceived and completed, and historians of Africa would be better served with an edition of this text, with all its faults and flaws. Not only that but the use of any of the texts that stems from the Ramusio version—that is, all that have been published until today—render questionable any study of the problems of hybridity that are so important in assessing the import of the *Description.* Obviously, the texts published during the sixteenth and seventeenth centuries are the ones to be considered in a study of the influence of Leo's text, and of the ways that it deeply transformed the view of Africa in early modern Europe. They are much less helpful, and sometimes truly misleading, if one wants to study the modalities in which this text was written on the frontiers of cultures. This study is fascinating in its own right, given the conditions in which the author has composed this text: it is a deeply hybrid work, in which Otherness is not just an abstract concept but a constant presence in the text, and even in the language that it uses. In a moment, I will try to show a few of the ways in which this presence manifests itself. For now, the concerns I have expressed explain why I have chosen excerpts from the manuscript to translate for this volume, since I am convinced of the necessity to go back to the original if one wishes fully to understand Leo's work and his place in cultural history.

Of course, despite the reservations expressed, it is through the text reworked by Ramusio that Leo has so greatly impacted on the European views of Africa. Among the texts contained in *Navigationi e Viaggi,* the *Description* in particular was given an enthusiastic reception. This is hardly surprising when one takes a look at the state of the knowledge of Africa in Europe just before its publication. The reading of influential geographers, such as Joannes Boemus or Sebastian Münster, shows that the core of the description of Africa in Europe until the first half of the sixteenth century goes back as far as Antiquity. The sources most often quoted in these texts, as well as in many others, are ancient authors, such as Strabo, Ptolemy, and Pliny. Their view of Africa had been transmitted through the Middle Ages via the work of influential authors such as Isidorus of Sevilla.[13] The information on Africa during the first half of the sixteenth century was only slightly renewed. Not only had Europeans very little knowledge of the more recent historical events in this continent but their texts conveyed a view of Africa that was based much more on myth than reality. Africa was generally considered a desert dominated by powerful wild animals. As for the people, they were portrayed as less than human, and sometimes as monstrous. It was thought that, in sharp contrast with Europe and Asia, they had built no civilizations and had no history.[14] Geographers and historians had only begun to take into account the new information brought by the Portuguese exploration of the Atlantic coast of Africa.

The *Description* brought a mass of information on North and West Africa, their people, their languages, their history, their trade, their economy, their cities, and their customs. In most of the texts about these regions written in the sixteenth and seventeenth centuries, the bulk of the modern information can be retraced to Leo's work. Later works do shed some new light, but mainly on contemporary history. Only colonization brought a serious renewal of the geographical description of Africa in Europe, though Jose Garcia Baquero found

Leo's influence still strong in maps designed in the late nineteenth century.[15] There is no doubt that Leo is the single most important figure to link North Africa and Europe between the years 1500 and 1800.[16] Moreover, the importance and the success of the *Description* is attested to by the fact that it is quoted in texts belonging to other fields of literature and knowledge from the political philosopher Jean Bodin[17] to the poet W. B. Yeats,[18] which helps to make Africa part of the European cultural horizon.

So the first European readers of Leo were quite ignorant of Africa. Leo is making a great effort to reach them by writing in a foreign language, and by using words and concepts that were familiar to them. If Leo has largely succeeded in transmitting a whole body of knowledge on Africa to Europe, he has nevertheless encountered failure in many smaller points. For example, one striking feature of the first book is the use of toponyms that belong to the European tradition. They designate generally big regions such as Numidia, Libya, and Barbaria. The first two originate from ancient texts while the last dates from the thirteenth century.[19] Later, Leo will also use "Ethiopia" to designate the Bilad al-Sudan, the Land of the Black People. Unfortunately, the correspondances between the European and Arab or Berber names are often inaccurate and have confused many early modern readers.[20]

But maybe the most remarkable example of the use of a European term is in the very first word of the text: "Africa." Leo gives the Arabic equivalent of "Ifrichia" (Ifriqiya), but at the time that Leo is writing, the equivalence is not correct.[21] And at the end of the paragraph, he explains it himself: in Arab geography, "Ifriqiya" designated only the region that corresponds roughly to modern Tunisia, the region that was called by the Romans *Africa proconsularis,* and is still called Africa Minor, Africa proper, or Little Africa by early modern European geographers. In fact, the concept of "continent" itself was foreign to Arab geographers, who followed another path inherited from Greek geography, and used the concept of *iqlim* (from the Greek "climate") to divide the world.[22] At the end of the chapter, Leo uses the expression "the western part," which is a literal translation of the word "Maghreb," thus juxtaposing Arab and European concepts, without explaining the discrepancy between the two.

Thus, Leo borrows the concept of continent from the European tradition. But his use of the derivative "African" may be confusing. He rarely uses it with the meaning of "inhabitant of the continent." Most often, "African" is opposed to "Arab," and designates the Berbers, the native people of North Africa, who have to be distinguished from their Arab invaders.[23] So Leo restricts the meaning of the word "African," which designates in his text only one people of the continent. One must note that the Arabic word "African" ("ifriqî") was not used in this sense in Arabic texts. At the same time, Leo's usage is contrary to the European usage and will confuse many readers. In other words, Leo here uses a European term but with a meaning that makes sense only in regard to the distinctions significant in his own culture. Thus we see the text shifting constantly between the two frames of reference; and the author had indeed a limited knowledge of one of them, the European, which leads him to a rather interesting blurring of the two in this case.

Through these few examples, we see that the efforts made by Leo were not entirely rewarding. If he had indeed contributed greatly to the transformation of the European image of Africa, it is not without fostering some mistakes. This loss comes in part from the very fact that he is writing in a foreign language. At the time, no European language was equipped to express notions and concepts elaborated on in the Islamic civilization. Leo here is trying to create a language that is capable of such expression. Sometimes he tries to find an equivalent, as when he translates "caliph" by "pontiff." This translation is deeply misleading, giving the illusion that the institution of the caliphate is equivalent to that of the Pontificate. Sometimes he simply translates an expression that has a very specific meaning in Islamic culture but no special resonance for Europeans. For instance, when he says that the pastoral tribes live with "no knowledge of the limits", when he talks about people who live with "no rules," he is making explicit reference to important concepts in Islamic theology. These phrases in particular allude to a fact that Leo will evoke elsewhere, that people in some regions, even though they were Muslims, did not abide by Islamic law in their lives, but rather followed customs that predate Islam. The translation falls far short of conveying the full meaning, and certainly does not convey its specificity to the European reader. The proof of this misunderstanding can be found in the Ramusio version: the expression "no knowledge of the limits" is actually omitted in the translation. Ramusio will consistently show a deep misunderstanding of the passages where Leo talks about law, justice, and its administration in Islamic countries, as well as those where he describes aspects of political life, such as the role of religious scholars ("ulama") and their link to the government. This is not surprising because there was little knowledge of Islamic culture at the time in Europe. These examples show that the Ramusio version of the text is not reliable; Ramusio has diminished Leo's effort to get across to the European readers concepts that are foreign to them and certainly betrayed the text written by Leo.

The ambiguities in Leo's text go very deep indeed, and have to be understood at the epistemological level when the author uses European concepts and mixes them with notions inherited from Arab geography and historiography. This work, to be fully understood, demands that one takes into account the two traditions to which it belongs and the efforts made by the author to make them coexist. These efforts were all the more difficult because not only did the readers have no other source of knowledge of the content but the author himself had no access to the books that could help him in writing. In the first book, he mentions that it has been about ten years since he had read a book on Arabic or African history and geography.

This lack of availability of his written sources has an important counterpart: the promotion of the character of Leo in the text. The particular conditions under which he wrote, removed from his sources, forced him to rely as much, and maybe more so, on his memory of his travels and of his own observations. After referring to Ibn al-Raqiq, he adds that he can also rely on his personal knowledge of the African people, which should compensate for his inability to refer to books. Thus his own experience becomes as much a source as the written tradition to which he has access only through his memory, and maybe more so. Memory, therefore, plays a vital role in this text, being at the same time the occasion of its elaboration and the most severe constraint on its writing. In particular, Leo has to insist on his own experience as a source of his text, since it is the one that will validate his authority as a geographer. And it is the only source, since he can quote but sketchily the texts that he would have liked to cite and, furthermore, that are unknown to his readers. Thus he cannot create his text's authority through a network of shared cultural references with his reader, sustained by evocative names. He is alone in making his work credible: he has to bring proof of his authority, to show that he is qualified and should be believed. To do so, he has to conjure up his own experience in the text, and to reiterate it again and again, since it is the sole source of his credibility.

Among the effects of this necessary promotion of Leo the character is the fact that the *Description* is often talked of as a travel account, belonging to the geographical genre that is closest to autobiography. In fact, as this first book shows, it is not a travel account, but a book of descriptive geography, belonging mainly to a long tradition of Arabic geography, the "Kingdoms and Routes."[24] This has not prevented many to read and present his text as a travel account. One reason is the fascination that many have felt for Leo the person, his remarkable life story, and his crossing of the frontiers of culture. This has led many authors at least to allude to aspects of his biography, his travels, and his conversion, before they quoted his text, which is far from common for a geographer. The most extensive tribute to Leo is certainly his novelized biography, written in French by the Lebanese author Amin Maalouf,[25] which was a best-seller in several countries. Presented as the autobiography of the fictional Leo, it constructs its character as a living link between civilizations and religions, an independent spirit and a fighter for more tolerance between peoples of different cultures. This unusual interest in the life of a geographer is linked to many aspects of Leo's text. The author himself stresses his extensive travels, as well as the large extent of his education, his knowledge of Arabic literature and philosophy, and of Islamic theology and law. Though that may reflect aspects of his character, it is made necessary by important issues of authority and credibility. The less he is able to quote his references, the more he has to mention his own experience.

Strangely, what he himself considered as a weakness (the lack of precise references to authors who had written before him), helps make his text so valuable for us, because it is less dependent on the conventions of the genre and more directly linked to his own experience. So Leo, by the insistence on his life story, has made himself a character in the text, and this character is a traveler; in the several hundreds of personal notations in the *Description,* the more numerous are the ones that mention his travels. After describing a place, he always mentions that he has visited it for personal reasons or as part of his work as a representative of the King. For example, in Book II, he gives a long account of an episode of his first trip to sub-Saharan countries when, as a teenager, he accompanied his ambassador uncle and was sent by him to visit a local lord of southern Morocco. He composed and read a poem to the lord, who was impressed by his talent and received him as an important emissary. In this episode, Leo gives a colorful anecdote, and also shows how well traveled he was, how, from an early age, he had contact with the elite of society, and how talented a writer he was. All the elements of the legitimation of his authorship are present.

The references to his travels are numerous in the text, and they usually have a legitimating function, which is linked to the fact that there is no cultural memory that is shared with the reader. Thus the text reminds us constantly of the conditions of its elaboration: Leo is so present in the text because the author has been led by circumstances to write in a foreign place and for a foreign audience.

Leo ends this first book with two stories in which he shows a great awareness of the complexities of his situation, and of the possible misunderstandings to which it may lead its readers. But the way in which Leo introduces the two stories and tries to anticipate and answer possible objections, asks the question on a different level. Here it becomes a moral question and comes fittingly after Leo has just illustrated a genre very common in Arabic culture, the listing of the qualities and flaws of people. This is as much an example of stylistic prowess as it gives instructive information on the way that classical Arabic culture viewed different nations.

These passages have attracted a lot of attention. In fact, the first commentary we have is a correction brought by Ramusio, who added an allusion to a disparaging opinion of Sallustius on the subject of the supposed untrustworthiness of the people of North Africa. First, Leo knows that he is the object as well as the subject of the discourse, and even more importantly he knows that the reader can't totally exempt him from it and tries to shield himself from attacks. Especially the story of the bird that avoids paying taxes to either king has attracted quite a few amused commentaries from sixteenth- and seventeenth-century writers. The story is simultaneously a striking example of the presence of Leo in the text and a way to show that the question of the hybridity of the text is also a moral and individual one. These factors combined account for a good part of the long-standing fascination exerted by this text and its author. He is at the same time the author of the *Description* and a character in it, and even the subject of its discourse. It is a text of descriptive geography but it is also in some ways a skewed autobiography. Moreover, the epistemological aspects of the crossing of the culture are also part and evidence of a personal crossing that gives the stories their particular resonance, which has struck so many readers. Leo has brought himself to live in the text by acknowledging the strangeness of his situation and the complexity of his own identity as it had been fashioned by circumstances, some of which he had chosen and some of which were not of his making. To these issues loaded with historical, epistemological and moral implications, his final answer, given through the story of the fish, brings a welcome touch of humor and playfulness.

Notes

1. It was not that unusual for people to cross the cultural frontiers between Islamic and Christian Mediterranean countries. Bartholomé and Lucille Benassar have studied the case of Christians living in North Africa and converting to Islam (*Les Chrétiens d'Allah* [Paris: Perrin, 1989]), and Ahmed Bouchareb the case of Muslims converted to Christianity and living in Spain and Portugal ("Les conséquences socio-culturelles de la conquête ibérique du littoral marocain," in *Relaciones de la Peninsula Iberica con el Magreb [Siglos XIII-XVI]* [Madrid, 1988]). What makes Leo

Africanus such a striking character is the fact that he has written a text reflecting his experience between two cultures, two religions, two civilizations.

2. See P. Curtin, S. Feierman, L. Thompson, J. Vansina, *African History* (New York: Longman, 1995); Charles-André Julien, *Histoire de l'Afrique du Nord* (Paris: Payot, 1986); Abdallah Laroui, *Histoire du Maghreb* (Paris: Maspéro, 1970); Fernand Braudel, *La Méditerranée et le Monde méditerranéen* (Paris: Armand Colin, 1982); Hernandez Pacheco, "Panorama historico-geographico de la Epoca de Leon el Africano," *Archivos del Instituto de Estudios Africanos* 30 (Madrid: Ares, 1954).
3. See Auguste Cour, *L'Etablissement des Dynasties des Chérifs au Maroc* (Paris: Leroux, 1903).
4. See Basil Davidson, *West Africa before the Colonial Era* (London: Longman, 1998): "Historians have called this one of the decisive battles of the world" (p. 58).
5. On this battle and its cultural impact in Europe and in Morocco, see Lucette Valensi, *Fables de la Mémoire* (Paris: Seuil, 1996).
6. Turin: Einaudi, 1978.
7. See entry "Leone Africano," *Enciclopedia Italiana di Scienze, Lettere ed Arti* 20 (Milan: Rizzoli, 1933).
8. It was published three times: 1556, 1559, 1632.
9. See Robert Brown, introduction to the 1896 edition of the English translation (London: Hakluyt Society).
10. The Italian text was republished in 1832 , then in 1978 (Turin: Einaudi); the French translation in 1830 and 1896; the English translation in 1896.
11. Paris: Maisonneuve, 1956.
12. Spanish, Russian, Arabic.
13. Sixth and seventh centuries.
14. On the mythical components of the European image of Africa, see Christopher Miller, *Blank Darkness* (Chicago: University of Chicago Press, 1985), and O. Zhiri, *L'Afrique au Miroir de l'Europe, Fortunes de Jean Léon l'Africain à la Renaissance* (Geneva: Droz, 1991).
15. See "Leon el Africano y la cartografia," *Archivos del Instituto de Estudios Africanos* 27 (Madrid: Ares, 1953).
16. See O. Zhiri, *L'Afrique au Miroir* and *Les Sillages de Jean Léon l'Africain du XVI° au XX° siècle* (Casablanca: Wallada, 1995).
17. *La République,* first published in 1575.
18. "Leo Africanus," *Yeats Annual* 1 (London: 1982).
19. On the history of the word, see Guy Turbet-Delof, *L'Afrique barbaresque dans la Littérature française* (Geneva: Droz, 1976).
20. In particular, this is the case for Numidia, which does not designate the same region at all in Europe and in Leo's text. He gives "Libia" as the equivalent of "Sahara," which is not the meaning the word had in European geography.
21. It has become since. Today "Ifriqiya" is used in Arabic to designate the continent as a whole.
22. *Iqlim* first designated a mathematical division of the world, but then came to mean simply region, sometimes administrative.
23. On the Berbers, see Michael Brett and Elizabeth Fentress, *The Berbers* (London: Blackwell, 1997).
24. On this question, see Louis Massignon, *Le Maroc dans les premières années du XVI° siècle, Tableau géographique d'après Jean Léon l'Africain* (Alger: Jourdan, 1906); Amando Melon, "Juan Leon Africano y su *Descripcion de Africa,*" *Archivos del Instituto de Estudios Africanos,* 29 (Madrid: Ares, 1954).
25. Paris: Lattès, 1986.

ABOUT THE CONTRIBUTORS

REBECCA CHUNG is a Ph.D. candidate in English Literature at the University of Chicago specializing in British Restoration and Eighteenth-Century Literature. She is preparing a literary critical edition of Lady Mary Wortley *Montagu's Travels of an English Lady (The Turkish Embassy Letters).*

MARY C. FULLER is Associate Professor of Literature at the Massachusetts Institute of Technology. Her current research focuses on colonial, mercantile, and military encounters between Renaissance England and other cultures. She is the author of *Voyages in Print: English Travel to America 1576–1624 (Cambridge, 1995).*

IVO KAMPS is Associate Professor of English at the University of Mississippi. He is the author of *Historiography and Ideology in Stuart Drama* (Cambridge, 1996) and editor of *Shakespeare Left and Right* (Routledge, 1991), *Materialist Shakespeare* (Verso, 1995). With Deborah Barker he has edited *Shakespeare and Gender* (Verso, 1995) and with Larry Danson *The Phoenix* for the Oxford edition of the complete works of Thomas Middleton (2001). Kamps is also the series editor for Early Modern Cultural Studies, 1500-1700 (Palgrave).

GERALD MACLEAN is Professor of English at Wayne State University, Detroit. His recent publications include *The Return of the King: An Anthology of English Poems Commemorating the Restoration of Charles II* (also online). Currently he is working on *The Rise of Oriental Travel: English Visitors to the Ottoman Empire, 1580–1720.*

SHANKAR RAMAN is the Class of 1957 Associate Professor in the Literature Faculty at the Massachusetts Institute of Technology. His recent book entitled *Framing "India": The Colonial Imaginary in Early Modern Culture,* under contract with Stanford University Press, is slated to appear in spring 2001. He has published on Christopher Marlowe's *Edward II (Deutsche Vierteljahresschrift)* and on John Fletcher's *The Island Princess (Renaissance Drama).*

JYOTSNA G. SINGH is Associate Professor in the English Department, Michigan State University. She is a co-author of *The Weyward Sisters: Shakespeare and Feminist Politics* (Blackwell, 1995) and the author of *Colonial Narratives/Cultural Dialogues: "Discoveries" of India in the Language of Colonialism* (Routledge, 1996).

GARY TAYLOR is Director of the Hudson Strode Program in Renaissance Studies at the University of Alabama, general editor of Oxford editions of The Complete Works of Shakespeare and Middleton, and author of *Reinventing Shakespeare: A Cultural History from the Restoration to the Present, Cultural Selection,* and (most recently) *Castration: Western Manhood from Jesus to the Posthuman.*

DANIEL VITKUS has a Ph.D. from Columbia University and taught for six years at the American University in Cairo. He has published articles on the representation of Islam during the Renaissance, and recently he edited an edition of *Three Turk Plays from Early Modern England.* He teaches in the English Department at Florida State University

OUMELBANINE ZHIRI is the author of *L'Afrique au Miroir de l'Europe, Fortunes de Jean Leon l'Africain à la Renaissance* (Droz, 1991), *Les Sillages de Jean Leon l'Africain du XVI° au XX° siecle* (Wallada, 1995) and *L'Extase et ses Paradoxes, Essai sur la Structure Narrative du Tiers Livre* (Champion, 1999). She is a Professor and teaches French and Comparative Literature at the University of California, San Diego.

INDEX

Page numbers printed in bold typeface designate entries found in the primary materials.